SALVAGE

CZECHOSLOVAK ACADEMY OF SCIENCES

THE CHEMICAL EQUILIBRIUM OF GASEOUS SYSTEMS

CZECHOSLOVAK ACADEMY OF SCIENCES

SCIENTIFIC EDITOR
Professor Dr. Ing. Eduard Hála, DrSc.

SCIENTIFIC ADVISER
Ing. Milan Šolc, CSc.

GRAPHIC DESIGN
Miroslav Houska

THE CHEMICAL EQUILIBRIUM OF GASEOUS SYSTEMS

ROBERT HOLUB

PETR VOŇKA

1976

D. REIDEL PUBLISHING COMPANY

DORDRECHT - HOLLAND / BOSTON - U.S.A.

Library of Congress Cataloging in Publication Data

Holub, Robert.

    The chemical equilibrium of gaseous systems.

    Based on Chemická rovnováha plynných reakcí, by R. Holub.
    Bibliography: p.
    1. Chemical reaction, Conditions and laws of. 2. Chemical equilibrium.
    I. Vonka, Petr, joint author. II. Holub, Robert. Chemická rovnováha plynných reakcí.
    III. Title.

QD501.H719   541'.392   75–34393
ISBN 90–277–0556–9

Sold and distributed in the U.S.A., Canada, and Mexico by D. Reidel Publishing Company, Inc.
Lincoln Building, 160 Old Derby Street, Hingham, Mass. 02043, U.S.A.

Published by D. Reidel Publishing Company, P.O. Box 17, Dordrecht, Holland, in co-edition
with ACADEMIA, Publishing House of the Czechoslovak Academy of Sciences, Prague

Preface

It is the purpose of this book to present a concise and sufficiently detailed description of the present state and possibilities of calculating chemical equilibria of gas mixtures. It is based on a book by one of the authors, published in Czech by the Publishing House Academia in Prague. The rapid development of the topic during the two years since publication of the Czech edition has made it necessary to revise practically all the sections in order to bring them up to the present level of knowledge.

One reason for writing this book was the practical requirement of contemporary industry, where a rational utilization of equilibrium composition calculations may provide valuable information concerning processes under study in all stages of their implementation. A second reason was the need of a text-book for studying this part of chemical thermodynamics in the scope as taught at the Institute of Chemical Technology, Prague.

These two basic motives determine the overall structure of the book, as well as the proportions and arrangement of the chapters. The book includes fundamental thermodynamic concepts as well as the mathematical apparatus needed to solve the problems involved, care being taken that the discussion should always lead to a practical procedure of performing equilibrium calculations in gas-phase systems of any degree of complexity whatever. Knowledge of chemical thermodynamics on the level of a fundamental university course is assumed.

The book is divided into seven chapters. A brief definition of the topic in Chapter 1 is followed by stoichiometric analysis of chemically reacting systems, leading up to an absolutely generalized description of the application of linear algebra methods.

Chapter 3 discusses the thermodynamic fundamentals of equilibrium behaviour, defines equilibrium criteria and indicates briefly the possibilities of their characterization.

Chapter 4 includes the calculation of chemical equilibria in simple systems in the ideal gas state, and the determination of the influence of simple reaction variables.

Chapter 5 is a review of methods employed to calculate complicated chemical equilibria in the ideal gas state. By means of detailed analysis, four very reliable and quite generalized procedures were selected. These have been practically tested and are described in complete detail up to the stage of block diagrams, from which prac-

5

tical programmes can be developed easily. Thus the procedures can readily be used to solve practical tasks in isothermic and adiabatic arrangements.

Chapter 6 describes the application of the individual methods for calculating equilibria in real gas systems.

Chapter 7 presents a critical survey of sources of thermochemical data, their calculation or estimation, method of tabulation, instructions for carrying out calculations with respect to maximum technological utilization of the results of calculations and an example of the treatment of a system of practical interest; errors, which may cause some uncertainty in the calculation, are briefly analyzed.

The book is amplified by a number of Appendices summing up some non-conventional mathematical procedures, proofs and some of the most important thermo chemical and physico-chemical data. In order to facilitate the understanding of illustrative problems and − at the same time − to stimulate the interest of the workers active in the industrial branches, we retained the traditional units (cal, K, atm).

We hope that this book will contribute to a more intensive application of methods of chemical equilibrium calculation, for which modern data processing equipment provides all necessary conditions. We also hope to have contributed a little towards making this branch of chemical thermodynamics as well known as it deserves to be in view of its practical application.

Finally, we should like to express our sincere thanks to Prof. Dr. Ing. Eduard Hála, Dr.Sc. and Prof. Dr. Ing. Jiří Pick, Dr.Sc. for valuable advice, Ing. Milan Šolc for carefully reading the manuscript and for helpful discussion, Dr. Ota Sofr for the translation and last not least, Mrs. Jaroslava Margoliusová for carefully typing the manuscript and drawing the pictures.

Prague, June 1974   Robert Holub
                                                               Petr Voňka

List of Symbols

a	constant of the relationship C_P as function of T
a_i	constant of the relationship $(C_P)_i$ as function of T
a, a_i	constant of the Beattie-Bridgman equation
a, a_i	constant of the Benedict-Webb-Rubin equation
a_i	activity of the constituent i
a_{ij}	constitution coefficient (denotes the number of atoms of the element j, present in the compound i)
A_i	chemical formula of the compound i
A_0	constant of the Beattie-Bridgman equation
A_k	constant of the power series K_a as function of T
b	constant of the relationship C_P as function of T
b_i	constant of the relationship $(C_P)_i$ as function of T
b, b_i	constant of the Beattie-Bridgman equation
b, b_i	constant of the Benedict-Webb-Rubin equation
B	constant
B, B'	second virial coefficient
B_{jk}	second virial coefficient of the interaction between constituents j and k
B_r	reduced second virial coefficient
B_j	chemical symbol of an element, or chemical formula of a basic particle species
B_k	constant of the relationship of the second virial coefficient as function of T
B_0	constant of the Beattie-Bridgman equation
B_0	constant of the Benedict-Webb-Rubin equation
c	constant of the relationship C_P as function of T
c_i	constant of the relationship $(C_P)_i$ as function of T
c, c_i	constant of the Beattie-Bridgman equation
c, c_i	constant of the Benedict-Webb-Rubin equation
c_i	constituent concentration
c_i	$= (G_i^\circ/RT) + \ln P)$
C, C'	constant
C_P	molar heat

$(C_P)_i$	molar heat of constituent i
C, C'	third virial coefficient
C_{jkl}	third virial coefficient of the interaction between the constituents j, k and l
C_k	constant of the relationship C_P as function of T
C_r	reduced third virial coefficient
d	total differential of the quantity denoted by the following symbol
d	bond length
D, D'	fourth virial coefficient
D_{jklp}	fourth virial coefficient of the interaction between the constituents j, k, l, and p
e_j	value of (G^o/RT) corresponding to the j-th element or basic particle species
E^o	standard electromotoric force
f	fugacity
f_i	fugacity of constituent i
f_r	function symbol
F	free energy (Helmholtz function)
F_i	free energy (Helmholtz function) of constituent i
F	Faraday constant
F	function symbol
G	free enthalpy (Gibbs function)
G_i	free enthalpy (Gibbs function) of constituent i
G_f	free enthalpy of formation
H	rank of a matrix of constitution coefficients
H	enthalpy
H_i	enthalpy of the constituent i
H_f	enthalpy of formation
H_r	enthalpy of chemical reaction (heat of reaction)
I_H	integration constant of the heat of reaction vs. temperature relationship
I_K	integration constant of the equilibrium constant vs. temperature relationship
K	Kelvin
K, K_a, K_r	equilibrium constant
K_f	equilibrium constant of reaction of formation
M	number of elements (basic particle species) in a closed system
M_i	molecular weight of compound i
n	overall number of moles in a system
n_i	number of moles of constituent i
N	number of compounds in a closed system
p_i	$= \sum_{k=1}^{M} a_{ik}\lambda_k - c_i$

P	number of linear combinations of columns of the matrix of constitution coefficients
P	overall pressure in a system
P_i	partial pressure of constituent i
P_c	critical pressure
P_r	reduced pressure
r_{jk}	coefficient of a system of linear equations
R	number of linearly independent reactions
R	gas constant
q_i	$= n_i + \sum_{r=H+1}^{N} v_{ri} n_r$
Q	function symbol $V, G/RT$
S	entropy
t	reciprocal value of the overall number of moles $(1/n)$
T	absolute temperature
T_c	critical temperature (K)
T_r	reduced temperature
u	function symbol
u	$= \left(n^{(p+1)}/n^{(p)} - 1\right)$
U	internal energy
v	molar volume
v_i	molar volume of the constituent i
V	overall volume
W	function symbol
y_i	molar fraction of constituent i
z	linking condition
z	compressibility factor
z_c	critical compressibility factor

Superscripts

e	quantity in equilibrium
M	quantity of mixing
o	quantity in standard state (or pure constituent)
0	initial value (zero-th approximation)
p	step (p-th approximation of the quantity denoted by the preceding symbol)
$'$	notation of alternative quantity, denoted by the preceding symbol
$-$	partial molar quantity
$*$	under conditions close to ideal ones

Subscripts

| c | critical state |

f	for reaction of formation
i	index of summation for compounds $(1, 2, ..., N)$
j	index of summation for elements (basic particle species) $(1, 2, ..., N)$
k	index of summation
l	index of summation
m	relating to mixture
p	index of summation
r	index of summation for reactions $(1, 2, ..., R)$
r	reduced state
s	index of summation
x	required value of quantity denoted by the preceding symbol

Greek letters

α, α_i	constant of the Benedic-Webb-Rubin equation
α	directional angle
β	functional expression in the Beattie-Bridgman equation
β_r	correction factor, replacing the reaction coordinate ξ_r
γ	functional expression in the Beattie-Bridgman equation
γ, γ_i	constant of the Benedict-Webb-Rubin equation
Γ	dimension-less function of free enthalpy (equal to G/RT, or $\sum_{i=1}^{N} f_i$)
∂	partial differential
δ	functional expression in the Beattie-Bridgman equation
δ_{jk}	Kronecker's delta (unit diagonal matrix of the dimension $j \times k$)
Δ	final difference, or deviation of quantity following the symbol
ε	arbitrarily small positive number
λ	Lagrangian multiplier
λ	magnitude of step
λ_i	arbitrary chosen vector, substituting a stoichiometric coefficient
μ_i	chemical potential of constituent i
v_{ri}	stoichiometric coefficient (denotes the number of moles of the i-th constituent taking part in the r-th reaction)
ξ	reaction coordinate (degree of conversion, extent of reaction)
ξ_r	reaction coordinate of the r-th reaction
$\prod_{i=1}^{N}$	denotes the product over i $(i = 1, 2, ..., N)$
σ	reciprocal molar volume $(1/v)$
$\sum_{i=1}^{N}$	denotes the sum over i $(i = 1, 2, ..., N)$
φ_i	fugacity coefficient of constituent i
Φ	function symbol

List of contents

12

1 Introduction

Chemical equilibrium can be established when a system contains constituents between which chemical conversion can take place. Determination of the products which can be formed by chemical conversion of the initial constituents and of the composition of the system in equilibrium is one of the frequent, important and often highly interesting problems which have to be solved in practice.

Up to recent times, when the chemical industry was still mainly based on coal, organic syntheses set out from relatively pure individual compounds e.g. acetylene, ethylene, benzene, etc., reacting systems could generally be described by means of a simple chemical reaction. With conversion of the chemical industry to oil and petrochemical processes, it becomes increasingly important to utilize multi-component mixtures as raw materials. This development is supported by economic aspects as well as progress in chemical technology, instrumental analysis and engineering. Separation of the components from petroleum or its fractions is difficult and expensive, while on the other hand, contemporary technology, utilizing modern analytical techniques is able to study even highly complex reaction systems. In this situation it is essential to be able to calculate the chemical equilibrium of systems in which several chemical reactions will be taking place.

There is no difficulty in calculating the chemical equilibrium of a system, in which a single chemical reaction takes place. The calculation, however, becomes increasingly difficult with the rising number of simultaneous reactions, until application of the same procedure to systems with more than three reactions proceeding simultaneously is practically impossible. Therefore, techniques have had to be worked out for more complicated chemically reacting systems, based on principles somewhat different from those of simple equilibrium calculation. The result are methods which allow equilibrium compositions to be calculated for systems of any degree of complexity whatever, in the ideal as well as real gas state.

A survey of the most important and generally applicable procedures, together with practical instructions for their application in solving actual problems is the topic of this book.

2 Stoichiometry of chemical reactions

In chemistry, stoichiometry is conventionally understood to mean the relationship between elements or fundamental particles and components in their mutual conversions. In the field of chemical equilibria, stoichiometry permits investigation of concentration changes as well as an accurate determination of the maximum number of reactions which may take place in a system, and allows the optimum combinations of these reactions to be selected. For this reason, equilibrium considerations proper must be preceded by a detailed stoichiometrical analysis of the system involved. This puprpose may well be achieved by utilization of linear algebra, and a closed system may be described formally as a system of linear algebraic equations.

2.1 DEFINITION OF A CHEMICAL REACTION

Let us consider a closed system consisting of M elements and N constituents. Let A_i stand for the i-th constituent and B_j for the j-th element. Then,

$$A_i = \sum_{j=1}^{M} a_{ij}B_j \quad i = 1, 2, ..., N \tag{2.1}$$

where a_{ij} is the constitution coefficient, denoting the number of gram atoms of the j-th element in the i-th constituent.

For conventional use this intuitive concept of a compound is sufficient, and there is no difficulty in extending it to a radical. Provided an electron is additionally defined as one of the set of elementary particles B_j, ions may also be considered to be independent constituents.

When every compound is described as a line vector of constitutional coefficients, a system of N constituents may be defined by a matrix of $N \times M$ elements. Clearly the elements of this matrix must be non-negative integers. With regard to the physical sense of the matrix, none of its rows or columns may be composed exclusively of zeros.

16

Example 1

Let us consider a system composed of three elements (C, H, O) and five constituents $(CH_4, H_2O, H_2, CO, CO_2)$. The matrix of constitution coefficients will take the form

	$j = 1$ C	$j = 2$ H	$j = 3$ O
(CH_4) $i = 1$	1	4	0
(H_2O) $i = 2$	0	2	1
(H_2) $i = 3$	0	2	0
(CO) $i = 4$	1	0	1
(CO_2) $i = 5$	1	0	2

A number of reactions may take place in a system defined between the given N constituents. Assuming the number of possible reactions in the system to be R, the chemical conversion of the initial substances to products may be expressed by the relation

$$\sum_{i=1}^{N} v_{ri} A_i = 0 \quad r = 1, 2, \ldots, R \tag{2.2}$$

where v_{ri} is the stoichiometric coefficient, the absolute value of which indicates the number of moles of the i-th constituent in the r-th reaction. Stoichiometric coefficients of products have a positive sign, while signs of stoichiometric coefficients of initial substances are negative. Although relation (2.2) characterises the course of specific chemical reactions, the way it is written is not yet sufficient for the quantitative description which is essential for chemical equilibrium investigations. The reason is that this relation does not include the absolute magnitude of the stoichiometric coefficients. Thus e.g. methanol synthesis may be expressed by the stoichiometric equations

$$CO + 2\,H_2 \;=\; CH_3OH$$

$$\tfrac{1}{2}CO + H_2 \;=\; \tfrac{1}{2}CH_3OH, \text{ etc.} \tag{2.3}$$

Since thermodynamic functions are extensive quantities, an additional condition must be prescribed to normalize the absolute magnitude of the stoichiometric coefficients. Several proposals were published[3,62,125] the following two seem to be best:

1. To select stoichiometric coefficients so as to obtain whole non-codivisible numbers only. In the example given by equation (2.3) this condition would be satisfied by the first-named equation.

2. Select the coefficients so as to satisfy the relation

$$\sum_{i=1}^{N} |v_{ri}| = 1 \quad r = 1, 2, ..., R \,. \tag{2.4}$$

This relation was proposed by Pings[125]: the absolute magnitude of the stoichiometric coefficients is determined by first writing every equation with arbitrary coefficients v'_{ri} and then determining the absolute magnitude from the relation

$$v_{ri} = \frac{v'_{ri}}{\displaystyle\sum_{s=1}^{N} |v'_{rs}|} \quad \begin{array}{l} i = 1, 2, ..., N \\ r = 1, 2, ..., R \end{array} \,. \tag{2.5}$$

In this case, methanol synthesis would be described by an equation of the form of

$$\tfrac{1}{4} CO + \tfrac{1}{2} H_2 = \tfrac{1}{4} CH_3OH \,. \tag{2.6}$$

As will be seen from the following, it is more advantageous to use stoichiometric coefficients in the form of mutually non-divisible integers.

Considering now every chemical reaction as a row vector of its stoichiometric coefficients, then the overall chemical conversion of the system is determined by a matrix of stoichiometric coefficients of the type $R \times N$. The elements of this matrix may be any rational numbers.

The system is uniquely defined by the two matrixes, the matrix of constitution coefficients and the matrix of stoichiometric coefficients. The matrix of constitution coefficients, however, is of fundamental significance because it describes the qualitative properties of the system and the matrix of stoichiometric coefficients may be derived from it.

2.2 DETERMINATION OF THE NUMBER OF LINEARLY INDEPENDENT REACTIONS

For every system containing a greater number of constituents, a set of stoichiometric equations may easily be constructed by means of a suitable combination of row vectors of the matrix of constitution coefficients. Only a certain number of these equations will, however, be mutually linearly independent. The other equations may then be expressed by a linear combination of the preceding reactions. It is essential for the following considerations to determine the maximum number of linearly independent stoichiometric equations, and to select out of all possible combinations those which describe a given system in the simplest possible manner.

The maximum number of linearly independent reactions may be derived as follows. Let us assume that at least one reaction may take place in the given system

$$\sum_{i=1}^{N} v_i A_i = 0 \,. \tag{2.7}$$

Let us now ask, how many linearly independent vectors $(v_1, v_2, \ldots, v_N)$ may be constructed. Substituting relation (2.1) into equation (2.7) we obtain

$$\sum_{i=1}^{N} v_i \left(\sum_{j=1}^{M} a_{ij} B_j \right) = 0 . \tag{2.8}$$

When the other of addition is altered, equation (2.8) converts to the form

$$\sum_{j=1}^{M} \left(\sum_{i=1}^{N} a_{ij} v_i \right) B_j = 0 . \tag{2.9}$$

Relation (2.9) may evidently be satisfied only, if

$$\sum_{i=1}^{N} a_{ij} v_i = 0 \quad j = 1, 2, \ldots, M . \tag{2.10}$$

Equation (2.10) forms a set of M equations for N unknowns $(v_1, v_2, \ldots, v_N)$. Let H be rank of the matrix of constitution coefficients. From linear algebra follows, that the dimension of the set of solutions — let us denote this dimension by the symbol $R_{\max}$ — of the set (2.10) is

$$R_{\max} = N - H \tag{2.11}$$

where $R_{\max}$ thus gives the maximum number of linearly independent solutions of the set (2.10).

Denoting the number of linearly independent reactions which take place in the system by R, we may write

$$R \leq R_{\max} = N - H . \tag{2.12}$$

Since there holds in general, that $H \leq N$, it is reasonable to demand moreover $N - H \neq 0$. Without this latter demand the investigation of the given system would be devoid of physical sense, since no chemical reaction could proceed in it. As example let us take the system $\{A_1 = CO, A_2 = CO_2\}$, where $N = H = 2$. For systems, in which at least one reaction can take place, there holds

$$0 < R \leq R_{\max} = N - H . \tag{2.13}$$

Relation (2.13) is called the Gibbs stoichiometric rule[26].

Example 2

In the system used in example 1 there holds, that $N = 5, M = 3, H = 3$. According to relation (2.13) $R_{\max} = 2$ and a maximum of two linearly independent reactions can take place in the system.

For chemical equilibrium purposes, the determination of $R_{\max}$ suffices. However, determination of this quantity is insufficient for judging the reaction mechanism or for kinetic studies, and the system must generally be measured by experimental means. This is due to the fact, that the number of dominating reactions, which determine the kinetics of the system involved, may be considerably less than the value of R_{max}.

Let a number R of linearly independent vectors $(v_{r1}, v_{r2}, ..., v_{rN})$, where $R \leq R_{max}$, $r = 1, 2, ..., R$ be the solution of the set (2.10). We shall call the matrix $\mathbf{N} = \{v_{ri}\}$, $r = 1, 2, ..., R$, $i = 1, 2, ..., N$ the matrix of stoichiometric coefficients. The elements of this matrix will form R × N rational numbers. From relation (2.10) follows, that there holds for an arbitrary matrix of stoichiometric coefficients

$$\sum_{i=1}^{N} v_{ri}a_{ij} = 0 \quad \begin{array}{l} j = 1, 2, ..., M \\ r = 1, 2, ..., R \end{array} \tag{2.14}$$

and, written in matrix form

$$\mathbf{N} \cdot \mathbf{A} = \mathbf{O} \tag{2.15}$$

where $\mathbf{A}$ is the matrix of constitution coefficients and $\mathbf{O}$ is a matrix of the type $R \times M$, composed of zeros only.

The rank of the matrix of constitution coefficients enables us to classify all constituents present in the system into two groups, fundamental (primary, basic) and derived (secondary). In every system there exists a certain minimum number of constituents by means of which the system is described in qualitative terms. These constituents are called fundamental ones (or also components) and their number is determined by the rank of the matrix of constitution coefficients. The maximum number of possible choices of a selected set of fundamental constituents is $\binom{N}{H}$, since the H rows of the matrix of constitution coefficients are not all linearly independent.

Example 3

Let us consider the system form example 1. There applies that the rank of the matrix of constitution coefficients is $H = 3$, since e.g. from the rows 1, 2, 3 a square matrix may be constructed which will have a non-zero determinant

$$\det \begin{vmatrix} 1 & 4 & 0 \\ 0 & 2 & 1 \\ 0 & 2 & 0 \end{vmatrix} = -2.$$

There likewise follows, that the constituents 1, 2, 3 may be selected as fundamental ones. The remaining two constituents may then be obtained by a combination of the constituent 1, 2, 3. The maximum possible number of choices of a set of fundamental constituents is $\binom{5}{3} = 10$. Since a square matrix constructed from an arbitrary selection of three rows of the matrix of constitution coefficients is regular, there actually exist 10 possibilities of chosing a set of fundamental constituents: $(1, 2, 3)$, $(1, 2, 4)$, $(1, 2, 5)$, $(1, 3, 4)$, $(1, 3, 5)$, $(1, 4, 5)$, $(2, 3, 4)$, $(2, 3, 5)$, $(2, 4, 5)$, $(3, 4, 5)$.

Generally, particularly in the case of more complicated systems, $(N \gg M)$ the rank of the matrix of constitution coefficients is given by the number of elements, i.e. $H = M$. In practical calculations, we usually have $M \leq 6$. When, however, there are P linear relationships between the columns of the matrix of constitution coefficients, the rank of this matrix will be $H = M - P$.

Example 4

Let us define a system composed of three elements (C, H, O) and three constituents (CO, H_2, CH_3OH). The matrix of the constitution coefficients will be

$$\begin{vmatrix} 1 & 0 & 1 \\ 0 & 2 & 0 \\ 1 & 4 & 1 \end{vmatrix}.$$

The rank of this matrix is $H = 2$, since the first and third columns are identical. Therefore this system has only two fundamental constituents and there holds that $R_{max} = N - H = 1$. This single equation is

$$CO + 2 H_2 = CH_3OH.$$

2.3 CALCULATION OF THE MATRIX OF STOICHIOMETRIC COEFFICIENTS

The matrix of stoichiometric coefficients may be determined from the matrix of constitution coefficients by means of the equations (2.10). Let us assume, without loss of generality, that the first H constituents are fundamental ones, i.e.

$$\det \begin{vmatrix} a_{11}, & a_{12}, & ..., & a_{1H} \\ a_{21}, & a_{22}, & ..., & a_{2H} \\ & \vdots & & \\ a_{H1}, & a_{H2}, & ..., & a_{HH} \end{vmatrix} \neq 0 \tag{2.16}$$

where H is the rank of the matrix of constitution coefficients. This may always be achieved by altering the order of the constituents. The set (2.10) may be rewritten in the form of

$$\sum_{i=1}^{H} a_{ij} v_i = - \sum_{i=H+1}^{N} a_{ij} v_i \quad j = 1, 2, ..., M. \tag{2.17}$$

From linear algebra theory follows, that a general solution of the system (2.17) will take the form of

$$v_s = \sum_{i=H+1}^{N} C_{si}\lambda_i \quad s = 1, 2, \ldots, H \tag{2.18}$$

$$v_s = \lambda_s \qquad s = H + 1, \ldots, N,$$

where $\lambda_{H+1}, \ldots, \lambda_N$ are arbitrary real numbers and the terms C_{si} are functions only of the constitution coefficients. The maximum number of linearly independent vectors $(v_1, v_2, \ldots, v_N)$, which can be constructed by suitable choice of the parameters $\lambda_{H+1}, \ldots, \lambda_N$ is $R_{max} = N - H$.

Example 5

Using data from examples 1 and 3, calculate the matrix of stoichiometric coefficients and write the corresponding chemical reactions.

When CH_4, H_2O and H_2 are taken to be the fundamental constituents, relation (2.17) converts to the specific form of

$$v_1 \qquad = -v_4 - v_5$$
$$4v_1 + 2v_2 + 2v_3 = 0$$
$$v_2 \qquad = -v_4 - 2v_5 .$$

The set of equations may be expressed in matrix form as

$$(v_1, v_2, v_3) \begin{pmatrix} 1 & 4 & 0 \\ 0 & 2 & 1 \\ 0 & 2 & 2 \end{pmatrix} = -(\lambda_4, \lambda_5) \begin{pmatrix} 1 & 0 & 1 \\ 1 & 0 & 2 \end{pmatrix},$$

where $v_4 = \lambda_4$, $v_5 = \lambda_5$. By modifying the set we obtain the general solution

$$v_1 = -\lambda_4 - \lambda_5$$
$$v_2 = -\lambda_4 - 2\lambda_5$$
$$v_3 = 3\lambda_4 + 4\lambda_5$$
$$v_4 = \lambda_4$$
$$v_5 = \lambda_5,$$

where λ_4 and λ_5 are arbitrary numbers. Converting back to the matrix form, the general solution will be given by $(v_1, v_2, v_3, v_4, v_5) = \lambda_4(-1, -1, 3, 1, 0) + \lambda_5(-1, -2, 4, 0, 1)$. Successive substitution of the values 0 and 1 for λ_4 and, conversely, 1 and 0 for λ_5 will give a set of stoichiometric coefficients of the first and second linearly independent reaction. Therefore the matrix of stoichiometric coefficients will take the form of

$$\begin{pmatrix} -1 & -1 & 3 & 1 & 0 \\ -1 & -2 & 4 & 0 & 1 \end{pmatrix}$$

and the corresponding stoichiometric equations are

$$CH_4 + H_2O = 3\,H_2 + CO$$

$$CH_4 + 2\,H_2O = 4\,H_2 + CO_2\,.$$

On substitution of $N - H$ concrete linearly independent vectors $(\lambda_{r,H+1}, \lambda_{r,H+2}, \ldots, \lambda_{r,N})$ $r = 1, 2, \ldots, N - H$ into equations (2.18) we evidently obtain $N - H$ linearly independent vectors $(v_{r1}, v_{r2}, \ldots, v_{rN})$ $r = 1, 2, \ldots, N - H$. Such a set of linearly independent vectors of stoichiometric coefficients is called the particular solution of the set (2.17) or (2.10). For practical calculations, it is useful to find a particular solution of such a type that the matrix of stoichiometric coefficients should contain a maximum number of zeros in every row. From linear algebra theory follows, that two cases must be distinguished:

1. The set of vectors of an arbitrarily selected number H of rows of the matrix of constitution coefficients is linearly independent, i.e. the set of basic constituents may be chosen by $\begin{pmatrix} N \\ H \end{pmatrix}$ means.

In this case every chemical reaction, and thus also every row of the matrix of stoichiometric coefficients must contain at least $H + 1$ non-zero terms. The reason is, that if e.g. a chemical reaction should involve only H constituents, then the rows of the matrix of constitution coefficients of these constituents would have to be linearly dependent, which disagrees with our assumption. Let us chose in relation (2.18) $\lambda_{r,H+i} = \delta_{r,i}$ (Kronecker's delta), where $\delta_{r,i} = 0$ $r \neq i$ and $\delta_{r,i} = 1$ for $r = i$, i.e. the r-th coordinate of the vector $(\lambda_{r,H+1}, \ldots, \lambda_{r,N})$ is equal to one and the rest are zeros. This kind of choice guarantees that every row of the matrix of stoichiometric coefficients will have exactly $H + 1$ non-zero terms, and a better result cannot be achieved. The condition that stoichiometric coefficients should be non-divisible numbers can be satisfied by multiplying every row of the matrix of stoichiometric coefficients by a suitable number. When we select a different set of basic constituents as basis of our calculation, the matrix of stoichiometric coefficients obtained will be formally different, but the number of zeros will be the same.

2. The case described sub 1 does not apply.

A general investigation of this case is difficult and time-consuming. Let us only stress the fact, that differing from the first case the number of zeros in the rows of the stoichiometric coefficient now depends on the selection of the set of basic constituents.

Evidently, there will exist reactions which will be composed of less than $H + 1$ constituents. The procedure is illustrated in the following example.

Example 6

Similarly to decomposition of methane by steam, a synthetic mixture may also be obtained by decomposition of ethane. Let us assume the summary scheme

$$C_2H_6 + H_2O \rightarrow C_2H_6, H_2O, CH_4, H_2, CO, CO_2$$

and let us determine the matrix of stoichiometric coefficients so as to express the result of the reaction in the stoichiometrically most advantageous form. If the order of constituents is kept as it is in the above scheme, then with elements in the order C, H, O the matrix of constitution coefficients will take the form of

$$\begin{vmatrix} 2 & 6 & 0 \\ 0 & 2 & 1 \\ 1 & 4 & 0 \\ 0 & 2 & 0 \\ 1 & 0 & 1 \\ 1 & 0 & 2 \end{vmatrix} \quad H = 3, R_{max} = 3$$

The number of basic constituents is equal to three, therefore there is a maximum of $\binom{6}{3} = 20$ possible combinations for their selection. It can be shown that actually there are only 19 possible combinations, because rows 1, 3, 4 form a singular matrix. Thus one reaction must exist which involves all three constituents, in other words there will exist precisely one row in the matrix of stoichiometric coefficients which has only three non-zero terms. This one reaction is a combination of the constituents 1, 3, 4.

$$C_2H_6 + H_2 = 2 CH_4 .$$

All other chemical reactions will contain at least four constituents. A total of thirteen possible reactions corresponds to all the possible choices of the set of basic constituents (disregarding the direction of the reaction). By means of these reactions the shift of the system from the initial state to equilibrium can be described. They are the following:

$$
\begin{aligned}
(1) \quad & C_2H_6 + H_2 && = \;\; 2 CH_4 \\
(2) \quad & C_2H_6 + 2 H_2O && = \;\; 5 H_2 + 2 CO \\
(3) \quad & C_2H_6 + 4 H_2O && = \;\; 7 H_2 + 2 CO_2 \\
(4) \quad & C_2H_6 + 2 CO_2 && = \;\; 3 H_2 + 4 CO \\
(5) \quad & C_2H_6 + 5 CO_2 && = \;\; 7 CO + 3 H_2O
\end{aligned}
$$

$$
\begin{array}{rll}
(6) & 2\,C_2H_6 + CO_2 &= 3\,CH_4 + 2\,CO \\
(7) & 3\,C_2H_6 + H_2O &= 5\,CH_4 + CO \\
(8) & 4\,C_2H_6 + 2\,H_2O &= 7\,CH_4 + CO_2 \\
(9) & CH_4 + CO_2 &= 2\,H_2 + 2\,CO \\
(10) & CH_4 + H_2O &= 3\,H_2 + CO \\
(11) & CH_4 + 2\,H_2O &= 4\,H_2 + CO_2 \\
(12) & CH_4 + 3\,CO_2 &= 2\,H_2O + 4\,CO \\
(13) & CO + H_2O &= CO_2 + H_2
\end{array}
$$

Selecting C_2H_6, H_2O and CH_4 as basic constituents we obtain the general solution by means of a procedure similar to the one used in example 5

$$
\begin{aligned}
v_1 &= \lambda_4 - 3\lambda_5 - 4\lambda_6 \\
v_2 &= - \lambda_5 - 2\lambda_6 \\
v_3 &= -2\lambda_4 + 5\lambda_5 + 7\lambda_6 \\
v_4 &= \lambda_4 \\
v_5 &= \lambda_5 \\
v_6 &= \lambda_6
\end{aligned}
$$

Selecting, successively, the vectors $(1, 0, 0)$, $(0, 1, 0)$ and $(0, 0, 1)$ for the vector $(\lambda_4, \lambda_5, \lambda_6)$ we obtain the matrix of stoichiometric coefficients

$$
\begin{vmatrix}
1 & 0 & -2 & 1 & 0 & 0 \\
-3 & -1 & 5 & 0 & 1 & 0 \\
-4 & -2 & 7 & 0 & 0 & 1
\end{vmatrix}
$$

the corresponding chemical reactions being

$$
\begin{aligned}
2\,CH_4 &= C_2H_6 + H_2 \\
3\,C_2H_6 + H_2O &= 5\,CH_4 + CO \\
4\,C_2H_6 + 2\,H_2O &= 7\,CH_4 + CO_2 .
\end{aligned}
$$

Table 1 shows a detailed stoichiometrical analysis of the system involved. It is advantageous for some methods of calculating chemical equilibria that the value of $\left| \sum\limits_{i=1}^{N} v_{ri} \right|$ $r = 1, 2, ..., R$ and the value $\sum\limits_{r=1}^{R} \left| \sum\limits_{i=1}^{N} v_{ri} \right|$ be minimised. These values are tabulated in the last two columns of Table 1 for each particular solution. Note, that when at least two of the constituents 1, 3, 4 are included in the basic constituents, then the equation $C_2H_6 + H_2 = 2\,CH_4$, i.e. an equation including only three constituents, is included among the respective stoichiometric equations.

Table 1. Stoichiometric analysis of a system including the constituents C_2H_6, H_2, CH_4, H_2O, CO and CO_2

Serial number of system	Selected basis constituents			Form of matrix of stoichiometric coefficients						Corresponding stoichiometric equations	$\lvert \sum\limits_{i=1}^{N} v_{ri} \rvert$			$\sum\limits_{r=1}^{R} \lvert \sum\limits_{i=1}^{N} v_{ri} \rvert$
											$r = 1$	$r = 2$	$r = 3$	
1	1	2	3	1	0	-2	1	0	0	$2\,CH_4 = C_2H_6 + H_2$	0	2	2	4
				-3	-1	5	0	1	0	$3\,C_2H_6 + H_2O = 5\,CH_4 + CO$				
				-4	-2	7	0	0	1	$4\,C_2H_6 + 2\,H_2O = 7\,CH_4 + CO_2$				
2	1	2	4	-1	0	2	-1	0	0	$C_2H_6 + H_2 = 2\,CH_4$	0	4	4	8
				-1	-2	0	5	2	0	$C_2H_6 + 2\,H_2O = 5\,H_2 + 2\,CO$				
				-1	-4	0	7	0	2	$C_2H_6 + 4\,H_2O = 7\,H_2 + 2\,CO_2$				
3	1	2	5	-3	-1	5	0	1	0	$3\,C_2H_6 + H_2O = 5\,CH_4 + CO$	2	4	4	10
				-1	-2	0	5	2	0	$C_2H_6'+ 2\,H_2O = 5\,H_2 + 2\,CO$				
				1	-3	0	0	-7	5	$3\,H_2O + 7\,CO = C_2H_6 + 5\,CO_2$				
4	1	2	6	-4	-2	7	0	0	1	$4\,C_2H_6 + 2\,H_2O = 7\,CH_4 + CO_2$	2	4	4	10
				-1	-4	0	7	0	2	$C_2H_6 + 4\,H_2O = 7\,H_2 + 2\,CO_2$				
				-1	3	0	0	7	-5	$C_2H_6 + 5\,CO_2 = 3\,H_2O + 7\,CO$				
5	1	3	4	Matrix of constitution coefficients of basic constituents is singular										
6	1	3	5	3	1	-5	0	-1	0	$5\,CH_4 + CO = 3\,C_2H_6 + H_2O$	2	0	2	4
				1	0	-2	2	0	0	$2\,CH_4 = C_2H_6 + H_2$				
				2	0	-3	0	-2	1	$3\,CH_4 + 2\,CO = 2\,C_2H_6 + CO_2$				
7	1	3	6	4	2	7	0	0	-1	$7\,CH_4 + CO_2 = 4\,C_2H_6 + 2\,H_2O$	2	0	2	4
				1	0	-2	1	0	0	$2\,CH_4 = C_2H_6 + H_2$				
				-2	0	3	0	2	-1	$2\,C_2H_6 + CO_2 = 3\,CH_4 + 2\,CO$				

Table 1 (cont.)

Serial number of system	Selected basic constituents	Form of matrix of stoichiometric coefficients	Corresponding stoichiometric equations	$\lvert \sum_{i=1}^{N} \nu_{ri} \rvert$			$\sum_{r=1}^{R} \lvert \sum_{i=1}^{N} \nu_{ri} \rvert$
				$r=1$	$r=2$	$r=3$	
8	1 4 5	$\begin{array}{rrrrrr} 1 & 2 & 0 & -5 & -2 & 0 \\ -1 & 0 & 2 & -1 & 0 & 0 \\ 1 & 0 & 0 & -3 & -4 & 2 \end{array}$	$2\,CO + 5\,H_2 = C_2H_6 + 2\,H_2O$ $C_2H_6 + H_2 = 2\,CH_4$ $4\,CO + 3\,H_2 = C_2H_6 + 2\,CO_2$	4	0	4	8
9	1 4 6	$\begin{array}{rrrrrr} 1 & 4 & 0 & -7 & 0 & -2 \\ -1 & 0 & 2 & -1 & 0 & 0 \\ -1 & 0 & 0 & 3 & 4 & -2 \end{array}$	$2\,CO_2 + 7\,H_2 = C_2H_6 + 4\,H_2O$ $C_2H_6 + H_2 = 2\,CH_4$ $C_2H_6 + 2\,CO_2 = 3\,H_2 + 4\,CO$	4	0	4	8
10	1 5 6	$\begin{array}{rrrrrr} -1 & 3 & 0 & 0 & 7 & -5 \\ -2 & 0 & 3 & 0 & 2 & -1 \\ -1 & 0 & 0 & 3 & 4 & -2 \end{array}$	$C_2H_6 + 5\,CO_2 = 7\,CO + 3\,H_2O$ $2\,C_2H_6 + CO_2 = 3\,CH_4 + 2\,CO$ $C_2H_6 + 2\,CO_2 = 3\,H_2 + 4\,CO$	4	2	4	10
11	2 3 4	$\begin{array}{rrrrrr} 1 & 0 & -2 & 1 & 0 & 0 \\ 0 & -1 & -1 & 3 & 1 & 0 \\ 0 & -2 & -1 & 4 & 0 & 1 \end{array}$	$2\,CH_4 = C_2H_6 + H_2$ $CH_4 + H_2O = CO + 3\,H_2$ $CH_4 + 2\,H_2O = CO_2 + 4\,H_2$	0	2	2	4
12	2 3 5	$\begin{array}{rrrrrr} 3 & 1 & -5 & 0 & 1 & 0 \\ 0 & -1 & -1 & 3 & 1 & 0 \\ 0 & -2 & -1 & 4 & 0 & 1 \end{array}$	$5\,CH_4 + CO = 3\,C_2H_6 + H_2O$ $CH_4 + H_2O = 3\,H_2 + CO$ $CH_4 + 2\,H_2O = 4\,H_2 + CO_2$	2	2	2	6
13	2 3 6	$\begin{array}{rrrrrr} 4 & 2 & -7 & 0 & 0 & -1 \\ 0 & -2 & -1 & 4 & 0 & 1 \\ 0 & 2 & -1 & 0 & 4 & -3 \end{array}$	$7\,CH_4 + CO_2 = 4\,C_2H_6 + 2\,H_2O$ $CH_4 + CO_2 = 4\,C_2H_6 + 2\,H_2O$ $CH_4 + 3\,CO_2 = 2\,H_2O + 4\,CO$	2	2	2	6

Table 1 (cont.)

Serial number system	Selected basic constituents	Form of matrix of stoichiometric coefficients	Corresponding stoichiometric equations	$\|\sum_{i=1}^{N} v_{ri}\|$ $r=1$	$r=2$	$r=3$	$\sum_{r=1}^{R}\|\sum_{i=1}^{N} v_{ri}\|$
14	2 4 5	1 2 0 −5 −2 0 0 1 1 −3 −1 0 0 −1 0 1 −1 1	$2\,CO + 5\,H_2 = C_2H_6 + 2\,H_2O$ $CO + 3\,H_2 = CH_4 + H_2O$ $CO + H_2O = H_2 + CO_2$	4	2	0	6
15	2 4 6	1 4 0 −7 0 −2 0 2 1 −4 0 −1 0 1 0 −1 1 −1	$2\,CO_2 + 7\,H_2 = C_2H_6 + 4\,H_2O$ $CH_4 + 2\,H_2O = 4\,H_2 + CO_2$ $CO_2 + H_2 = CO + H_2O$	4	2	0	6
16	2 5 6	1 −3 0 0 −7 5 0 −2 1 0 −4 3 0 −1 0 1 −1 1	$7\,CO + 3\,H_2O = C_2H_6 + 5\,CO_2$ $4\,CO + 2\,H_2O = CH_4 + 3\,CO_2$ $CO + H_2O = H_2 + CO_2$	4	2	0	6
17	3 4 5	1 0 −2 1 0 0 0 1 1 −3 −1 0 0 0 1 −2 −2 1	$2\,CH_4 = C_2H_6 + H_2$ $CO + 3\,H_2 = CH_4 +{'}H_2O$ $2\,CO + 2\,H_2 = CH_4 + CO_2$	0	2	2	4
18	3 4 6	1 0 −2 1 0 0 0 2 1 −4 0 −1 0 0 −1 2 2 −1	$2\,CH_4 = C_2H_6 + H_2$ $CO_2 + 4\,H_2 = CH_4 + 2\,H_2O$ $CH_4 + CO_2 = 2\,H_2 + 2\,CO$	0	2	2	4
19	3 5 6	2 0 −3 0 −2 1 0 2 −1 0 4 −3 0 0 −1 2 2 −1	$3\,CH_4 + 2\,CO = 2\,C_2H_6 + CO_2$ $CH_4 + 3\,CO_2 = 2\,H_2O + 4\,CO$ $CH_4 + CO_2 = 2\,H_2 + 2\,CO$	2	2	2	6
20	4 5 6	1 0 0 −3 −4 2 0 1 0 −1 1 −1 0 0 1 −2 −2 1	$3\,H_2 + 4\,CO = C_2H_6 + 2\,CO_2$ $H_2 + CO_2 = H_2O + CO$ $2\,H_2 + 2\,CO = CH_4 + CO_2$	4	0	2	6

2.4 EXPRESSION OF THE MASS BALANCE

In a closed system with N constituents and M elements, in which R reactions take place $(R \leq N - H)$, numbers of moles of the individual constituents are not independent variables, as they are linked by relationships which we call mass balance equations. There are two fundamental ways of expressing the mass balance: by means of constitution coefficients or of stoichiometric coefficients. We shall show that the two ways are equivalent under the assumption $R = R_{\max}$.

The mass balance equation constructed by means of the constitution coefficients has the form

$$\sum_{i=1}^{N} a_{ij} n_i = b_j \quad j = 1, 2, \ldots, M , \tag{2.19}$$

where b_j is the overall number of gramatoms of the j-th element in the system and n_i is the number of moles of the i-th constituent. As far as the rank of H is less than M (which is a non-typical case) then $M - H$ equations in the relation (2.19) are linear combinations of the other equations, and these $M - H$ equations may be left out.

By means of stoichiometric coefficients, the mass balance equations are expressed in the form of

$$n_i = n_i^{\circ} + \sum_{r=1}^{R} v_{ri} \xi_r \quad i = 1, 2, \ldots, N , \tag{2.20}$$

where ξ_r is called reaction coordinate and n_i° is the number of moles of the i-th constituent in the initial mixture. We may convince ourselves easily by means of the relation (2.10), that substitution from relations (2.20) into equation (2.19) will satisfy the equations (2.19). The reverse need not be true. This is so, because the dimension of the solution of the set (2.19) is $N - H = R_{\max}$, while the dimension of the set (2.20) is R, where R is in general less or equal to $R_{\max}$. In practical calculations R is always assumed equal to $R_{\max}$, in order to obtain a maximum of information about the system. Detail about the strategy of calculation will be found in Chapter 7.

In this manner a chemical equilibrium is formulated as the problem of simultaneously solving a set of N algebraic equations. This set may be made up of two groups, depending on the nature of the constituents, their numbers of moles in state of equilibrium being unknown quantities. The first group forms H balance relations of the type (2.19). The second, i.e. the equilibrium group, expresses one degree of freedom of the system for every derived constituent, and thus also the possibility for one linearly independent chemical reaction to proceed.

Exercises:

1. Construct matrixes of constitution coefficients for the following systems, and determine their rank:

a) CO, H_2, CH_3OH, $(CH_3)_2O$, H_2O, CH_4, CH_2O

b) C_2H_5OH, H_2S, C_2H_5SH, $(C_2H_5)_2S$, C_2H_4, H_2O, $(C_2H_5)_2O$, C_2H_4O, C_2H_6, H_2

c) CH_4, O_2, C_2H_2, CO, CO_2, H_2O, H_2, C_2H_4, C_3H_4, C_4H_4, C_4H_6

d) SO_2, H_2O, H_2S, CO, CO_2, CH_4, COS, CS_2, S_2

e) C_3H_6, NH_3, N_2, O_2, $CH_2{:}CH.CN$, CH_3CN, HCN, H_2O, CO, CO_2, $CH_2{:}$ $:CH.CO.NH_2$

2. Determine the number of linearly independent reactions in each of the following systems

$$
\begin{aligned}
\text{a)} \quad 2\,C_2H_4 + O_2 &= 2\,C_2H_4O \\
C_2H_4 + 2\,O_2 &= 2\,CO + 2\,H_2O \\
C_2H_4 + 3\,O_2 &= 2\,CO_2 + 2\,H_2O \\
2\,C_2H_4O + 3\,O_2 &= 4\,CO + 4\,H_2O \\
2\,C_2H_4O + 5\,O_2 &= 4\,CO_2 + 4\,H_2O \\
CO + H_2O &= CO_2 + H_2
\end{aligned}
$$

$$
\begin{aligned}
\text{b)} \quad CH_4 + H_2O &= CO + 3\,H_2 \\
CH_4 + 2\,H_2O &= CO_2 + 4\,H_2 \\
CH_4 + CO_2 &= 2\,H_2 + 2\,CO \\
CH_4 + 3\,CO_2 &= 2\,H_2O + 4\,CO \\
CO + H_2O &= CO_2 + H_2
\end{aligned}
$$

$$
\begin{aligned}
\text{c)} \quad 4\,NH_3 + 5\,O_2 &= 4\,NO + 6\,H_2O \\
4\,NH_3 + 7\,O_2 &= 4\,NO_2 + 6\,H_2O \\
4\,NH_3 + 3\,O_2 &= 2\,N_2 + 6\,H_2O \\
4\,NH_3 + 6\,NO &= 5\,N_2 + 6\,H_2O \\
2\,NO + O_2 &= 2\,NO_2 \\
2\,NO &= N_2 + O_2 \\
N_2 + 2\,O_2 &= 2\,NO
\end{aligned}
$$

$$
\begin{aligned}
\text{d)} \quad 2\,CO + S_2 &= 2\,COS \\
2\,COS &= CO_2 + CS_2 \\
CO_2 + H_2S &= COS + H_2O \\
COS + H_2S &= CS_2 + H_2O
\end{aligned}
$$

$$CO + H_2O = CO_2 + H_2$$
$$CO + H_2S = COS + H_2$$

e) $CH_3COOH = CO_2 + CH_4$
$$2(CH_3COOH) = (CH_3COOH)_2$$
$$CH_3COOH = CH_2CO + H_2O$$
$$CH_4 + H_2O = CO + 3H_2$$
$$CO + H_2O = CO_2 + H_2$$
$$CH_4 + CO_2 = 2H_2 + 2CO$$
$$CH_4 + 3CO_2 = 2H_2O + 4CO$$

3. Determine the most advantageous matrix of stoichiometric coefficients for the systems given sub 2., and write down the corresponding sets of stoichiometric equations.

3 Chemical equilibrium of a system

3.1 GENERAL CONSIDERATIONS

The dependence of thermodynamic quantities on variables of state and the number of moles of individual constituents is

$$dU = T\,dS - P\,dV + \sum_{i=1}^{N} \left(\frac{\partial U}{\partial n_i}\right)_{S,V,n_{j\neq i}} dn_i \qquad (3.1a)$$

$$dH = T\,dS + V\,dP + \sum_{i=1}^{N} \left(\frac{\partial H}{\partial n_i}\right)_{S,P,n_{j\neq i}} dn_i \qquad (3.1b)$$

$$dF = -S\,dT - P\,dV + \sum_{i=1}^{N} \left(\frac{\partial F}{\partial n_i}\right)_{T,V,n_{j\neq i}} dn_i \qquad (3.1c)$$

$$dG = -S\,dT + V\,dP + \sum_{i=1}^{N} \left(\frac{\partial G}{\partial n_i}\right)_{T,P,n_{j\neq i}} dn_i \;. \qquad (3.1d)$$

From equations (3.1a-d) follows, that

$$\left(\frac{\partial U}{\partial n_i}\right)_{S,V,n_{j\neq i}} = \left(\frac{\partial H}{\partial n_i}\right)_{S,P,n_{j\neq i}} = \left(\frac{\partial F}{\partial n_i}\right)_{T,V,n_{j\neq i}} = \left(\frac{\partial G}{\partial n_i}\right)_{T,P,n_{j\neq i}} \;. \qquad (3.2)$$

Gibbs called the differential quotients of equation (3.2) the chemical potential μ. In the following, let us only consider the Gibbs function G, which is best suited for expressing equilibrium relationships with respect to pressure and temperature as independent variables. The differential of the Gibbs function may be written in the form of

$$dG = -S\,dT + V\,dP + \sum_{i=1}^{N} \mu_i\,dn_i \;. \qquad (3.3)$$

Let us now consider a closed system, in which for the sake of simplicity a single chemical reaction is taking place. We shall write this reaction in the form

$$\sum_{i=1}^{N} \nu_i A_i = 0 \;, \qquad (3.4)$$

where A_i denotes corresponding compounds in the sense of equation (2.1). The mass balance equation may be expressed in the form (see section 2.4)

$$n_i = n_i^{\circ} + v_i \xi \quad i = 1, 2, \ldots, N \tag{3.5}$$

and thus

$$\mathrm{d}n_i = v_i \, \mathrm{d}\xi . \tag{3.6}$$

At a given temperature T and pressure P, the Gibbs function is solely function of the variable ξ, and there follows from (3.1d) and (3.6)

$$\mathrm{d}G = \left(\sum_{i=1}^{N} v_i \mu_i \right) \mathrm{d}\xi . \tag{3.7}$$

From the differential relationship (3.7) follows

$$\frac{\mathrm{d}G}{\mathrm{d}\xi} = \sum_{i=1}^{N} v_i \mu_i . \tag{3.8}$$

Chemical equilibrium is achieved in the point $\xi = \xi_0$, in which the function $G(\xi)$ attains a minimum, i.e. in the point for which there holds

$$\frac{\mathrm{d}G}{\mathrm{d}\xi} = 0 \tag{3.9}$$

$$\frac{\mathrm{d}^2 G}{\mathrm{d}\xi^2} > 0 . \tag{3.10}$$

In a homogenous system, the inequality (3.10) is always satisfied under the assumption that equation (3.9) is valid (conditions of thermodynamic stability), and the equation

$$\sum_{i=1}^{N} v_i \mu_i = 0 \tag{3.11}$$

will then be an expression of the condition of chemical equilibrium of the system, in which chemical conversion can be characterised by a single chemical reaction. In an ideal mixture of ideal or of real gases, the inequality (3.10) may actually be proved to be true for all values of ξ, for which the relation $n_i = n_i^{\circ} + v_i \xi \geqq 0$, $i = 1, 2, \ldots, N$ hold (see section 3.6). The relatively rare case of a heterogenous gaseous system, in which the inequality (3.10) need not necessarily follow from equation (3.9), shall be excluded from our considerations for the present. Note in relation (3.11), that formally its form is the same as that of the stoichiometric equation (3.4), with the exception that for every constituent the chemical formula is replaced by the chemical potential.

A completely analogue consideration may be used to derive the equilibrium criterium of a system in which R linearly independent reaction are taking place. Expressing them in the shape of

$$\sum_{i=1}^{N} v_{ri}A_i = 0 \quad r = 1, 2, ..., R,$$ (3.12)

then the mass balance equations are expressed by the relations

$$n_i = n_i^o + \sum_{r=1}^{R} v_{ri}\xi_r \quad i = 1, 2, ..., N$$ (3.13)

and, therefore,

$$dn_i = \sum_{r=1}^{R} v_{ri}\,d\xi_r \quad i = 1, 2, ..., N.$$ (3.14)

Substituting relations (3.14) into the equation (3.3) we obtain, for a given temperature and pressure,

$$dG = \sum_{r=1}^{R} \left(\sum_{i=1}^{N} v_{ri}\mu_i \right) d\xi_r,$$ (3.15)

from which there follows

$$\frac{\partial G}{\partial \xi_r} = \sum_{i=1}^{N} v_{ri}\mu_i \quad r = 1, 2, ..., R.$$ (3.16)

Since, in equilibrium, the function $G = G(\xi_1, \xi_2, ..., \xi_R)$ becomes minimum, the following conditions must apply in equilibrium

$$\sum_{i=1}^{N} v_{ri}\mu_i = 0 \quad r = 1, 2, ..., R$$ (3.17)

and

$$\sum_{s=1}^{R} \sum_{p=1}^{R} \frac{\partial^2 G}{\partial \xi_s \, \partial \xi_p} h_s h_p > 0$$ (3.18)

for every vector $(h_1, h_2, ..., h_R)$, $\sum_{i=1}^{R} h_i^2 \neq 0$. The inequality (3.18) is abbreviated thus:

$$d^2G > 0.$$ (3.19)

In respect of the relationship between equation (3.17) and inequality (3.18), the system may be stated as for the system in which a single chemical reaction takes place.

3.2 THE EQUILIBRIUM CONSTANT AND ΔG° OF A REACTION

Since the following relationship applies between the chemical potential and activity

$$\mu_i = \mu_i^\circ + RT \ln a_i \tag{3.20}$$

where μ_i° is the chemical potential of the i-th constituent in the standard state, the equilibrium criterion expressed by equation (3.9) may be modified to read

$$\frac{\mathrm{d}G}{\mathrm{d}\xi} = \sum_{i=1}^{N} v_i \mu_i = \sum_{i=1}^{N} v_i \mu_i^\circ + RT \ln \prod_{i=1}^{N} a_i^{v_i} = 0 . \tag{3.21}$$

Denoting

$$K_a = \prod_{i=1}^{N} a_i^{v_i} , \tag{3.22}$$

$$\Delta G^\circ = \sum_{i=1}^{N} v_i \mu_i^\circ \tag{3.23}$$

and

$$\Delta G = \frac{\mathrm{d}G}{\mathrm{d}\xi} , \tag{3.24}$$

the equilibrium criterion may be rewritten in the form of

$$\Delta G = \Delta G^\circ + RT \ln K_a = 0 \tag{3.25}$$

or,

$$\Delta G^\circ = -RT \ln K_a \tag{3.26}$$

where K_a is called equilibrium constant of the reaction and ΔG^O the standard change of free enthalpy. In equilibrium, concentrations of individual constituents in a reaction mixture stabilize in such a manner that the expression composed of activities of the right-hand side of the relation (3.22) remains unchanged, irrespective of the initial composition of the system. This is so, because the value of ΔG° is independent of composition, and is constant at a given temperature and pressure.

This consideration may again be extended to systems in which R independent reactions take place. Equation (3.17) must apply to each of them

$$\frac{\partial G}{\partial \xi_r} = \sum_{r=1}^{N} v_{ri} \mu_i = \sum_{i=1}^{N} v_{ri} \mu_i^\circ + RT \ln \prod_{i=1}^{N} a_i^{v_{ri}} = 0 \quad r = 1, 2, \ldots, R . \tag{3.27}$$

Denoting, similarly

$$(K_a)_r = \prod_{i=1}^{N} a_i^{v_{ri}} \quad r = 1, 2, \ldots, R \tag{3.28}$$

$$\left(\Delta G^\circ\right)_r = \sum_{i=1}^{N} \nu_{ri}\mu_i^\circ \quad r = 1, 2, ..., R \tag{3.29}$$

$$\left(\Delta G\right)_r = \frac{dG}{d\xi_r} \quad r = 1, 2, ..., R \tag{3.30}$$

the equilibrium criteria may be rewritten in the form

$$\left(\Delta G\right)_r = \left(\Delta G^\circ\right)_r + RT \ln \left(K_a\right)_r = 0 \quad r = 1, 2, ..., R \tag{3.31}$$

or,

$$\left(\Delta G^\circ\right)_r = -RT \ln \left(K_a\right)_r \quad r = 1, 2, ..., R \tag{3.32}$$

where $\left(\Delta G^\circ\right)_r$ is the standard change of free enthalpy of the r-th reaction and $(K_a)_r$ is the equilibrium constant of the r-th reaction.

3.3 SELECTION OF THE STANDARD STATE

Relations (3.22) and (3.28) apply quite in general. However, the selection of the standard state must always be defined in order that it should be possible to express the equilibrium composition by means of convential quantities of concentration. According to the definition of activity

$$a_i = \frac{f_i}{f_i^\circ} \tag{3.33}$$

where f_i is the fugacity of the i-th constituent in the mixture and f_i° is the fugacity of the i-th constituent in the standard state, the numerical value of activity for each participating constituent depends on its fugacity in the standard state. This relationship is also transferred to the equilibrium constant, the value of which is useless without a definition of the standard state.

In principle, there is no limit to the selection of standard states. For practical reasons it is advantageous to make use of one of the following possibilities.

a) For gaseous constituents in nearly ideal systems it is best to select for the standard state so-called unit fugacities, i.e. the pure constituent in the state of an ideal gas, at the temperature of the system and pressure of 1 atm. Then, there applies

$$a_i = \frac{f_i}{1} = \varphi_i y_i P = \varphi_i P \frac{n_i}{n} \quad i = 1, 2, ..., N , \tag{3.34}$$

where φ_i is the fugacity coefficient of the i-th constituent in the mixture and y_i is the molar fraction of the i-th constituent. In ideal systems there hold

$$a_i = y_i P = P_i \quad i = 1, 2, ..., N , \tag{3.35}$$

where P_i is the partial pressure of the i-th constituent.

On substituting the relation (3.35) into the expression for the equilibrium constant, we obtain

$$(K_a)_r = \prod_{i=1}^{N} a_i^{\nu_{ri}} = \prod_{i=1}^{N} f_i^{\nu_{ri}} = P^{\nu_r} \prod_{i=1}^{N} \varphi_i^{\nu_{ri}} \prod_{i=1}^{N} y_i^{\nu_{ri}} =$$

$$= \left(\frac{P}{n}\right)^{\nu_r} \prod_{i=1}^{N} \varphi_i^{\nu_{ri}} \prod_{i=1}^{N} n_i^{\nu_{ri}} = (K\varphi)_r \cdot \left(\frac{P}{n}\right)^{\nu_r} \prod_{i=1}^{N} n_i^{\nu_{ri}} \qquad r = 1, 2, \ldots, R, \qquad (3.36)$$

where the term

$$(K\varphi)_r = \prod_{i=1}^{N} \varphi_i^{\nu_{ri}} \qquad r = 1, 2, \ldots, R, \qquad (3.37)$$

includes a correction for non-ideal behaviour of the gas mixture. For ideal gaseous systems these expressions are reduced to

$$(K_a)_r = \left(\frac{P}{n}\right)^{\nu_r} \prod_{i=1}^{N} n_i^{\nu_{ri}} = \prod_{i=1}^{N} P_i^{\nu_{ri}} \qquad r = 1, 2, \ldots, R, \qquad (3.38)$$

where $\nu_r = \sum_{i=1}^{N} \nu_{ri}$ is the change of the mole number of the r-th reaction. With a standard state thus selected, the value of ΔG° (and therefore also the equilibrium constant) is independent of pressure.

b) In strongly non-ideal systems the pure constituent at the temperature and pressure of the system is selected as standard state of the system. There applies

$$a_i = \frac{f_i}{f_i^\circ} = \frac{\varphi_i y_i P}{\varphi_i^\circ P} = \frac{\varphi_i}{\varphi_i^\circ} \cdot \frac{n_i}{n} \qquad i = 1, 2, \ldots, N. \qquad (3.39)$$

where φ_i° is the fugacity coefficient of the pure constituent at the temperature and pressure of the system. The equilibrium constant may be expressed by the expressions

$$(K_a)_r = \prod_{i=1}^{N} a_i^{\nu_{ri}} = \prod_{i=1}^{N} \left(\frac{\varphi_i}{\varphi_i^\circ}\right)^{\nu_{ri}} \left(\frac{1}{n}\right)^{\nu_r} \prod_{i=1}^{N} n_i^{\nu_{ri}} \qquad r = 1, 2, \ldots, R. \qquad (3.40)$$

For an ideal mixture of real gases $\left(f_i = y_i f_i^\circ\right)$ we have $\varphi_i = \varphi_i^\circ$. There hold the relation

$$(K_a)_r = \left(\frac{1}{n}\right)^{\nu_r} \prod_{i=1}^{N} n_i^{\nu_{ri}} = \prod_{i=1}^{N} y_i^{\nu_{ri}} \qquad r = 1, 2, \ldots, R. \qquad (3.41)$$

c) Sometimes, particularly in connection with chemical kinetics, the standard state employed is that of the pure constituent in the ideal gas state at the temperature of the system and molarity equal to one. The following applies,

$$f_i = \varphi_i y_i P = \varphi_i y_i \frac{nRT}{V} = \varphi_i RT \frac{n_i}{V} = \varphi_i RT c_i$$

$$f_i^\circ = RT, \qquad (3.42)$$

where c_i is the concentration of the i-th constituent. The equilibrium constant can be expressed in one of the following forms

$$(K_a)_r = \left(\frac{1}{RT}\right)^{v_r} \prod_{i=1}^{N} f_i^{v_{ri}} = \prod_{i=1}^{N} \varphi_i^{v_{ri}} \prod_{i=1}^{N} c_i^{v_{ri}} = (K\varphi)_r \prod_{i=1}^{N} c_i^{v_{ri}} \quad r = 1, 2, \ldots, R. \quad (3.43)$$

As long as the gas mixture behaves in the ideal manner, we have

$$(K_a)_r = \prod_{i=1}^{N} c_i^{v_{ri}} = \left(\frac{1}{RT}\right)^{v_r} \prod_{i=1}^{N} P_i^{v_{ri}} \quad r = 1, 2, \ldots, R. \quad (3.44)$$

Note: There follows from relations (3.28) and (3.33), that the equilibrium constant is a dimension-less quantity, depending solely on the choice of the standard state. The relationship a) would make it appear that the dimension of the equilibrium constant K_a is that of pressure. This is not true, because the value of f_i° is replaced by 1 atm for the case that the standard state of unit fugacity is chosen.

Example 7

It was found by experimental means in the dehydrogenation of ethane to ethylene, which proceeds according to the equation

$$C_2H_6(g) \;=\; C_2H_4(g) + H_2(g)$$

that the equilibrium mixture contains 17 molars % ethylene at 873 K and 0.84 atm. Assuming ideal behaviour in all constituents present, calculate the equilibrium constant of the reaction for these three standard states:

a) pure constituent at unit fugacity
b) pure constituent at the temperature and pressure of the system
c) pure constituent at unit molarity

a) $\quad K_a = \dfrac{P_{H_2} P_{C_2H_4}}{P_{C_2H_6}} = \dfrac{y_{H_2} y_{C_2H_4}}{y_{C_2H_6}} P = \dfrac{0.17 \times 0.17}{0.66} \times 0.84 = 3.68 \times 10^{-2}$

b) $\quad K_a = \dfrac{y_{H_2} y_{C_2H_4}}{y_{C_2H_6}} = \dfrac{0.17 \times 0.17}{0.66} = 4.38 \times 10^{-2}$

c) $\quad K_a = \dfrac{P_{H_2} P_{C_2H_4}}{P_{C_2H_6}} \dfrac{1}{RT} = \dfrac{y_{H_2} y_{C_2H_4}}{y_{C_2H_6}} \dfrac{P}{RT} = \dfrac{0.17 \times 0.17 \times 0.84}{0.66 \times 0.082 \times 873} = 5.14 \times 10^{-4}.$

The standard state of unit fugacity is prevalently used for chemical equilibria of gas mixtures, since in this case the equilibrium constant depends on temperature only. As long as the standard state will not be specified in the following text, the standard state of unit fugacity will always be used.

3.4 DEPENDENCE OF THE EQUILIBRIUM CONSTANT ON VARIABLES OF STATE

3.4.1 Dependence on temperature

The temperature coefficient of the equilibrium constant

The relationship between temperature and the equilibrium constant can be expressed from equation (3.26), which re-write in the form of

$$\frac{\Delta G^{\circ}}{T} = -R \ln K_a . \tag{3.45}$$

The temperature coefficient of the function $\Delta G^{\circ}/T$ can then be expressed as

$$\left(\frac{\partial\left(\dfrac{\Delta G^{\circ}}{T}\right)}{\partial T}\right)_P = \frac{T\left(\dfrac{\partial \,\Delta G^{\circ}}{\partial T}\right)_P - \Delta G^{\circ}}{T^2} = \frac{-T\,\Delta S^{\circ} - \Delta G^{\circ}}{T^2} = -\frac{\Delta H^{\circ}}{T^2} . \tag{3.46}$$

Evidently, equation (3.46) converts to the form

$$\left(\frac{\partial \ln K_a}{\partial T}\right)_P = \frac{\Delta H^{\circ}}{RT^2} \tag{3.47}$$

which is usually called van't Hoff reaction isobar. In analogy, the following relation can be derived for changes at constant volume

$$\left(\frac{\partial \ln K_a}{\partial T}\right)_V = \frac{\Delta U^{\circ}}{RT^2} , \tag{3.48}$$

where ΔU° is the standard change of internal energy.

Dependence of the equilibrium constant on temperature at $\Delta H^{\circ} = const.$

In order to elucidate the temperature dependence of the equilibrium constant between two temperatures, equation (3.47) must be integrated. To this purpose an assumption must be made about the dependence of the heat of reaction on temperature. When it is required to convert the equilibrium constant in a relatively narrow temperature range, it may be assumed that ΔH° is independent of temperature. Limited integration of equation (3.47) then leads to

$$\ln \frac{(K_a)_{T_2}}{K_{(a)}_{T_1}} = -\frac{\Delta H^{\circ}}{R}\left(\frac{1}{T_2} - \frac{1}{T_1}\right). \tag{3.49}$$

while unlimited integration under the same assumption gives

$$\ln{(K_a)_T} = -\frac{\Delta H°}{RT} + C,$$ (3.50)

where C is an integration constant.

Relations (3.49) and (3.50) show, that in this case $\log K_a$ is linear function of reciprocal absolute temperature

$$\log{(K_a)_T} = \frac{B}{T} + C',$$ (3.51)

where B and C' are constants.

Equation (3.51) is very useful, since it can be used in many cases for graphic interpolation of equilibrium constant values. Relatively good agreement is evident from Fig. 1, in which experimentally determined equilibrium constants of methanol synthesis[111] are plotted, which in a temperature interval of 553 to 611 K satisfy the relationship

$$\log{(K_a)_T} = \frac{6691}{T} - 15.42.$$ (3.52)

Very frequently, the reverse problem must be solved, i.e. the heat of reaction is to be determined from the temperature dependence of the equilibrium constant. The graphic procedure is used with advantage, the heat of reaction being obtained from the relation

$$\Delta H° = -2.303RB$$ (3.53)

where B is the slope of the straight line corresponding to equation (3.51). Although we have reason to fear that the calculated values may be subject to a great error as they are obtained by derivating a measured quantity, this method is to be recommended particularly for those cases, where measurement of the equilibrium is more accurate than direct calorimetric measurement or, in cases where the reaction is not accessible to calorimetric measurement.

Dependence of the equilibrium constant on temperature at $\Delta H° = f(T)$

When it is required to calculate the temperature dependence of the equilibrium constant in a wider temperature range, the temperature dependence of the heat of reaction must be taken into account when integrating the equation (3.47).

The temperature dependence of the heat of reaction is expressed by the Kirchhoff equation

$$\left(\frac{\partial \Delta H°}{\partial T}\right)_P = \sum_{i=1}^{N} v_i C_{P_i}°,$$ (3.54)

40

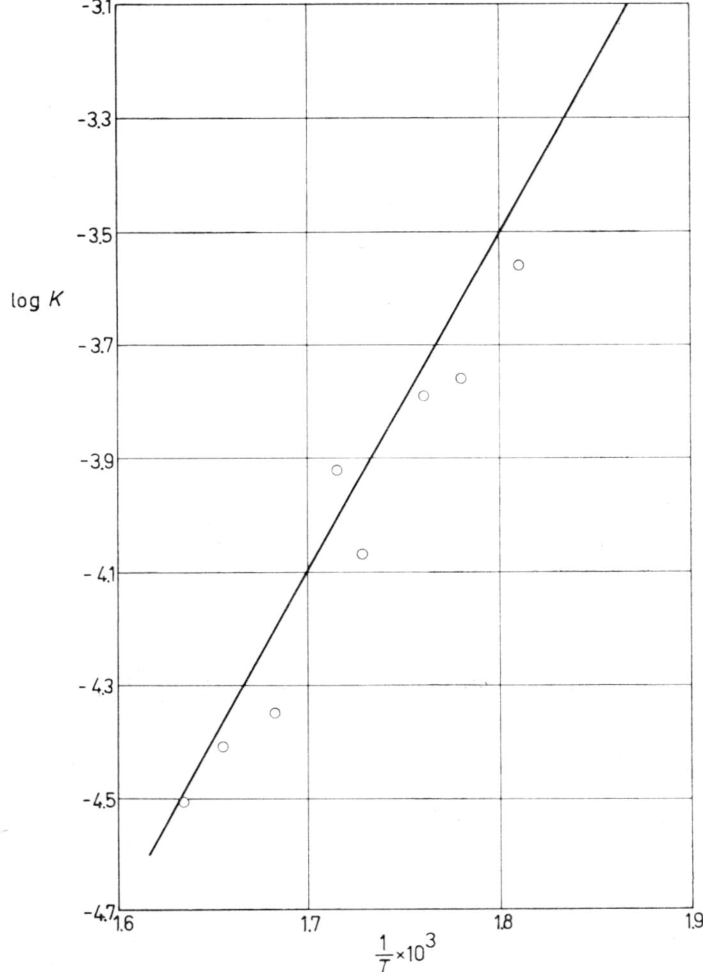

Fig. 1. Dependence of equilibrium constant of methanol synthesis on temperature in the 553 to 611 K range.

(Points — experimental values, curve — dependence according to equation (3.52)).

where $C^{\circ}_{P_i}$ is the molar heat of the i-th constituent in the standard state. Substituting relation (3.54) into equation (3.47) we obtain

$$\ln (K_a)_T = \int \frac{1}{RT^2} \left[\int (\sum_{i=1}^{N} v_i C^{\circ}_{P_i}) \, dT + I_H \right] dT + \ln I_K , \qquad (3.55)$$

where I_H and I_K are integration constants. The equation can be calculated numerically if the molar heats of all constituents are known. When the temperature range is not too broad, the temperature dependence of the molar heats can be expressed rather

41

well by means of the empirical second-power relation

$$C_{P_i}^{o} = a_i + b_i T + c_i T^2 \quad i = 1, 2, \ldots, N \tag{3.56}$$

where a_i, b_i, c_i are constants.

Then,

$$\Delta H_T^o = \int (a + bT + cT^2)\, dT + I_H ,$$

$$\Delta H_T^o = aT + \tfrac{1}{2}bT^2 + \tfrac{1}{3}cT^3 + I_H , \tag{3.57}$$

where

$$a = \sum_{i=1}^{N} v_i a_i ,$$

$$b = \sum_{i=1}^{N} v_i b_i ,$$

$$c = \sum_{i=1}^{N} v_i c_i .$$

Substituting from (3.57) into (3.47) and integrating, we arrive at

$$R \ln (K_a)_T = a \ln T + \frac{1}{2} bT + \frac{1}{6} cT^2 - \frac{I_H}{T} + R \ln I_K . \tag{3.58}$$

For practical application of equation (3.58) we must finally determine the constants I_H and I_K, to which purpose either the equilibrium constants must be known at two temperatures or, at least at one temperature and additionally one value of the heat of reaction.

The procedure will be seen from the following example:

Example 8

When methane is decomposed with steam, the following reaction takes place among others

$$CO(g) + H_2O(g) = CO_2(g) + H_2(g)$$

Determine the dependence of the equilibrium constant on temperature and calculate the equilibrium constant of this reaction at 1000 K from the following data:

$(\Delta H^o)_{700K} = -9051$ cal

$(K_a)_{700K} = 9.0188$

$CO : C_P^o = 6.34 + 1.84 \times 10^{-3}T - 2.8 \times 10^{-7}T^2 \ (\text{cal K}^{-1} \text{ mol}^{-1})$

$H_2O : C_P^o = 7.19 + 2.37 \times 10^{-3}T + 2.08 \times 10^{-7}T^2 \ (\text{cal K}^{-1} \text{ mol}^{-1})$

$$CO_2 : C_P^o = 6.34 + 10.19 \times 10^{-3}T - 35.33 \times 10^{-7}T^2 \; (\text{cal K}^{-1} \text{mol}^{-1})$$

$$H_2 : C_P^o = 6.95 - 0.20 \times 10^{-3}T - 4.80 \times 10^{-7}T^2 \; (\text{cal K}^{-1} \text{mol}^{-1}) \,.$$

According to equation (3.57):

$$\Delta H_T^o = -0.19T + 2.89 \times 10^{-3}T^2 - 13.14 \times 10^{-7}T^3 + I_H$$

$$I_H = -9051 + 0.19 \times 700 - 2.89 \times 10^{-3} \times 700^2 + 13.14 \times 10^{-7} \times 700^3$$

$$I_H = -9883.4$$

and from equation (3.58):

$$R \ln (K_a)_{700} = 0.19 \times 2.303 \times 2.84510 + 2.89 \times 10^{-3} \times 700 -$$

$$- 6.52 \times 10^{-7} \times 700^2 - \frac{9883.4}{700} + R \ln I_K$$

$$R \ln I_K = 18.03191 \,.$$

The general temperature dependence of the equilibrium constant is:

$$\log (K_a)_T = -0.096 \log T + 0.631 \times 10^{-3}T - 1.4246 \times 10^{-7}T^2 -$$

$$- \frac{2159.5}{T} + 3.93997 \,.$$

Then,

$$\log (K_a)_{1000K} = 1.98101 \,,$$

$$(K_a)_{1000K} = 95.73 \,.$$

Validity of equation (3.58) is limited to the interval in which the empirical relationship for molar heat are valid, the precision of the correlation of molar heat values being decisive for the precision of calculated equilibrium constant data. For practical purposes this precision is usually satisfactory. For illustration, Fig. 2 shows the course of the temperature dependence of equilibrium constants for some reactions which are of significance in industry.

In cases where experimental data are available for the equilibrium constant at different temperatures, the temperature coefficient of the equilibrium constant can be calculated directly by solving a corresponding number of equations. This is best done by means of a relationship which is identical with equation (3.58)

$$\log (K_a)_T = A_1 + \frac{A_2}{T} + A_3 \log T + A_4 T + A_5 T^2 \,, \tag{3.59}$$

where A_i are constants.

Where more than five equilibrium constant data are available, it is recommended to calculate the coefficients by means of one of the statistical procedures (e.g. the least squares methods, etc.) using all the data available, rather than to use higher polynomials. On the other hand, with less than five data one of the simplified forms of equation (3.59) is used, in which the respective coefficient (e.g. A_5) is eliminated. Note, however, that on gradual simplification of equation (3.59) it is generally

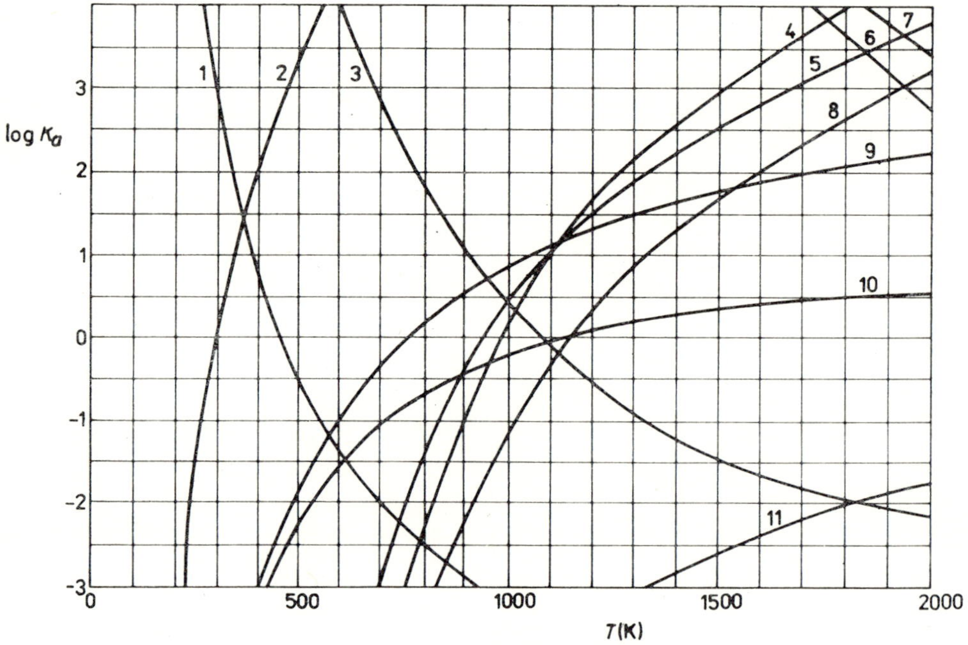

Fig. 2. Temperature dependence of equilibrium constant of several reactions that have technological significance.

1. $\frac{1}{2}N_2 + \frac{3}{2}H_2 = NH_3$
2. $N_2O_2 = 2NO$
3. $SO_2 + \frac{1}{2}O_2 = SO_3$
4. $CO_2 + C = 2CO$
5. $C + H_2O = CO + H_2$
6. $CO + \frac{1}{2}O_2 = CO_2$
7. $H_2 + \frac{1}{2}O_2 = H_2O$
8. $CaCO_3 = CaO + CO_2$
9. $NO_2 = NO + \frac{1}{2}O_2$
10. $H_2 + CO_2 = H_2O + CO$
11. $\frac{1}{2}N_2 + \frac{1}{2}O_2 = NO$.

necessary to narrow down the temperature range of its validity in order to maintain the required degree of precision.

3.4.2 Dependence on pressure

The dependence of the equilibrium constant is a function of the choice of the standard state. For the conventionally employed standard state of unit fugacity,

44

the equilibrium constant is independent of pressure, as the value of $\Delta G^\circ = \sum_{i=1}^{N} v_i \mu_i^\circ$ depends solely on temperature in the case of this standard state.

Let us now discuss the pressure dependence of the equilibrium constant in cases where the standard state selected is the pure constituent at the temperature and pressure of the system. From relation (3.22) follows

$$\left(\frac{\partial \Delta G^\circ}{\partial P} \right)_T = \Delta V^\circ ,$$

$$\Delta V^\circ = \sum_{i=1}^{N} v_i \bar{v}_i , \tag{3.60}$$

where $\bar{v}_i$ is the partial molar volume of the i-th constituent. Substitution from relation (3.60) into relation (3.26) gives the following expression for the pressure dependence of the equilibrium constant

$$\left(\frac{\partial \ln K_a}{\partial P} \right)_T \approx - \frac{\Delta V^\circ}{RT} . \tag{3.61}$$

Let us assume that the expression $\bar{v}_i \approx v_i^\circ = RT/P$ applies for the partial molar volume of the i-th constituent, v_i° being the molar volume of the pure constituent at the temperature and pressure of the system. Substituting this approximation into relation (3.60) we obtain

$$\left(\frac{\partial \ln K_a}{\partial P} \right)_T \approx - \frac{v}{P} , \tag{3.62}$$

where $v = \sum_{i=1}^{N} v_i$, is usually called change of the molar number of the reaction involved.

3.5 DEPENDENCE OF THE EQUILIBRIUM CONSTANT ON THE ABSOLUTE MAGNITUDE OF THE STOICHIOMETRIC COEFFICIENTS

Let us assume a closed system in which one chemical reaction is taking place

$$\sum_{i=1}^{N} v_i A_i = 0 . \tag{3.63}$$

In order to be able to elucidate the influence of the absolute magnitude of the stoichiometric coefficients on the value of the equilibrium constant we re-write equation (3.63) in the form

$$\sum_{i=1}^{N} (\alpha v_i) A_i = 0 , \tag{3.64}$$

where α is an arbitrary non-zero number. Denoting the standard change of free enthalpy of the reactions (3.63) and (3.64) ΔG_1^o and ΔG_2^o, resp. there holds

$$\Delta G_2^o = \alpha \, \Delta G_1^o \tag{3.65}$$

and, therefore,

$$K_2 = K_1^\alpha , \tag{3.66}$$

where K_1 and K_2 is the equilibrium constant of the reactions (3.63) and (3.64), resp.

Example 9

For ammonia synthesis, taking place according to the reaction

$$\tfrac{1}{2} N_2(g) + \tfrac{3}{2} H_2(g) \;=\; NH_3(g)$$

an equilibrium constant value of 0.0090 is stated in the literature[5] for 700 K at the standard state of unit fugacity. Determine the equilibrium constant of the following reaction for the same temperature

$$2 NH_3(g) \;=\; N_2(g) + 3 H_2(g) .$$

The coefficient α is equal to -2 and, therefore,

$$K_2 = K_1^{-2} = (0.009)^{-2} \approx 0.123 \times 10^5 .$$

3.6 PROPERTIES OF THE FREE ENTHALPY OF A SYSTEM

From the analysis presented in section 3.1 follows that the shift of a closed single-phase gaseous system from the initial to the final state, as described by the respective stoichiometric relations, can be expressed by means of the value of overall free enthalpy, the minimum point of this function corresponding to the required state of chemical equilibrium.

Let us assume an ideal system of ideal gases, in which only one chemical reaction is taking place

$$\sum_{i=1}^{N} v_i A_i = 0 . \tag{3.67}$$

The mass balance equation will take the form

$$n_i = n_i^o + v_i \xi . \tag{3.68}$$

Let us assume, without loss of generality, that constituent number one is the key constituent, i.e. the one which would be the first to react completely in the system. Let us define the degree of conversion x by means of the expression

$$x = \frac{n_1^o - n_1}{n_1^o} . \qquad (3.69)$$

Then, there evidently holds that

$$n_i = n_i^o - n_1^o \frac{v_i}{v_1} x . \qquad (3.70)$$

Substitution leads to

$$-\frac{n_1^o}{v_1} x = \frac{n_1^o}{|v_1|} x = \xi \qquad (3.71)$$

i.e. back to the relation (3.68). The advantage of introducing the quantity x consists in the fact, that the degree of conversion always lies in the interval $\langle 0, 1 \rangle$, since from relation (3.69) follows

$$n_1 = n_1^o(1 - x) . \qquad (3.72)$$

Let us now turn to the function $G = G(x)$. Employing the standard state of unit fugacity, we have

$$G = \sum_{i=1}^{N} n_i \mu_i = \sum_{i=1}^{N} n_i(\mu_i^o + RT \ln (y_i P)) = \sum_{i=1}^{N} n_i G_i^* + RT \sum_{i=1}^{N} n_i \ln \left(\frac{n_i}{n}\right), \quad (3.73)$$

where $G_i^* = G_i^o + RT \ln P = \mu_i^o + RT \ln P$ is the molar free enthalpy of the pure i-th constituent in the state of the ideal gas at the temperature and pressure of the system. We shall now write relation (3.73) in the form of

$$G = \sum_{i=1}^{N} n_i G_i^* + G^M , \qquad (3.74)$$

G^M being the free enthalpy of mixing. The term G^M is always negative (we stress the assumption of an ideal system). Substitution of the relation (3.68) into equation (3.74) gives expressions for the first and second derivative of the Gibbs function

$$\frac{dG}{d\xi} = \sum_{i=1}^{N} v_i G_i^* + RT \sum_{i=1}^{N} v_i \ln \left(\frac{n_i}{n}\right) \qquad (3.75)$$

$$\frac{d^2 G}{d\xi^2} = \frac{d^2 G^M}{d\xi^2} = RT \left(\sum_{i=1}^{N} \frac{v_i^2}{n_i} - \frac{v^2}{n} \right) > 0 , \qquad (3.76)$$

where $v = \sum_{i=1}^{N} v_i$. Proof of the validity of the inequality contained in the relation

(3.76) is presented in Appendix 3 for all values of ξ which satisfy the relation $n_i \equiv$ $\equiv \left(n_i^o + v_i\xi\right) > 0$ $i = 1, 2, \ldots, N$. From the substitution (3.71) follows

$$\frac{\mathrm{d}G}{\mathrm{d}x} = \frac{n_1^o}{|v_1|}\frac{\mathrm{d}G}{\mathrm{d}\xi}\,, \tag{3.77}$$

$$\frac{\mathrm{d}^2G}{\mathrm{d}x^2} = \left(\frac{n_1^o}{|v_1|}\right)^2\frac{\mathrm{d}^2G}{\mathrm{d}\xi^2} > 0\,. \tag{3.78}$$

By combination of relation (3.78) with the validity of the limits

$$\lim_{x \to 0_+}\frac{\mathrm{d}G}{\mathrm{d}x} = -\infty \quad \lim_{x \to 1_-}\frac{\mathrm{d}G}{\mathrm{d}x} = \infty \tag{3.79}$$

we obtain an important property of the function $G = G(x)$ as well as $G = G(\xi)$. In an ideal system of ideal gases, in which one chemical reaction is taking place, the

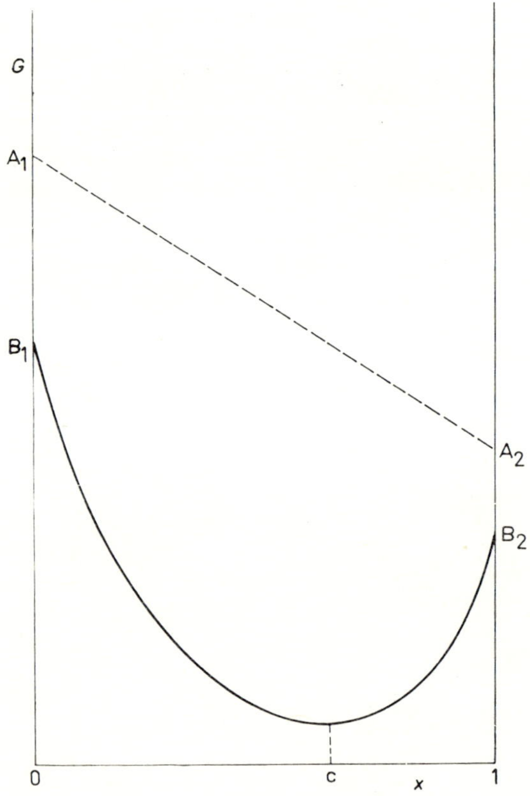

Fig. 3. Course of the overall free enthalpy of a system, in which a single chemical reaction is taking place, in dependence on the number of moles of the initial constituent 1.

48

function G is a convex function of the variable x or ξ, which for $x \in (0, 1)$ has precisely one minimum.

The first term on the right-hand side of equation (3.74) gives the value of free enthalpy of the system before mixing of the constituents. The second term expresses the decrease of free enthalpy of the system, caused by mixing of the free constituents. The free enthalpy of the system of initial constituents before their mixing $\left(\sum_{i=1}^{N} n_i^o G_i^* \right)$ is represented by point A_1 in Fig. 3. When these initial constituents are mixed (the reaction proper has not yet started), the free enthalpy of the system decreases to point $B_1 \left(\sum_{i=1}^{N} n_i^o G_i^* + RT \sum_{i=1}^{N} n_i^o \ln \left(n_i^o / n^o \right) \right)$. Then the chemical reaction takes place and the free enthalpy of the closed system continues to decrease, until it achieves a minimum value in the point $x = C$ and chemical equilibrium is established. A similar procedure may be followed starting from the system of reaction products. As long as there is only one initial constituent $(n_1^o \neq 0, n_2^o = \ldots = = n_N^o = 0)$ point A_1 is identical with point B_1. Similarly, with only one reaction products, A_2 becomes identical with B_2.

The above consideration may be extended to systems in which R linearly independent reactions are taking place. In this case the free enthalpy of the system is a function of R variables $\xi_1, \xi_2, \ldots, \xi_R$. Convexity of the function $G = G(\xi_1, \xi_2, \ldots, \xi_R)$ in an ideal gas system is proved in Appendix 6 on a set Ω formed by all points $(\xi_1, \xi_2, \ldots, \ldots, \xi_R)$ for which the relation

$$n_i = n_i^o + \sum_{r=1}^{R} v_{ri} \xi_r > 0 \quad i = 1, 2, \ldots, N \tag{3.80}$$

applies.

3.7 PROCEDURE OF CALCULATING A CHEMICAL EQUILIBRIUM

The preceding sections show that there are in principle two approaches to the solution of the chemical equilibrium of a system.

The first, up to now more conventional approach sets out from a description of the overall chemical conversion of the given system by means of a set of chemical reactions, and the corresponding standard changes of free enthalpy or equilibrium constants. Assuming R linearly independent reactions and ideal behaviour of the gas mixture, this procedure converts to the solution of a non-linear set of R equations for R unknown variables $\xi_1, \xi_2, \ldots, \xi_R$

$$(K_a)_r = \left(\frac{P}{n} \right)^{v_r} \prod_{i=1}^{N} n_i^{v_{ri}} \quad r = 1, 2, \ldots, R \tag{3.81}$$

where we substitute for n_i

$$n_i = n_i^{\circ} + \sum_{r=1}^{R} v_{ri}\xi_r \quad i = 1, 2, \ldots, N$$

$$n = \sum_{i=1}^{N} n_i = n^{\circ} + \sum_{r=1}^{R} v_r\xi_r \, . \tag{3.82}$$

In the second approach, the system is considered as a whole, its thermodynamic properties being expressed by the overall free enthalpy of the system. Calculation of the chemical equilibrium is then equivalent to the problem of finding the minimum of free enthalpy of the system in dependence on the variables ξ_r or n_i maintaining validity of the balance relations (2.20) and (2.19).

It can of course be proved easily that the two methods are equivalent. From relation (3.82) follows

$$dn_i = \sum_{r=1}^{R} v_{ri} \, d\xi_r \quad i = 1, 2, \ldots, N \tag{3.83}$$

for the total differential of overall free enthalpy of the system holds

$$dG = \sum_{i=1}^{N} \mu_i \, dn_i \, . \tag{3.84}$$

Substituting from equation (3.83) into relation (3.84) and inverting the order of addition we obtain

$$dG = \sum_{r=1}^{R} \left(\sum_{i=1}^{N} \mu_i v_{ri} \right) d\xi_r \, . \tag{3.85}$$

The expression in brackets is quite simply ΔG_r, whereby relation (3.85) converts to

$$dG = \sum_{r=1}^{R} \Delta G_r \, d\xi_r \, . \tag{3.86}$$

Since the equation $dG = 0$ must apply when equilibrium is established, all values of ΔG_r must likewise be zero and conversely.

Exercises

1. The following types of chemical reactions are given:

$$
\begin{aligned}
\text{a)} \quad & -A_1 + A_2 = 0 \\
\text{b)} \quad & -A_1 + 2A_2 = 0 \\
\text{c)} \quad & -2A_1 + A_2 = 0 \\
\text{d)} \quad & -A_1 - A_2 + A_3 = 0 \\
\text{e)} \quad & -A_1 - A_2 + 2A_3 = 0 \\
\text{f)} \quad & -A_1 - A_2 + A_3 + A_4 = 0 \, .
\end{aligned}
$$

Assuming, for the sake of simplicity, that in all cases reactions at 500 K will be accompanied by a change in standard free enthalpy equal to -1000 cal mol^{-1}, plot for every reaction the course of free enthalpy of the system in dependence on the loss of initial constituents, and show that this relationship is of convex character, having a minimum identical with that which can be calculated from the relation

$$RT \ln K_a = -\Delta G^{\circ}.$$

Assume in the calculation that the initial constituents are ideal gases in the standard state.

2. Consider the following reactions:

$$-A_1 - 2 A_2 + A_3 = 0$$

$$-A_1 - 3 A_2 + 2 A_3 = 0.$$

How great must be the change of standard free enthalpy in the two cases for the conversion at least 10% at 800 K:

    a)   at a pressure of 1 atm

    b)   at a pressure of 450 atm .

3. Calculate the equilibrium constant of the reation

$$SO_2(g) + \tfrac{1}{2} O_2(g) = SO_3(g)$$

at 1000 K and 1 atm, if all constituents exhibit ideal behaviour and the following data are known for them:

Consti- tuent	$\Delta H_f^{\circ}(298.15\ \mathrm{K})$ (cal mol^{-1})	$S^{\circ}(298.15\ \mathrm{K})$ (cal $\mathrm{K}^{-1}\ \mathrm{mol}^{-1}$)	$C_P^{\circ} = a + bT + cT^2$		
			a	$b \times 10^3$	$c \times 10^6$
SO_2	$-70\,960$	59.40	7.116	9.512	-3.511
O_2	0	49.00	6.095	3.253	-1.017
SO_3	$-94\,450$	61.24	6.077	23.537	-9.687

4. For the case of hydrogenation of ethylene to ethane according to the equation

$$C_2H_4(g) + H_2(g) = C_2H_6(g)$$

the dependence of the equilibrium constant on temperature in the temperature range of 300 to 1000 K was found to be

$$\log K_a = \frac{6806}{T} - 2.981 \log T + 7.668 \times 10^{-4}T - 1.764 \times 10^{-7}T^2 + 1.81 \,.$$

Calculate: a) the temperature dependence of the heat of reaction in the temperature range stated,

b) the value of the heat of reaction at 800 K.

4 Chemical equilibrium of simple systems in the ideal gas state

4.1 REACTION COORDINATE

It was mentioned already in the preceding sections, that the initial state of every closed system is unambigously determined by the number of moles of the initial constituents. This determines the overall number of gramatoms of the individual basic constituents, which must remain constant in the course of chemical conversions (provided no nuclear processes take place).

The linkage between individual constituents can then be characterized by a newly selected quantity ξ_r for every linearly independent reaction: this is called reaction coordinate (also extent of reaction, in German Reaktionslaufzahl). This selected quantity is a quantitative measure of the depth to which chemical conversion, corresponding to a given stoichiometric description, has proceeded. Together with balance relationships the reaction coordinate elucidates the actual concentration of individual constituents with absolute accuracy.

For some of the simples cases, the mutual relationship between reacting constituents may be expressed by graphical means:

Let us first consider the simples chemical conversion

$$v_2 A_2 - v_1 A_1 = 0 . \tag{4.1}$$

On introduction of the reaction coordinate ξ the actual concentration of the two components will be defined by the equations

$$n_1 = n_1^\circ + v_1 \xi$$
$$n_2 = v_2 \xi . \tag{4.2}$$

Chemical conversion of the pure component A_1 to the pure product A_2 can be illustrated by the straight line $\overline{MN}$ in Fig. 4. Clearly, any equilibrium composition must lie on this straight line, irrespective of whether the chemical equilibrium is more favourable, due to the numerical value of the equilibrium constant, to the formation of products or vice versa.

Similarly, the chemical reaction can be described for a three-constituent system. Such a system can be described in two ways: either two constituents react to form a third one or, one constituent forms the other two, according to the schemes

$$-v_1 A_1 - v_2 A_2 + v_3 A_3 = 0 \,,$$

or

$$-v_1 A_1 + v_2 A_2 + v_3 A_3 = 0 \,. \tag{4.3}$$

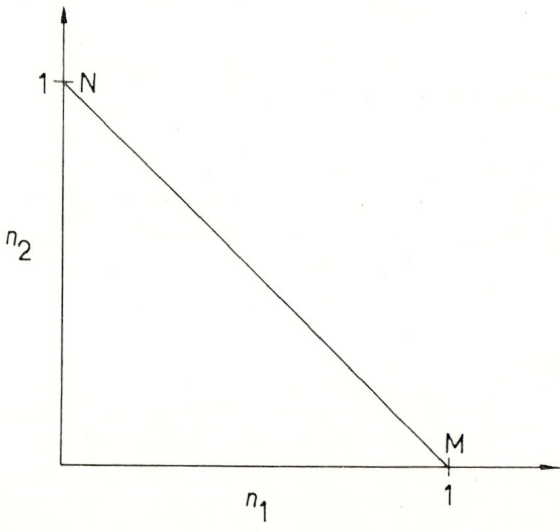

Fig. 4. Diagram of chemical conversion corresponding to the stoichiometric equation
$$-A_1 + A_2 = 0.$$

Again, the concentration of the individual constituents can be expressed in the first case by the relations

$$n_1 = n_1^o + v_1 \xi \,,$$

$$n_2 = n_2^o + v_2 \xi \,,$$

$$n_3 = v_3 \xi \,; \tag{4.4}$$

and in the second

$$n_1 = n_1^o + v_1 \xi \,,$$

$$n_2 = v_2 \xi \,,$$

$$n_3 = v_3 \xi \,. \tag{4.5}$$

Chosing, for the sake of simplicity, $n_1^o = n_2^o = 1$ and considering that $v_i = 1$, then conversion of the pure constituent (or, of the pure constituents) to products is characterized by the straight line $\overline{XY}$ for the case of one initial constituent in Fig. 5a and the line $\overline{VW}$ for the case of two initial constituents in Fig. 5b.

Finally, Fig. 6 illustrates the position of possible equilibrium compositions for the simples case, in which two reactions are taking place in a system. These can be written as consecutive reactions,

$$-v_1 A_1 + v_2 A_2 = 0,$$
$$-v_2 A_2 + v_3 A_3 = 0, \qquad (4.6)$$

or parallel reactions

$$-v_1 A_1 + v_2 A_2 = 0,$$
$$-v_1 A_1 + v_3 A_3 = 0. \qquad (4.7)$$

In both cases, two reaction coordinates must be employed, ξ_1 and ξ_2. Considering again for simplicity that $n_1^0 = 1$ and all $v_i = 1$, then the possible equilibrium compositions will fill the area MNO.

Although the possibilities of graphical representation are herewith exhausted, it is yet clear that with a system of N constituents and R linearly independent reactions, possible equilibrium compositions will fill an R-dimensional area in an N-dimensional

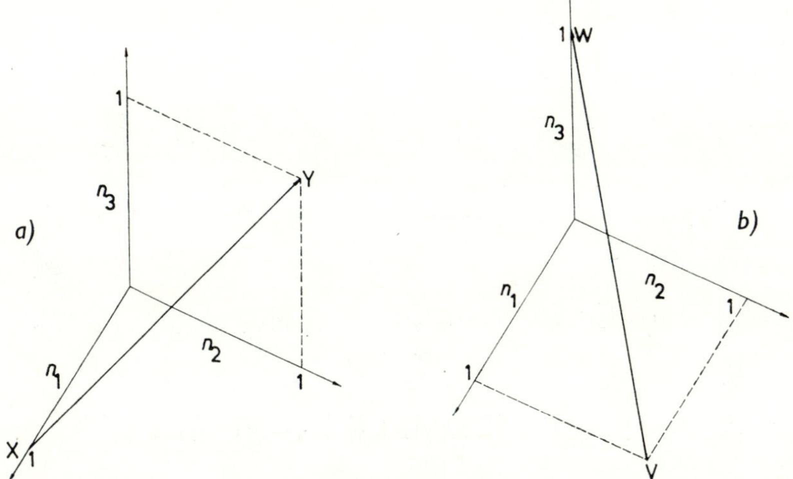

Fig. 5. Diagram of chemical conversion in a system of three constituents with a single chemical reaction.

(a — according to stoichiometric equation $-A_1 - A_2 + A_3 = 0$
b — according to stoichiometric equation $-A_1 + A_2 + A_3 = 0$)

space. Shifting of the system from the initial state to equilibrium will be determined by the vector

$$\xi = (\xi_1, \xi_2, \ldots, \xi_R) \qquad (4.8)$$

and the mutual concentration ratio of constituents in the equilibrium mixture is defined by the relative ratio of individual ξ_r values, taking into account the shape of the stoichiometric equations.

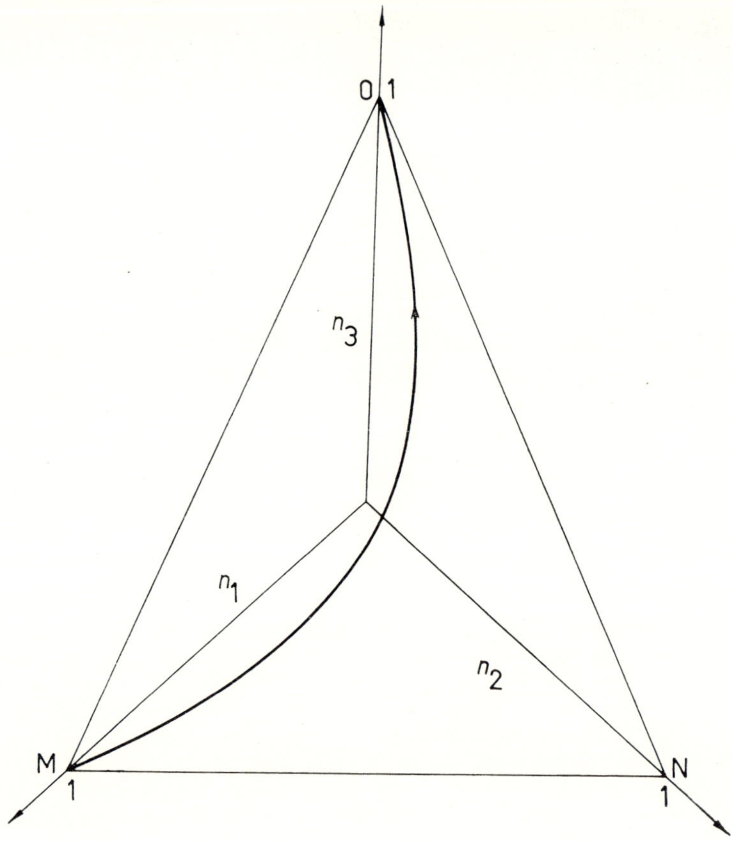

Fig. 6. Diagram of chemical conversion in a system of three constituents with two chemical reactions.

4.2 CALCULATION OF SIMPLE EQUILIBRIA

Use of the quantity ξ allows rapid conversion of equations which describe the chemical equilibrium of simple systems, into forms which are relatively easy to solve.

When only one chemical reaction takes place in a system, substitution of individual numbers of moles in equation (3.38) by the expressions

$$n_i = n_i^o + v_i\xi \quad i = 1, 2, \ldots, N \tag{4.9}$$

leads to the form

$$K = \left(\frac{P}{n^o + v\xi}\right)^v \prod_{i=1}^{N} (n_i^o + v_i\xi)^{v_i}, \tag{4.10}$$

which is an algebraic equation with one unknown, frequently of higher order. The equation can be solved by analytical means up to the fourth order, while for higher orders one of the methods of numerical mathematic[44,96,140] must be used (see Appendix 1).

Example 10

Determine the equilibrium composition of a system in which the reaction

$$CH_4(g) + H_2O(g) = CO(g) + 3 H_2(g)$$

is taking place at 900 K and 1 atm. The initial constituents are present in an equimolar ratio, the system exhibits ideal behaviour. Thermochemical data for the individual constituents are listed in the following table

	$-(G° - H_0°/T)$ (cal K^{-1} mole^{-1})	$\Delta H_0°$ (cal mole^{-1})
CH_4	46.470	$-15\ 937$
H_2O	46.120	$-57\ 104$
H_2	32.004	0
CO	48.097	$-27\ 202$

The mass balance of the individual constituents in equilibrium is determined by the following relations (when $n_1^o = n_2^o = 1$ in order that $\xi \in (0, 1)$)

$$n_{CH_4} = 1 - \xi ,$$

$$n_{H_2O} = 1 - \xi ,$$

$$n_{H_2} = 3\xi ,$$

$$n_{CO} = \xi .$$

Substituting the balance relations into equation (4.10) and calculating the equilibrium constant from the given thermochemical data we obtain the algebraic equation

$$2.053 = \frac{\xi(3\xi)^3}{(1 - \xi)^2 (2 + 2\xi)^2} .$$

In this case the solution is very simple. The equation may be converted to the form

$$\frac{2.053 \times 4}{27} = \frac{\xi^4}{(1 - \xi^2)^2} ,$$

and, since ξ must be an element of the interval $(0, 1)$, the square root of the entire equation can be drawn. After modification we have

$$\xi^2(1 - Z) = Z, \quad \text{where} \quad Z = \sqrt{\frac{4 \times 2.053}{27}},$$

so that $\xi = 0.596$.

The resulting equilibrium composition will be

$$n_{CH_4} = 0.404$$
$$n_{H_2O} = 0.404$$
$$n_{H_2} = 1.788$$
$$n_{CO} = 0.596$$
$$\overline{\quad\quad\quad\quad\quad}$$
$$n = 3.192$$

Using an excess of one component instead of the equimolar composition (e.g. 2 moles water per 1 mole CH_4), the following relation results

$$2.053 = \frac{\xi(3\xi)^3}{(1 - \xi)(2 - \xi)(3 + 2\xi)^2},$$

which cannot be solved so easily. The most expedient solution will then be the following. We leave the relation in its original form, making it equal to zero by transposing the equilibrium constant to the right-hand side. Using the regula falsi method we then arrive very quickly at an accurate result.

We select

$$\xi^{(1)} = 0.7, \quad \text{then} \quad \frac{0.7 \times 2.1^3}{0.3 \times 1.3 \times 4.4^2} - 2.053 = -1.1941,$$

and

$$\xi^{(2)} = 0.8, \quad \text{then} \quad \frac{0.8 \times 2.4^3}{0.2 \times 1.2 \times 4.6^2} - 2.053 = 0.12469.$$

Narrowing the interval down and repeating the trial calculation we obtain

$$n_{CH_4} = 0.206$$
$$n_{H_2O} = 1.206$$
$$n_{H_2} = 2.382$$
$$n_{CO} = 0.794$$
$$\overline{\quad\quad\quad\quad\quad}$$
$$n = 4.588.$$

When two independent reactions are proceeding in a system, the balance relationship can be expressed sufficiently rapidly in the form of

$$n_i = n_i^o + v_{1i}\xi_1 + v_{2i}\xi_2 \quad (i = 1, 2, ..., N),$$ (4.11)

these relations then being substituted into the expressions

$$K_1 = \prod_{i=1}^{N} n_i^{v_{1i}} \left(\frac{P}{n}\right)^{v_1},$$ (4.12)

$$K_2 = \prod_{i=1}^{N} n_i^{v_{2i}} \left(\frac{P}{n}\right)^{v_2}.$$ (4.13)

The result are two equations with the unknown variables ξ_1 and ξ_2. As far as they are of higher order, a sufficiently rapid solution is obtained by trial substitution of several values of one unknown variable and plotting the calculated pairs of the other variable. When the respective points are connected, the point of intersection represents the solution (see Fig. 7).

Example 11

When methane is decomposed by steam, as in the case of Example 10, CO_2 is always formed as a side product. The system is then described by two chemical reactions, the most advantageous combination of which being (according to Example 6) two subsequent reactions

$$CH_4(g) + H_2O(g) = CO(g) + 3\,H_2(g),$$
$$CO(g) + H_2O(g) = CO_2(g) + H_2(g).$$

Calculate the equilibrium composition of the system, using data from Example 10, the respective data being

$$-\left(\frac{G^o - H_0^o}{T}\right)_{CO_2,\,900K} = 53.047 \text{ cal K}^{-1} \text{ mole}^{-1}$$

$$\text{and} \quad (\Delta H_0^o)_{CO_2} = -93.969 \text{ cal mole}^{-1}.$$

The ratio of initial constituents is $n_{CH_4}^o : n_{H_2O}^o = 1$. With two reaction coordinates, the balance relations will take the form of

$$n_{CH_4} = 1 - \xi_1$$
$$n_{H_2O} = 1 - \xi_1 - \xi_2$$
$$n_{H_2} = 3\xi_1 + \xi_2$$
$$n_{CO} = \xi_1 - \xi_2$$
$$n_{CO_2} = \xi_2$$

$$\overline{\quad n \quad = 2 + 2\xi_1\,.}$$

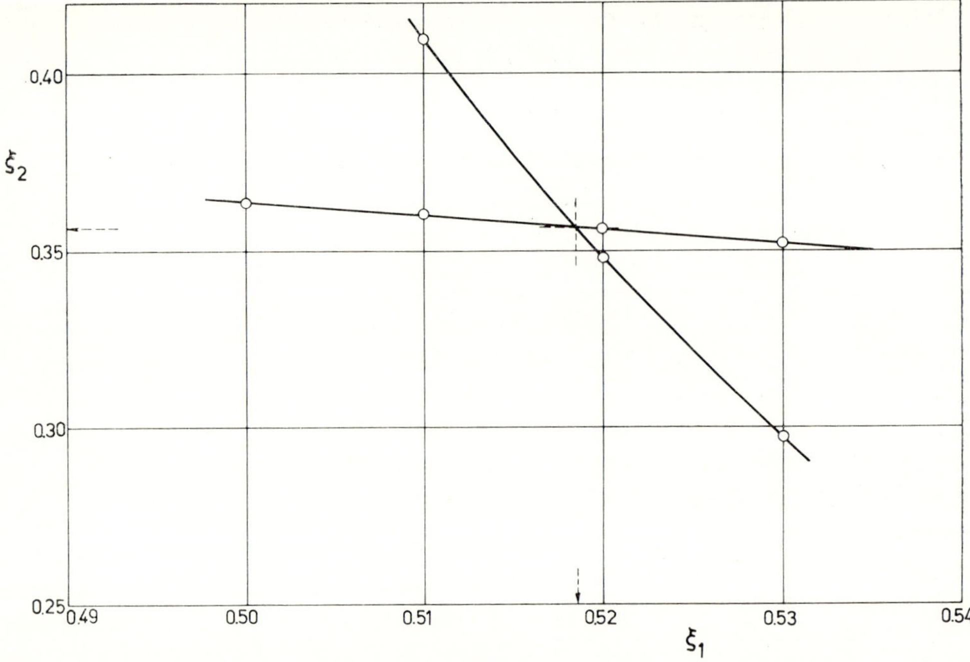

Fig. 7. Graphical determination of the equilibrium composition of a system in which two independent reactions are taking place.

The respective equilibrium relations are:

$$\frac{(\xi_1 - \xi_2)(3\xi_1 + \xi_2)^3}{(1 - \xi_1)(1 - \xi_1 - \xi_2)(2 + 2\xi_1)^2} \cdot \frac{1}{} = 2.053 \, ,$$

$$\frac{\xi_2(3\xi_1 + \xi_2)}{(\xi_1 - \xi_2)(1 - \xi_1 - \xi_2)} = 3.000 \, .$$

The two equations must be solved simultaneously. We again chose a trial-and-error procedure, modifying the second equilibrium relation, which is a quadratic equation, into the usual form. The corresponding pair of values ξ_2, which were calculated from the two equilibrium relations by substitution of ξ_1 values chosen as trial approximations are listed in the following table:

ξ_1	ξ_2^{I}	ξ_2^{II}
0.50	0.598	0.363
0.51	0.410	0.360
0.52	0.348	0.356
0.53	0.297	0.352

60

The solution is now obtained from Fig. 7 by reading values of the point of intersection of the two lines which connect correspoding values of ξ_2; $\xi_1 = 0.519$, $\xi_2 = 0.356$.

Therefore the equilibrium composition is

$$n_{CH_4} = 0.481$$

$$n_{H_2O} = 0.125$$

$$n_{H_2} = 1.913$$

$$n_{CO} = 0.163$$

$$n_{CO_2} = 0.356$$

$$\overline{\phantom{n_{CO_2} = 0.356}}$$

$$n = 3.038 .$$

Hougen, Watson and Ragatz[77] report a procedure by means of which this method can be applied even to systems in which three independent reactions are proceeding. The procedure is illustrated in Fig. 8. The solution is again obtained by trial approximations. Several values are selected for the coordinate ξ_3, and three values of each of the coordinates ξ_1 and ξ_2 are calculated with its use from the equilibrium relations. The results are plotted and points corresponding to individual equilibrium relations are connected. The intersection represents the value of the coordinates ξ_1 and ξ_2. When these values are substituted into any of the equilibrium relations, the value of ξ_3 is obtained. This solution, however, is quite laborious.

A case is frequently encountered, in which the chemical conversion in a system is described by more than three linearly independent reactions, but yet the resulting set of equilibrium relations can be simplified to such a degree that one of the above-described procedures can be applied. In principle, two kinds of systems may be involved:

a) Systems containing isomers only, in which case the resulting set is composed of linear equations only, or systems in which some of the equilibrium relations are linear.

b) Systems, in which the equilibrium constant for one of the reactions will differ in order of magnitude so that it can be neglected. This procedure, however, can only be employed on condition that a simplified stoichiometric consideration is used for the system. The reason is that we must eliminate the corresponding component, which moreover may not be involved as a product in any other thermodynamically preferred reaction.

In the general case, more complicated systems must be solved by a different method, which is described in Chapter 5.

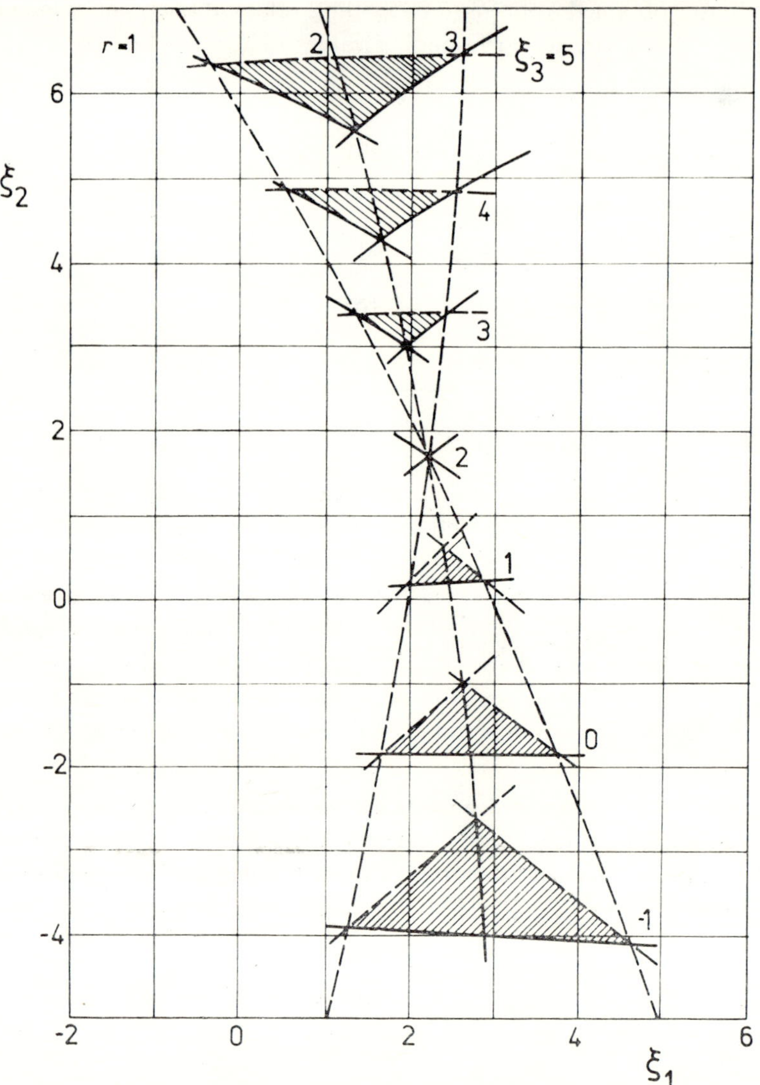

Fig. 8. Graphical determination of the equilibrium composition of a system in which three independent reactions are taking place.

4.3 DEPENDENCE OF THE DEGREE OF CONVERSION ON INDIVIDUAL REACTION CONDITIONS

In general the degree of conversion depends on temperature, pressure, the numerical value of the equilibrium constant and composition of the initial mixture $n^o =$

$= (n_1^o, n_2^o, \ldots, n_N^o)$. The dependence on temperature is implicitly contained in the numerical value of the equilibrium constant. We may therefore write

$$\xi = F(K_a(T), P, \mathbf{n}^o) \,. \tag{4.14}$$

For the sake of simplicity, let us first investigate the dependence of the degree of conversion on temperature and pressure in a closed homogenous system, in which one chemical reaction is taking place. In this case, free enthalpy of the system is a function of temperature, pressure and the degree of conversion ξ. In equilibrium, there holds

$$\left(\frac{\partial G}{\partial \xi}\right)_{T,P} = 0 \,. \tag{4.15}$$

Let us denote

$$\left(\frac{\partial G}{\partial \xi}\right)_{T,P} = G_1(T, P, \xi) \,, \tag{4.16}$$

and investigate the dependence of the degree of conversion ξ on temperature and pressure. Relation (4.15) may be written as

$$G_1(T, P, \xi(T, P)) = 0 \,. \tag{4.17}$$

Differentiating the left- and right-hand side of the equation (4.17) with respect to temperature at fixed pressure, we have

$$\left(\frac{\partial G_1}{\partial T}\right)_{P,\xi} + \left(\frac{\partial G_1}{\partial \xi}\right)_{T,P}\left(\frac{\partial \xi}{\partial T}\right)_P = 0 \,. \tag{4.18}$$

In equilibrium, the thermodynamic relationship applies

$$\left(\frac{\partial G}{\partial \xi}\right)_{T,P} = \left(\frac{\partial H}{\partial \xi}\right)_{T,P} - T\left(\frac{\partial S}{\partial \xi}\right)_{T,P} = 0 \,. \tag{4.19}$$

Joining together relations (4.18), (4.15) and (4.19) we obtain

$$T\left(\frac{\partial \xi}{\partial T}\right)_P = \frac{\left(\dfrac{\partial H}{\partial \xi}\right)_{T,P}}{\left(\dfrac{\partial^2 G}{\partial \xi^2}\right)_{T,P}} \,. \tag{4.20}$$

Similarly, differentiation of both sides of equation (4.17) with respect to pressure at fixed temperature leads to

$$\left(\frac{\partial \xi}{\partial P}\right)_T = - \frac{\left(\dfrac{\partial V}{\partial \xi}\right)_{T,P}}{\left(\dfrac{\partial^2 G}{\partial \xi^2}\right)_{T,P}} \,. \tag{4.21}$$

63

Since the expression $(\partial^2 G/\partial\xi^2)_{T,P} > 0$, applies in equilibrium, equations (4.19) and (4.20) can be interpreted unambiguously, as Le Chatelier did in qualitative terms as early as 1885[98]. The principle named after him can be put in simple terms thus: "A change of any of the quantities which determine the state of a system in equilibrium, will cause a shift of equilibrium in that direction in which the effect of this change is compensated most."

In quantitative terms the dependence of the degree of conversion on individual reaction variables can be studied by analysing the equation

$$K_a = \left(\frac{P}{n}\right)^v \prod_{i=1}^{N} n_i^{v_i} , \tag{4.22}$$

or, best, its logarithmic form

$$\ln K_a = v \ln P - v \ln n + \sum_{i=1}^{N} v_i \ln n_i , \tag{4.23}$$

where

$$n_i = n_i^\circ + v_i\xi \quad i = 1, 2, \ldots, N$$

and

$$n = \sum_{i=1}^{N} n_i = n^\circ + v\xi . \tag{4.24}$$

4.3.1. Dependence of the degree of conversion on temperature

The influence of temperature on the equilibrium composition may be judged from equation (4.20). The actual form of the relationship is obtained by derivating equation (4.23) by temperature at fixed pressure, i.e.

$$\frac{\Delta H^\circ}{RT^2} = -\frac{v^2}{n}\left(\frac{\partial\xi}{\partial T}\right)_P + \sum_{i=1}^{N}\frac{v_i^2}{n_i}\left(\frac{\partial\xi}{\partial T}\right)_P . \tag{4.25}$$

Modification of relation (4.25) leads to

$$\left(\frac{\partial\xi}{\partial T}\right)_P = \frac{\Delta H^\circ}{RT^2\left(\sum\limits_{i=1}^{N}\dfrac{v_i^2}{n_i} - \dfrac{v^2}{n}\right)} . \tag{4.26}$$

The denominator in relation (4.26) is always positive (see Appendix 3). Therefore it is clear that the dependence of the degree of conversion on temperature is given by the sign and absolute magnitude of the heat of reaction ΔH°. When the reaction is exothermic, $(\Delta H^\circ < 0)$, the value of the equilibrium constant as well as of the degree of conversion (with other variables being constant) will decrease with rising temperature. On the other hand, with endothermic reaction $(\Delta H^\circ > 0)$ the magnitude of the equilibrium constant and of the degree of conversion will rise with rising temperature. The absolute magnitude of the heat of reaction determines the slope of the degree of conversion vs. temperature relationship. This influence is illustrated in the following example.

Example 12

Dehydrogenation of ethane to ethylene takes place according to the equation

$$C_2H_6(g) \; = \; C_2H_4(g) \; + \; H_2(g).$$

Calculate the degree of conversion of this reaction for the temperature range 600 to 1500 K and 1 atm pressure from the following data (standard state of unit fugacity):

T (K)	600	700	800	900	1000
K_a	3.594×10^{-6}	2.101×10^{-4}	4.547×10^{-3}	5.021×10^{-2}	3.441×10^{-1}

T (K)	1100	1200	1300	1400	1500
K_a	1.671	6.219	18.93	48.95	112.07

Clearly, the degree of conversion is equal to

$$\xi = \sqrt{\frac{K_a + 1}{K_a}}.$$

The result is given in the following table for individual temperature values and respective values of K_a as shown in the above table. For better illustration, the heat of reaction is also given in the last column.

T (K)	ξ	ΔH_r^{o} (cal mole^{-1})
600	0.0019	$-\;9\,293$
700	0.0145	$-10\,259$
800	0.0673	$-11\,063$
900	0.2186	$-11\,718$
1000	0.5060	$-12\,225$
1100	0.7909	$-12\,630$
1200	0.9282	$-12\,910$
1300	0.9746	$-13\,160$
1400	0.9899	$-13\,330$
1500	0.9956	$-13\,450$

4.3.2 Dependence of the degree of conversion on pressure

Differentiating equation (4.23) with respect to pressure at fixed temperature, we obtain the expression

$$0 = \frac{v}{P} - \frac{v^2}{n}\left(\frac{\partial \xi}{\partial P}\right)_T + \sum_{i=1}^{N} \frac{v_i^2}{n_i}\left(\frac{\partial \xi}{\partial P}\right)_T. \tag{4.27}$$

Since the standard state of unit fugacity was chosen, the equilibrium constant is independent of pressure. Modification of equation (4.27) gives

$$\left(\frac{\partial \xi}{\partial P}\right)_T = -\frac{v}{P\left(\sum\limits_{i=1}^{N} \dfrac{v_i^2}{n_i} - \dfrac{v^2}{n}\right)}. \tag{4.28}$$

The denominator on the right-hand side of equation (4.28) is always positive (see Appendix 3). The sign of the quantity $v = \sum\limits_{i=1}^{N} v_i$, which is the sum of stoichiometric coefficients of the reaction, also called change of the mole number of the reaction, is therefore decisive for the influence of pressure on the degree of conversion. With $v < 0$ the degree of conversion rises with rising pressure. With $v = 0$, the degree of conversion is independent of pressure, in the case of $v > 0$ the degree of conversion decreases with increasing pressure. In some cases this pressure influence is very important, as it enables technologic utilization of reactions which are of little interest at normal pressure.

Example 13

Ammonia is prepared according to the reaction

$$\tfrac{3}{2}\,H_2(g) + \tfrac{1}{2}\,N_2(g) = NH_3(g).$$

The value of the equilibrium constant is $K_a = 0{,}30076$ at 500 K. Find, for this temperature, the dependence of the degree of conversion on pressure in the 1 to 1000 atm range for a stoichiometric ratio of initial constituents, disregarding the real behaviour of the mixture.

We shall set out from equation (4.10). The equilibrium relation will be

$$K_a = \frac{\xi}{(0.5 - 0.5\xi)^{0.5}\,(1.5 - 1.5\xi)^{1.5}}\left[\frac{P}{(2 - \xi)}\right]^{-1},$$

converting, after modification, to

$$PK_a \times 0.5^2 \times 3^{1.5} = \frac{\xi(2 - \xi)}{(1 - \xi)^2}$$

and finally

$$\xi = 1 - \frac{1}{\sqrt{(1 + F)}}, \quad \text{where} \quad F = PK_a \times 0.5^2 \times 3^{1.5}$$

$$F = P \times 0.39070 .$$

The dependence then follows from this table:

P (atm)	ξ
1	0.152
2	0.250
5	0.418
10	0.549
20	0.663
50	0.779
100	0.842
200	0.888
500	0.928
1000	0.949

Pressure may influence the equilibrium composition in one more sense, i.e. in the correction for non-ideal behaviour, as seen from equation (3.36). This influence may, particularly at high pressures, achieve large values, so that it generally cannot be neglected. Because of the importance of calculating the real behaviour of systems, this dependence will be dealt with more comprehensively in Chapter 6.

4.3.3 Dependence of the degree of conversion on the composition of the initial mixture

The degree of conversion may likewise be influenced by a change in the initial composition (see Appendix 5). It was believed for a long time, that the degree of conversion always rises with the rising concentration of one of the initial constituents. This phenomenon which, as may be shown need not apply generally[138], was employed in those cases where prices of the raw materials differed. The philosophy behind this approach involves the attempt to use an excess of easily available initial constituents in order to achieve more perfect utilization of an economically valuable initial constituent. Since, however, the above phenomenon need not always apply, the reaction must first be analysed and the composition of the initial mixture must then be chosen accordingly.

Let us, for the sake of simplicity, investigate the following special case: $v_1 < 0$, $v_2 < 0$, $v_i > 0$ $(i = 3, 4, ..., N)$ i.e. a case in which products are formed from two initial substances. Let us now study the relationship $\xi = \xi(n_2^o)$ at fixed temperature, pressure and input number of moles n_1^o of the first constituent. Differentiating the left- and right-hand sides of the equation (4.23) by n_2^o we obtain

$$0 = -\frac{v}{n}\left(1 + v\,\frac{d\xi}{dn_2^o}\right) + \sum_{i=1}^{N} \frac{v_i^2}{n_i}\,\frac{d\xi}{dn_2^o} + \frac{v_2}{n_2}\,. \tag{4.29}$$

Eliminating the expression $d\xi/dn_2^o$ we get

$$\frac{d\xi}{dn_2^o} = \frac{\dfrac{v}{n} - \dfrac{v_2}{n_2}}{\displaystyle\sum_{i=1}^{N} \frac{v_i^2}{n_i} - \frac{v^2}{n}}\,, \tag{4.30}$$

the value of the denominator on the right-hand side of equation (4.30) always being positive (see Appendix 3). Substitution of the relations

$$n_2 = n_2^o + v_2\xi$$
$$n = n^o + v\xi = n_1^o + n_2^o + v\xi\,, \tag{4.31}$$

into the numerator of the right-hand side of equation (4.30) shows, that in the case $(v_2 - v) > 0$ the function $\xi = \xi(n_2^o)$ passes through an extreme in the point n_2^o, for which there holds

$$n_2^o = n_1^o\,\frac{v_2}{v - v_2}\,. \tag{4.32}$$

In order to be able to investigate the relationship $\xi = \xi(n_2^o)$ in more detail, let us examine the asymptotic behaviour of the equation (4.22) for $n_2^o \to \infty$.

Evidently there holds

$$K_a = \left(\frac{P}{n}\right)^v \prod_{i=1}^{N} n_i^{v_i} \approx (n_2^o)^{v_2 - v}\,\frac{\xi}{n_1^o + v_1\xi} \qquad n_2^o \to \infty\,. \tag{4.33}$$

Validity of the expression (4.33) is proved in Appendix 4. Since the equilibrium constant K_a is a number which does not depend on the magnitude of n_2^o, it is clear that we must distinguish three cases

a) $(v_2 - v) < 0$.

From relation (4.33) follows

$$\lim_{n^o{}_2 \to \infty} \xi(n_2^o) = -\frac{n_1^o}{v_1} = \frac{n_1^o}{|v_1|} \tag{4.34}$$

and from relation (4.30) we see, that the function $\xi = \xi(n_2^o)$ is a rising function of the argument n_2^o. Conversion of the first initial constituent increases with the rising amount of the second initial constituent, the first constituent reacting completely when $n_2^o \to \infty$.

b) $\left(v_2 - v\right) = 0$.

In this case there applies

$$\lim_{n^o{}_2 \to \infty} \xi(n_2^o) = \xi_0 \,, \quad \xi_0 \neq 0, \xi_0 \neq \frac{n_1^o}{|v_1|} \tag{4.35}$$

and from relation (4.30) there follows, that the function $\xi = \xi(n_2^o)$ is likewise a growing function of the argument n_2^o. With the increasing amount of the second initial constituent the conversion of the first constituent rises, but the first constituent will not react completely even when $n_2^o \to \infty$.

c) $\left(v_2 - v\right) > 0$.

In this last variant, the relation

$$\lim_{n^o{}_2 \to \infty} \xi(n_2^o) = 0 \,, \tag{4.36}$$

becomes valid and the function $\xi = \xi(n_2^o)$ attains a maximum value in the point for which equation (4.32) applies.

The following table summarises our results. A normalised degree of conversion $x = \left(|v_1|/n_1^o\right) \xi$ is employed. We assume, that the first constituent is the key constituent in the sense of the definition given in section (3.6).

$v_2 - v$	$\lim\limits_{n^o{}_2 \to \infty} x(n_2^o)$	Property of function $x(n_2^o)$
< 0	1	rising function
$= 0$	$\neq 0, \neq 1$	rising function
> 0	0	maximum in point n_2^o from relation (4.32)

Example 14

Ammonia synthesis takes place according to the equation

$$\tfrac{3}{2} H_2(g) + \tfrac{1}{2} N_2(g) \;=\; NH_3(g) \,.$$

Determine the number of moles of nitrogen which must be added to 1.5 moles hydrogen in order that the maximum possible number of moles hydrogen should react. Assume ideal behaviour in the gas mixture.

There holds for the reaction involved, that $v_2 - v = -0.5 - (-1) = 0.5$, and, according to relation (4.32), $n_2^o = n_1^o$. Therefore, we must add the same amount of nitrogen moles to n_1^o moles hydrogen at any given temperature and pressure, in order to achieve a maximum yield from the reaction. The equilibrium relation takes the form of

$$K_a P = \frac{\left(n_2^o + 1.5 - \xi\right)\xi}{\left(1.5 - 1.5\xi\right)^{1.5}\left(n_2^o - 0.5\xi\right)^{0.5}} .$$

For 700 K, we have $\log K = -2.044$. The table shows the dependence of the degree of conversion ξ (in this case, $\xi = x$) on n_2^o for pressures of 150, 200, 300 atm. In practical calculations a correction for non-ideal behaviour must be taken. Due to the illustrative nature of this example, we shall not take this correction into account.

n_2^o	0.5	1	1.5	2	5	10	100	1000
ξ 150 atm	0.398	0.443	0.449	0.447	0.407	0.359	0.181	0.007
ξ 200 atm	0.453	0.503	0.510	0.507	0.467	0.417	0.223	0.009
ξ 300 atm	0.530	0.585	0.592	0.588	0.550	0.5	0.292	0.129

It will be seen from Example 14. that for $n_{N_2} > 1$ the relationship $\xi = \xi(n_{N_2}^o)$ varies very slowly. From the asymptotic relationship (4.33) follows, that in the case c) the slope of the relationship $\xi = \xi(n_2^o)$ is proportional to the magnitude of the difference $v_2 - v$. Assuming the stoichiometric coefficients to be mutually non-divisible integers, there applies in Example 14 $(v_2 - v) = 1$. Let us chose an example in which the difference $(v_2 - v)$ is considerably greater.

Example 15

Hydrogenation of naphtalene to trans-dekaline proceeds according to the equation

$$5\,H_2(g) + C_{10}H_8(g) \;\rightarrow\; C_{10}H_8(g) .$$

At 530 K, we have $\log K = -0.275$. The reaction takes place at atmospheric pressure. Let us assume, that the reaction begins with a mixture of 5 moles hydrogen and $n_2^o \geqq 1$ moles trans-dekaline. The equilibrium relation is

$$K_a = \frac{\xi(5 + n_2^o - 5\xi)^5}{(5 - 5\xi)^5 (n_2^o - \xi)} .$$

70

The following table shows the relationship between the degree of conversion and the initial number of $C_{10}H_8$ moles. The maximum is achieved in the point $n_2^o = 5 \times \times (-1)/(-4) = 1.25$.

n_2^o	1	1.25	1.5	2	5	10	20	30
ξ	0.155	0.159	0.155	0.144	0.069	0.021	0.003	0.005

Let us now take an example, in which the degree of conversion rises with the increasing amount of the second constituent, but will not achieve the maximum value in the limiting case of $n_2^o \to \infty$ i.e. the entire amount of the first constituent will not react.

Example 16

Consider the reaction

$$SO_2(g) + \tfrac{1}{2}O_2(g) = SO_3(g),$$

which takes place at atmospheric pressure with an equilibrium constant, the decadic logarithm of which is 1.502 at 800 K $(K_a = 31.77)$. In this case, $v_2 - v = 0$ and the equilibrium relation is $(n_1^o = 1)$

$$K_a = \frac{\xi}{1-\xi} \times \sqrt{\left(\frac{1 + n_2^o - 0.5\xi}{n_2^o - 0.5\xi}\right)}.$$

Evidently, there applies

$$\xi \to \frac{K_a}{1 + K_a} \quad \text{for} \quad n_2^o \to \infty .$$

The relationship $\xi = \xi(n_2^o)$ is given in the following table

n_2^o	0.5	1	2	5	10	50	100	1000
ξ	0.883	0.95	0.962	0.967	0.968	0.9695	0.9695	0.9695

The following example demonstrates a case, in which the degree of conversion rises with the amount of the second initial constituent, achieving a maximum value in the limiting case of $n_2^o \to \infty$. An example similar to Example 14 was intentionally chosen in order to stress the significance of the order of constituents.

Example 17

Let us again consider ammonia synthesis, writing the reaction however in a form which differs from that of Example 14:

$$\tfrac{1}{2} N_2(g) + \tfrac{3}{2} H_2(g) \;=\; NH_3(g),$$

and let us investigate the influence of an excess of hydrogen. Now we have $v_2 - v = -0.5$. Assuming $n_1^o = 0.5$, the relationship $\xi = \xi(n_2^o)$ takes a form which is summarized in the following table.

n_2^o	1.5	2	5	10	50	100	1000
ξ 150 atm	0.398	0.487	0.757	0.881	0.978	0.989	0.999

In this case, the equilibrium relation is

$$K_a P = \frac{\xi(0.5 + n_2^o - \xi)}{(0.5 - 0.5\xi)^{0.5}\,(n_2^o - 1.5\xi)^{1.5}}.$$

The results presented in this section appear to differ from Le Chatelier's principle, since addition of a large number of moles of the second constituent need not necessarily increase the conversion of the first constituent. This apparent discrepancy may be explained by the fact, that the increasing amount of the second constituent at the same time dilutes the mixture, and the phenomen described in section 4.3.4 sets in. With $v < 0$ dilution of the mixture acts in a direction opposed to conversion of the initial constituents. With $v < v_2$ this effect prevails over all other influences.

4.3.4 Dependence of the degree of conversion on inert constituents

The term inert constituents is understood to include all those constituents, which do not take part in the reaction proper. They influence the degree of conversion by diluting the reaction mixture, i.e. by increasing the value of the overall number of moles n on the right-hand side of equation (4.22). Thus it is clear, that the influence of inert constituents on the degree of conversion is reciprocal to the influence of pressure. With $v > 0$, the reaction yield is increased by the presence of inert constituents, and vice versa. When $v = 0$, the presence of an inert constituent does not influence the degree of conversion.

In practical work, inert constituents are frequently used intentionally either in the form of a protective atmosphere or, because they enable work at atmospheric pressure even though this is disadvantageous from the thermodynamic point of view. Particularly in petrochemistry, dilution with an inert constituent is preferred over work at decreased pressure for reasons of safety. This relationship is illustrated in the following example.

Example 18

Propylene can be prepared by dehydrogenation of propane according to the equation

$$C_3H_8(g) \ = \ C_3H_6(g) \ + \ H_2(g) \, .$$

Pressure will evidently have a negative influence on the degree of conversion. Work in vacuo is doubtless advantageous in thermodynamic respects but, on the other hand, it involves the risk of explosive mixtures being formed due to possible leaks in the apparatus. To enable the reaction to be carried out at atmospheric pressure, the pressure effect is eliminated by dilution of the initial constituent with steam.

Determine the dependence of the degree of conversion ξ as function of the number of moles of steam per 1 mole initial propane at 700 K and 1 atm pressure, assuming ideal behaviour of all constituents if the equilibrium constant value is $K_a = 0.02485$ at 700 K. The equilibrium relation will take the form of

$$K_a = \frac{\xi^2}{(1 - \xi)(1 + n^o_{H_2O} + \xi)} \, ,$$

$$\xi = \frac{-n_{H_2O}K_a + \sqrt{[(n^o_{H_2O})^2 + 4(1 + K_a)K_a(1 + n^o_{H_2O})]}}{2(1 + K_a)} \, .$$

For several initial steam concentrations the result is presented in the following table. For the sake of completeness the third line includes the corresponding pressure which would be needed, if the reaction were to take place in the absence of steam.

$n^o_{H_2O}$	0	1	2	3	5	7
ξ	0.156	0.208	0.247	0.277	0.326	0.346
P	760	418	291	227	159	123

$n^o_{H_2O}$	10	12	15	18	20
ξ	0.409	0.435	0.467	0.495	0.511
P	94	91	68	58	51

4.4 DEPENDENCE OF THE EQUILIBRIUM COMPOSITION ON INDIVIDUAL REACTION VARIABLES IN THE CASE OF MORE COMPLICATED SYSTEMS

In a closed system, in which R linearly independent reactions are taking place, the dependence of the degree of conversion $\xi_1, \xi_2, ..., \xi_R$ on temperature and pressure can be derived in a manner similar to that employed in section 4.3. Introducing, in agreement with equation (4.16), a notation corresponding to the case in which a single reaction is taking place,

$$\left(\frac{\partial G}{\partial \xi_j}\right)_{\xi_{i \neq j}, T, P} = G_{1,j}(T, P, \xi_1, \xi_2 ..., \xi_R) \quad j = 1, 2, ..., R \tag{4.37}$$

we may rewrite the equilibrium conditions in the form of

$$G_{1,r}(T, P, \xi_1(T, P), ..., \xi_R(T, P)) = 0 \quad r = 1, 2, ..., R . \tag{4.38}$$

Differentiating the left- and right-hand sides of equation (4.38) with respect to temperature at fixed, pressure, we can derive the relations

$$\sum_{j=1}^{R} \frac{\partial^2 G}{\partial \xi_r \partial \xi_j} \frac{\partial \xi_j}{\partial T} = T \frac{\partial H}{\partial \xi_r} \quad r = 1, 2, ..., R \quad [P] \tag{4.39}$$

and by analogy, differentiating both sides of equation (4.38) with respect to pressure at fixed temperature we obtain

$$\sum_{j=1}^{R} \frac{\partial^2 G}{\partial \xi_r \partial \xi_j} \frac{\partial \xi_j}{\partial P} = - \frac{\partial V}{\partial \xi_r} \quad r = 1, 2, ..., R \quad [T] . \tag{4.40}$$

Since the relation $d^2 G > 0$ applies in equilibrium, the determinant of the sets (4.39) and (4.40) must differ from zero. From relations (4.39) and (4.40) follows, that the general dependence of the degree of conversion on temperature and pressure is highly complicated: it will be shown in the following sections, that this applies to other reaction variables too. Due to the great variety of reaction schemes and possibility of the occurence of factors acting in opposite directions it is difficult to elucidate the prevailing effect.

4.4.1 Temperature dependence

Investigation of the temperature dependence of the degree of conversion in a system, in which a minimum of two linearly independent reactions are taking place, is rendered complicated by the fact, that the relationship involved depends on the ratios of heats of reaction as well as on the signs of the heats of reaction. For the sake of simplicity, let us consider two reactions between the isomers A_1, A_2, A_3.

$$-A_1 + A_2 = 0$$
$$-A_1 + A_3 = 0, \tag{4.41}$$

with one mole of the constituent A_1 at the start. The mass balance equations are

$$n_1 = 1 - \xi_1 - \xi_2$$
$$n_2 = \xi_1$$
$$n_3 = \xi_2, \tag{4.42}$$

the equilibrium equations have the form

$$K_1 = \frac{\xi_1}{1 - \xi_1 - \xi_2}$$

$$K_2 = \frac{\xi_2}{1 - \xi_1 - \xi_2}. \tag{4.43}$$

Expressing the degrees of conversion ξ_1, ξ_2 from equations (4.43) we obtain, after simple modification

$$\xi_1 = \frac{K_1}{1 + K_1 + K_2}$$

$$\xi_2 = \frac{K_2}{1 + K_1 + K_2}. \tag{4.44}$$

Differentiating the left- and right-hand sides of the equations (4.44) with respect to temperature (in this case the degree of conversion is independent of pressure) we obtain

$$\frac{d\xi_1}{dT} = \frac{\xi_1}{RT^2(1 + K_1 + K_2)} \left[\Delta H_1^\circ(1 + K_2) - K_2\Delta H_2^\circ\right]$$

$$\frac{d\xi_2}{dT} = \frac{\xi_2}{RT^2(1 + K_1 + K_2)} \left[\Delta H_2^\circ(K_1 + 1) - K_1\Delta H_2^\circ\right]. \tag{4.45}$$

In this simple case, a complete analysis of the degrees of conversion in the system of reactions (4.41) can be carried out with the use of the equations (4.45). For illustration, let us perform this analysis for the degree of conversion ξ_1, i.e. the amount of the isomer A_2 in the equilibrium mixture. Four basic cases are possible in the temperature interval studied $\langle T_1, T_2 \rangle$ each having several possible alternatives.

$$\text{I.} \quad \Delta H_1^\circ > 0, \quad \Delta H_2^\circ > 0 \quad T \in \langle T_1, T_2 \rangle.$$

If there applies in the entire temperature interval $\langle T_1, T_2 \rangle$, that the value of $(1 +$

$+ K_2)/K_2$ is greater or lesser than the value $\Delta H_2^o/\Delta H_1^o$, then the degree of conversion ξ_1 will rise or decrease with the rising temperature, respectively. When, in a point $T_0 \in \langle T_1, T_2 \rangle$ the value of $(1 + K_2)/K_2$ is equal to the quotient of heats of reaction, the degree of conversion ξ_1 will achieve an extreme value at this temperature (we assume that the second derivation differs from zero).

$$\text{II.} \quad \Delta H_1^o > 0, \quad \Delta H_2^o < 0 \quad\quad T \in \langle T_1, T_2 \rangle$$

in this case the degree of conversion ξ_1 will always be a rising function of temperature

$$\text{III.} \quad \Delta H_1^o < 0, \quad \Delta H_2^o > 0 \quad\quad T \in \langle T_1, T_2 \rangle \,.$$

In this case the degree of conversion ξ_1 will always be a decreasing function of temperature

$$\text{IV.} \quad \Delta H_1^o < 0, \quad \Delta H_2^o < 0 \quad\quad T \in \langle T_1, T_2 \rangle \,.$$

When, in the entire temperature interval $\langle T_1, T_2 \rangle$ the value of $(1 + K_2)/K_2$ is greater or lesser than the value $\Delta H_2^o/\Delta H_1^o$, the degree of conversion ξ_1 will be a decreasing or increasing function of temperature, respectively. By analogy to the first case, the degree of conversion ξ_1 will achieve an extreme value in the point $T = T_0 \in \langle T_1, T_2 \rangle$, in which the right-hand side of the first equation (4.45) becomes equal to zero.

We have assumed in this analysis, that the standard heats of reaction ΔH_1^o or ΔH_2^o maintain the same sign in the entire temperature interval $\langle T_1, T_2 \rangle$. If this assumption were not valid, the interval $\langle T_1, T_2 \rangle$ would have to be divided into several smaller intervals, the analysis being then carried out separately for each sub-interval.

Subsequently, the function $\xi_2 = \xi_2(T)$, i.e. the dependence of the amount of the isomer A_3 in the equilibrium mixture on temperature, should now be studied by a procedure analogical to that employed for the relationship $\xi_1 = \xi_1(T)$.

From the preceding example follows, that the relationship between degrees of conversion or, equilibrium composition and temperature is complicated even in the case of relatively simple systems of chemical reactions. With more complicated systems the temperature relationship usually cannot be predicted, and the dependence of the amounts of individual constituents in the equilibrium mixture on temperature can usually be obtained after the chemical equilibrium has been calculated for several temperatures and the results plotted.

Let us now consider several examples of graphical investigation of the dependence of the degree of conversion on equilibirium constants. Let us assume, that the equilibrium constant of the first reaction varies in the interval $(0.01-100)$ in dependence on temperature, the equilibrium constant of the second reaction varying in the interval $(0.01-100)$. Fig. 9 shows the dependence of the reaction coordinates ξ_1, ξ_2 on K_2 for different values of K_1 for the reaction system (4.41). Similarly, for the system

$$-A_1 + 2A_2 = 0$$

$$-A_1 + A_3 = 0, \tag{4.46}$$

this relationship is shown in Fig. 10, for the system

$$-A_1 + 2A_2 = 0$$

$$-A_1 + 2A_3 = 0,\tag{4.47}$$

in Fig. 11 and for the system

$$-A_1 + A_2 + A_3 = 0$$

$$-A_2 + A_3 + A_4 = 0,\tag{4.48}$$

in Fig. 12.

4.4.2 Pressure dependence

Dependence on pressure is determined by the overall stoichiometric coefficient of each of the individual reactions $\left(v_r = \sum_{i=1}^{N} v_{ri}\ r = 1, 2, ..., R\right)$. When the overall stoichiometric coefficients of all reactions are zeros, chemical equilibrium of the system will not be influenced by pressure (unless, of course, we consider real behaviour of the mixture). When at least one of the reactions has a non-zero overall stoichiometric coefficient, the system is dependent on pressure. For illustration, the dependence of degrees of conversion on pressure is plotted in Fig. 13 for some relatively simple systems of two chemical reactions for the case that both equilibrium constants are equal to one.

4.4.3 Dependence on the composition of the initial mixture

It has been shown in section 4.3.3, that this is a complicated problem in the case where a single reaction is taking place in a system. Where two or more linearly independent reactions are proceeding in a system, the influence of composition of the initial mixture cannot be elucidated beforehand.

4.4.4 Dependence on the amount of inert constituents

Similarly to simple systems, the influence of inert constituents is reciprocal to the influence of pressure. For practical applications, therefore, the alternative choice between decreased pressure and dilution of the initial mixture should be considered.

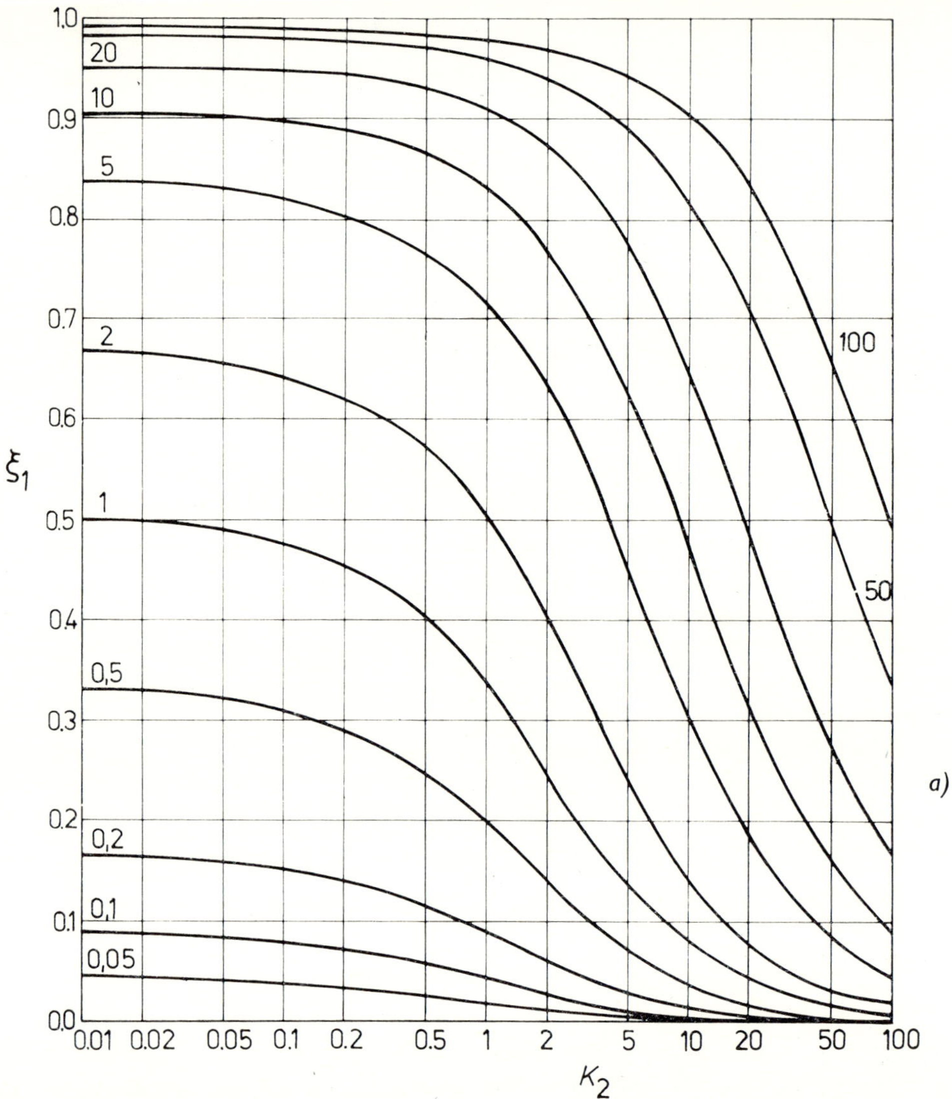

Fig. 9a. Dependence of reaction coordinates on the numerical value of equilibrium constants in the range K_1, $K_2 \in (0.01 - 100)$ in a system of three constituents and two chemical reactions according to stoichiometric equations (4.41) — dependence of ξ_1.

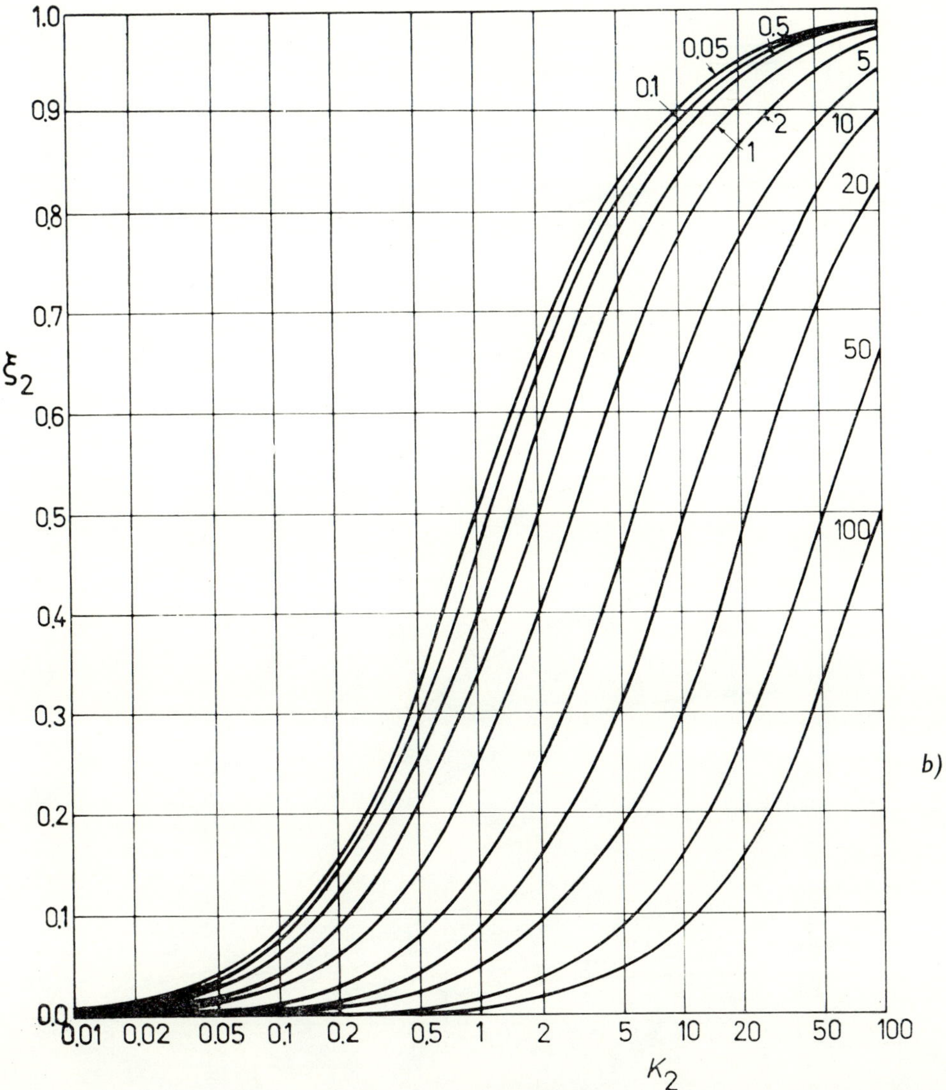

Fig. 9b — dependence of ξ_2.

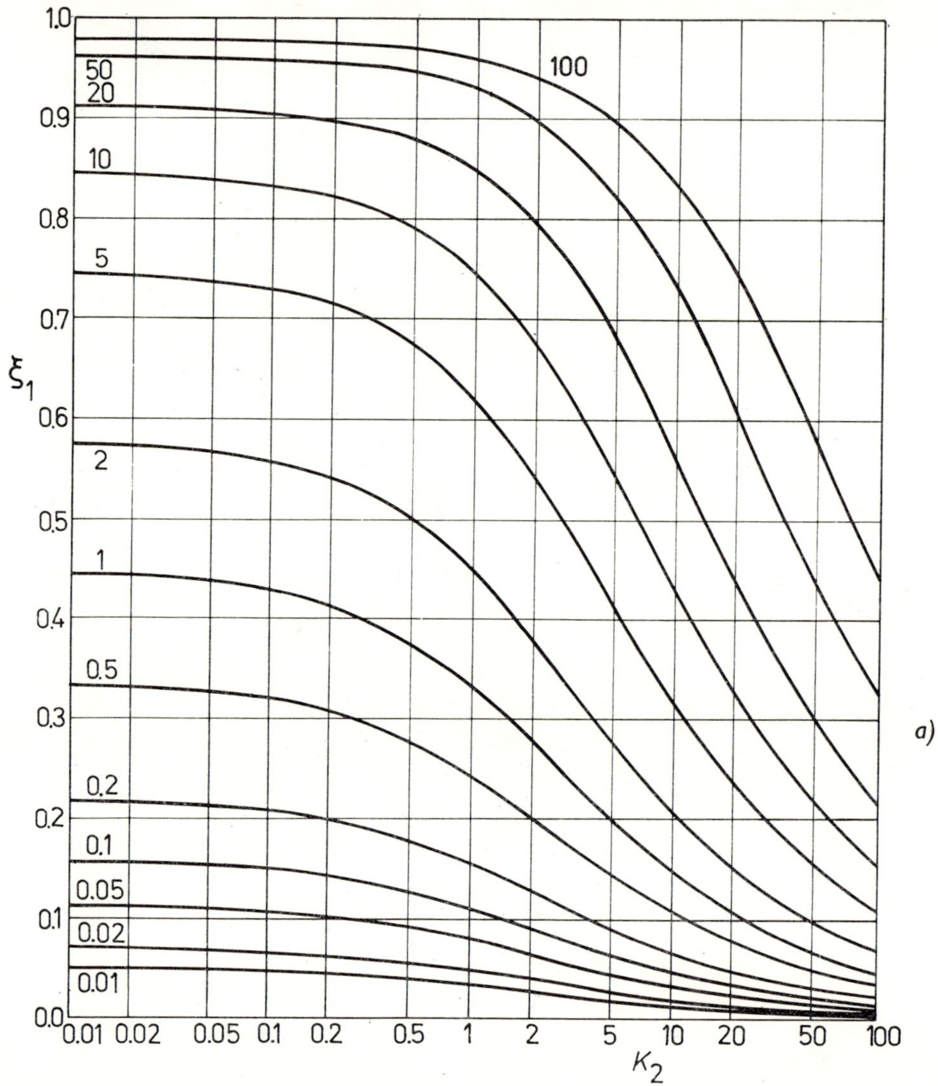

*Fig. 10*a. Dependence of reaction coordinates on the numerical value of equilibrium constants in the range K_1, $K_2 \in (0.01-100)$ in a system of three constituents and two chemical reactions according to stoichiometric equations (4.46) — dependence of ξ_1.

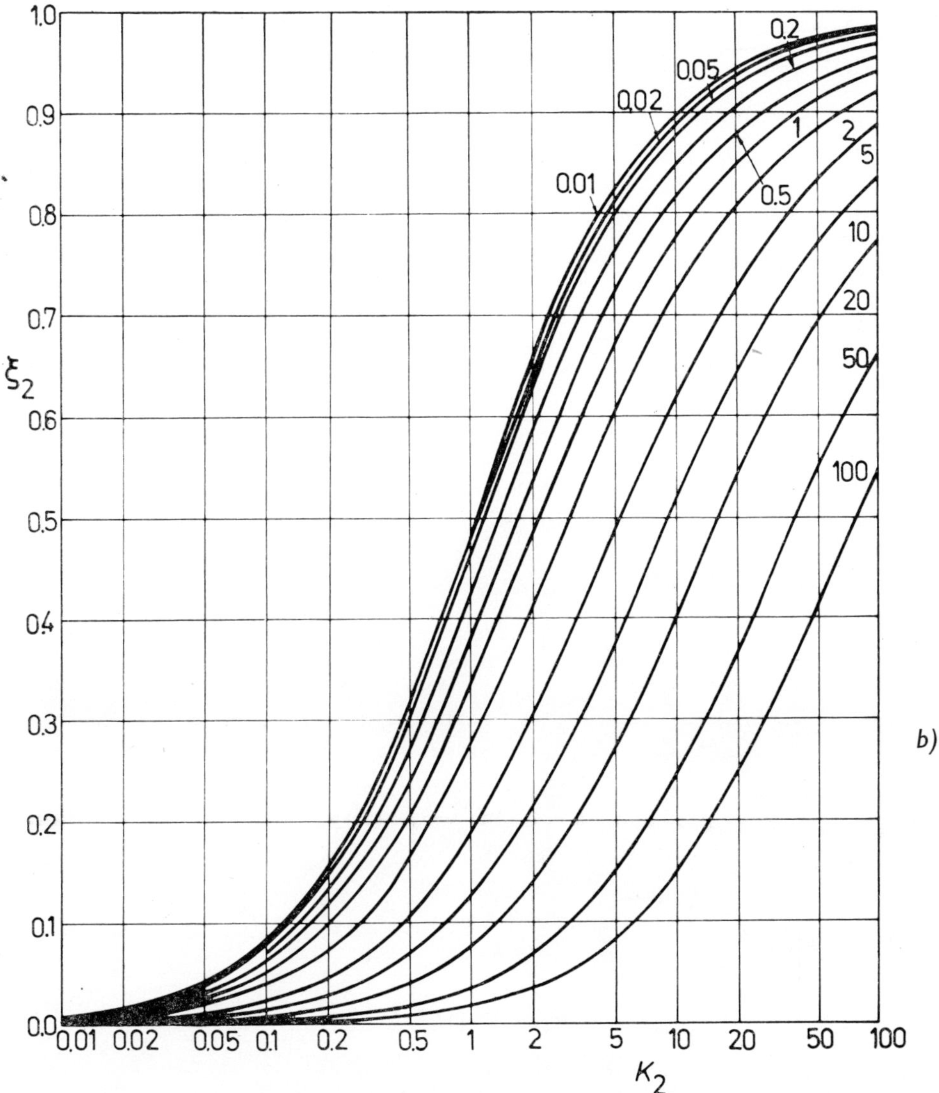

*Fig. 10*b — dependence of ξ_2.

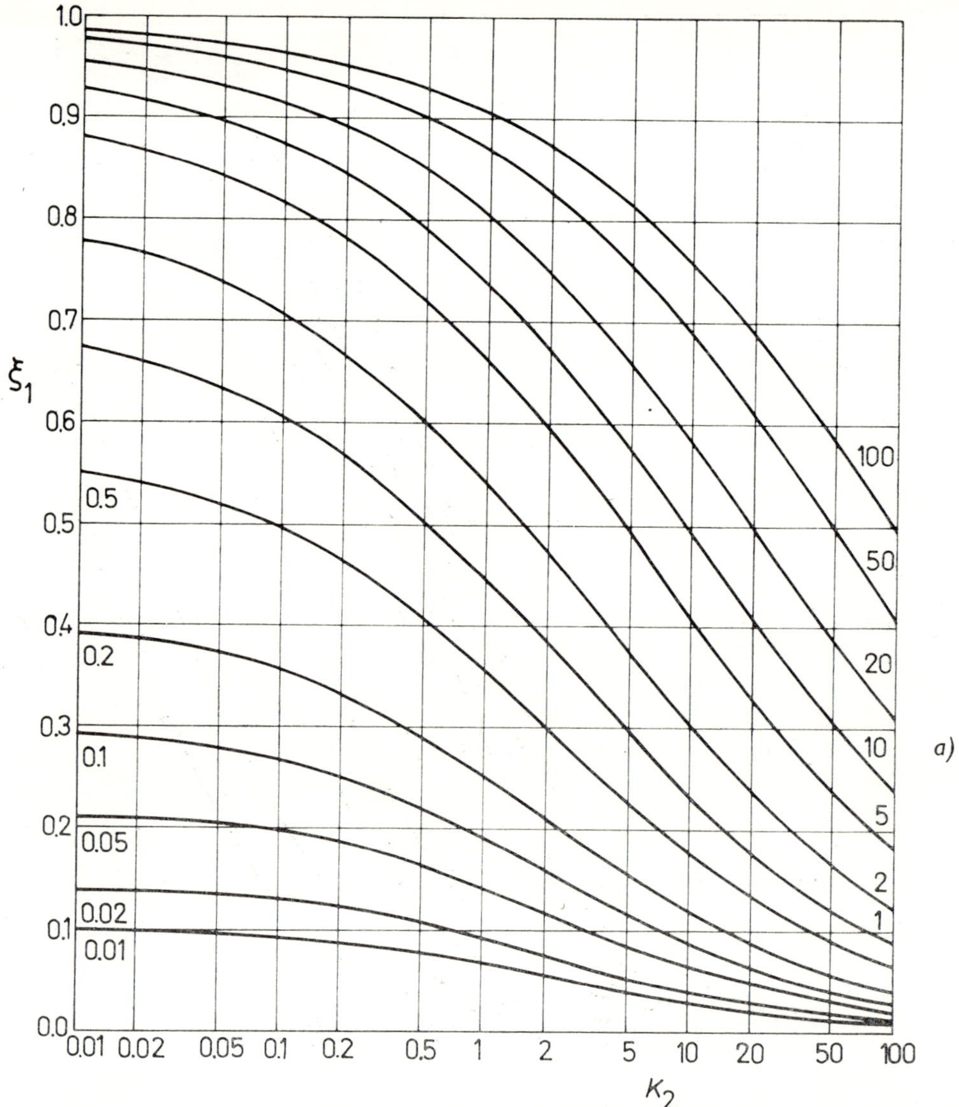

*Fig. 11*a. Dependence of reaction coordinates on the numerical value of equilibrium constants in the range $K_1, K_2 \in (0.01-100)$ in a system of three constituents and two chemical reactions according to stoichiometric equations (4.47) — dependence of ξ_1.

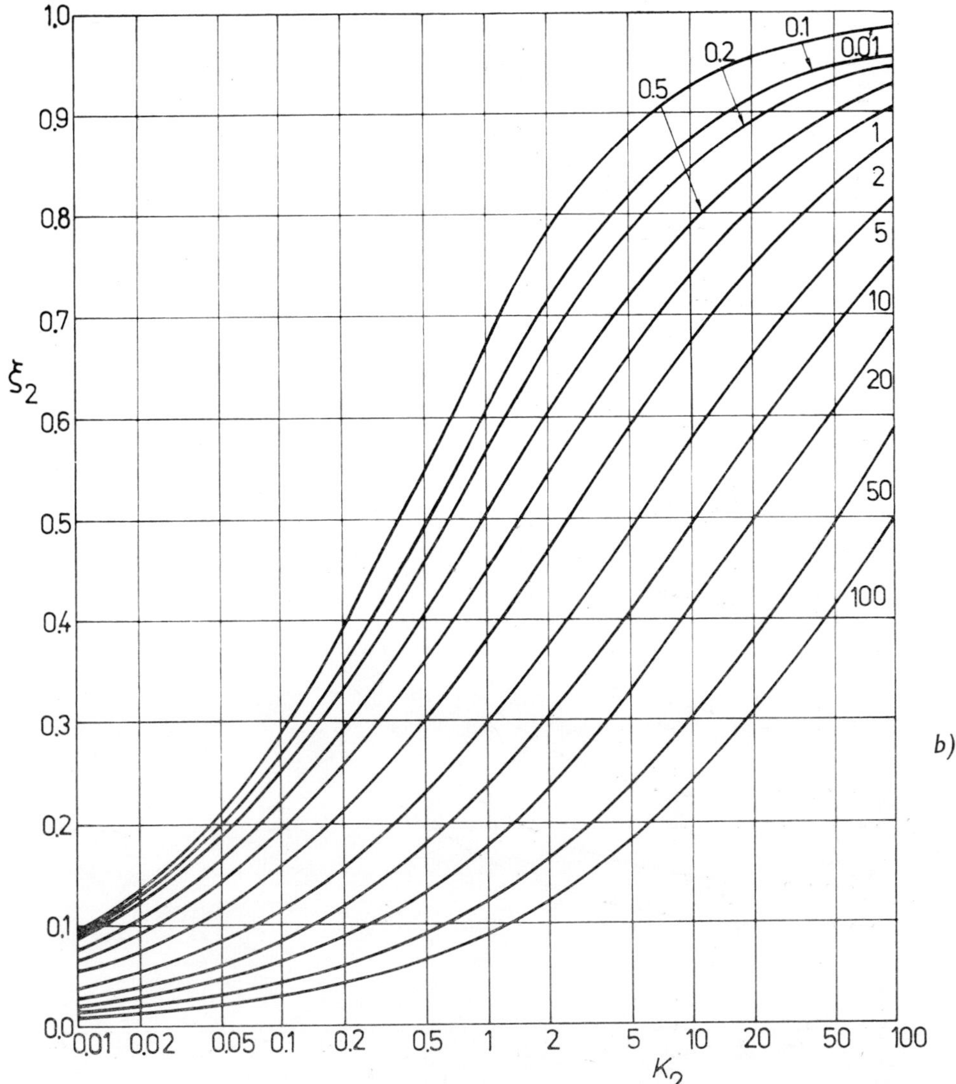

*Fig. 11*b — dependence of ξ_2.

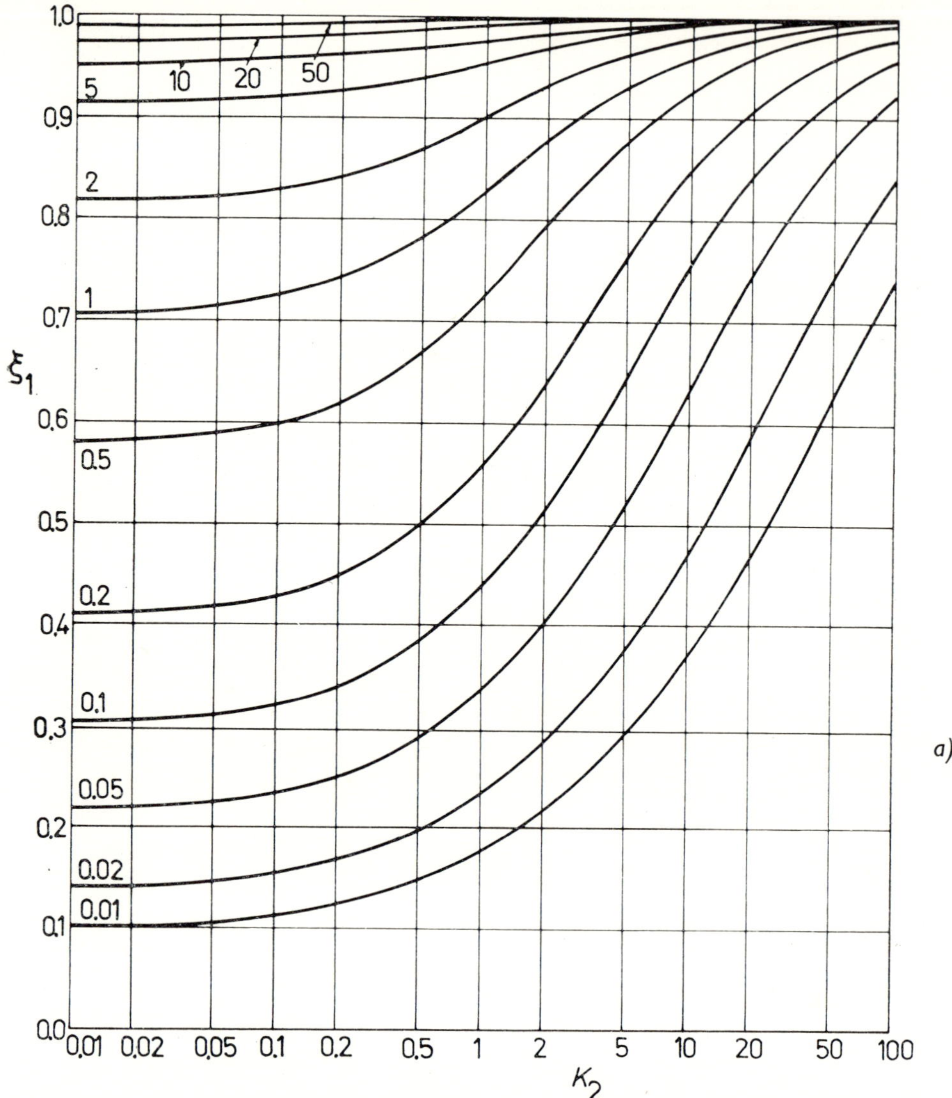

*Fig. 12*a. Dependence of reaction coordinates on the numerical value of equilibrium constants in the range $K_1 . K_2 \in (0.01 - 100)$ in a system of three constituents and two chemical reactions according to stoichiometric equations (4.48) — dependence of ξ_1.

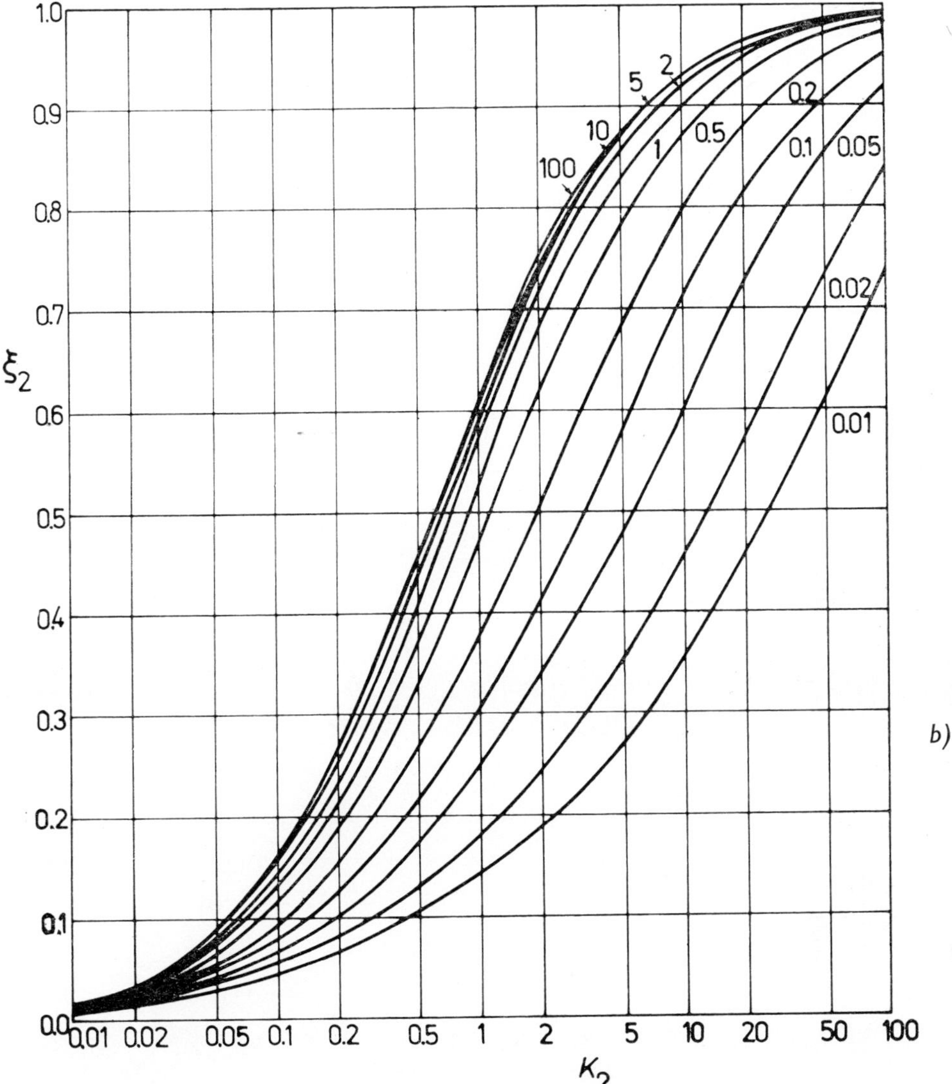

*Fig. 12*b — dependence of ξ_2.

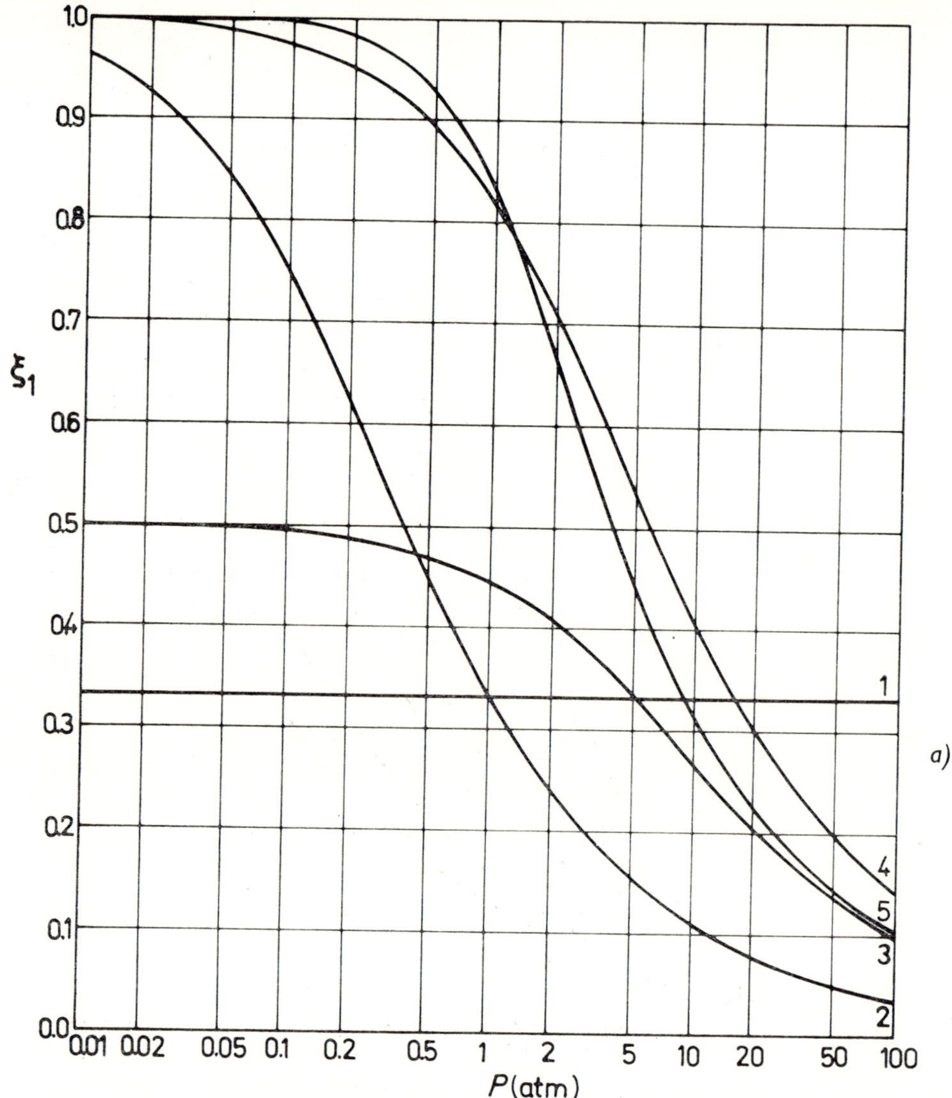

*Fig. 13*a. Dependence of reaction coordinates on pressure for the case of $K_1, K_2 = 1$ in the range of $P \in (0.01 - 100)$ atm — dependence of ξ_1.

1.	$-A_1 + A_2 = 0;$		$-A_1 + A_3 = 0$	
2.	$-A_1 + 2A_2 = 0;$		$-A_1 + A_3 = 0$	
3.	$-A_1 + 2A_2 = 0;$		$-A_1 + 2A_3 = 0$	
4.	$-A_1 + A_3 + A_4 = 0;$		$-A_1 + 2A_2 = 0$	
5.	$-A_1 + A_2 + A_3 = 0;$		$-A_2 + A_4 + A_3 = 0$	

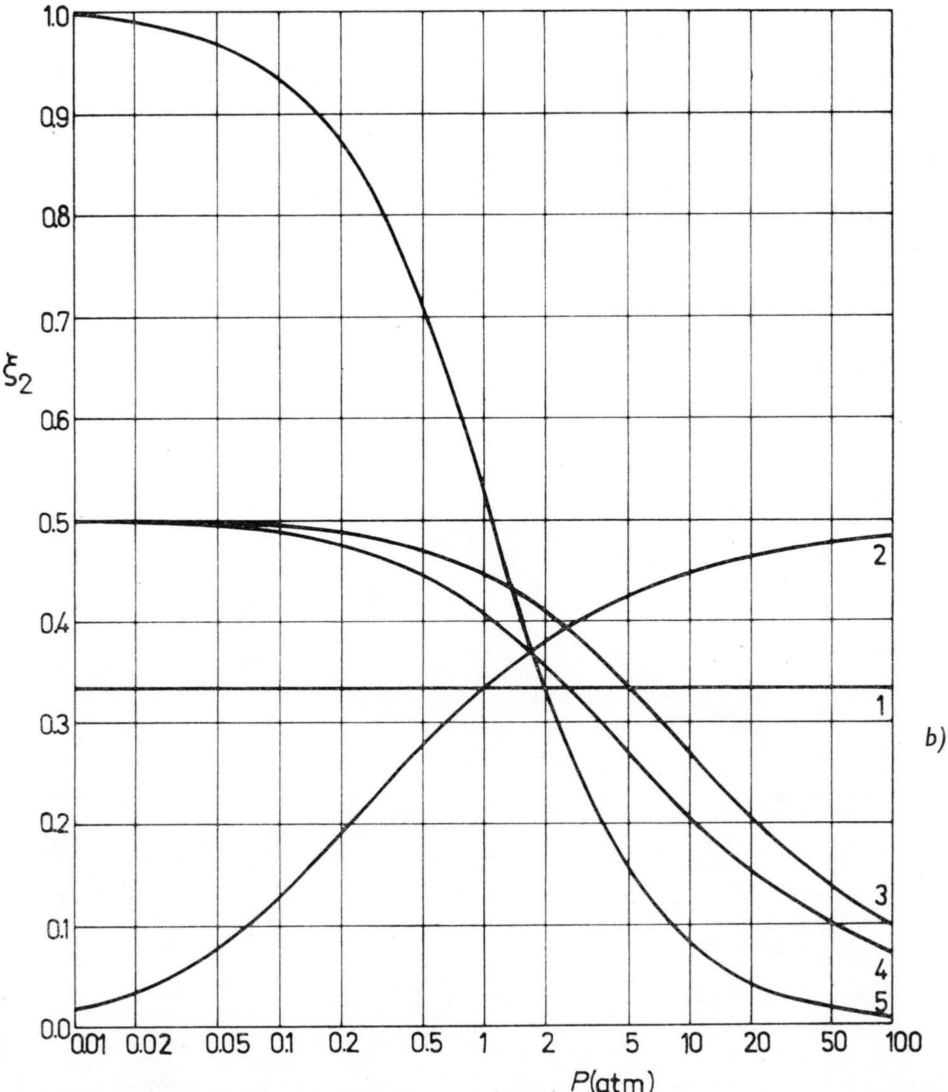

*Fig. 13*b — dependence of ξ_2.

Exercises:

1. Derive the general formulae for calculating the reaction coordinates under the assumption, that $K_2 = 0.1 K_1$ for the following reaction schemes:

a) $-A_1 + 2A_2 = 0$

      $-A_1 + A_3 \ = 0$

b) $-A_1 + 2A_2 = 0$

      $-A_1 + 2A_3 = 0$

c) $-A_1 + A_4 + 2A_3 + A_4 \ = 0$

      $-A_1 + A_4 + 2A_3 + A_5 \ = 0$

d) $-A_1 + A_2 + A_3 \ + A_4 \ = 0$

               $- A_3 \ + A_5 \ = 0$

e) $-A_1 - A_2 + A_3 \ + A_4 \ = 0$

               $- A_3 \ + 2A_5 = 0$

f) $-A_1 - A_2 + A_3 \ + A_4 \ = 0$

      $-A_3 - A_2 + A_5 \ + A_4 \ = 0$

g) $-A_1 - A_2 + A_3 \ + 2A_4 = 0$

      $-A_3 - A_2 + A_5 \ + A_4 \ = 0$

h) $-A_1 - A_2 - A_6 \ + A_3 + 2A_4 + A_7 = 0$

      $-A_3 - A_2 - A_6 \ + A_5 + A_4 \ + A_7 = 0$

i) $-A_1 - A_2 - A_6 \ + A_3 + 3A_4 + A_7 = 0$

      $-A_3 - A_2 - A_6 \ + A_5 + A_4 \ + A_7 = 0$

2. Write down possible chemical conversions corresponding to the schemes in Exercise 1, and as far as possible derive by stoichiometric analysis further matrixes of stoichiometric coefficients and corresponding chemical reactions for each system.

3. Using the formulae from Exercise 1, calculate the equilibrium composition of the mixture at 1 atm pressure. Assume $K_1 = 5$, ideal behaviour of the mixture and, with multi-component initial mixtures, assume equimolar composition and unit fugacity as the standard state.

4. Products of butane dehydrogenation may be 1-butene, 2-butene, cis and 2-butene, trans. The thermochemical data of the individual constituents for various temperatures are:

	$-(G° - H_0°)/T$ (cal K^{-1} mole^{-1})						$\Delta H_0°$ (cal mole^{-1})
T (K)	600	700	800	900	1000	1100	
butane	72.05	75.95	79.69	83.28	86.73	90.03	−23 332
1-butene	71.56	75.08	78.42	81.61	84.66	87.58	5 158
2-butene, cis	69.94	73.19	76.30	79.29	82.17	84.95	3 794
2-butene, trans	68.84	72.27	75.53	78.64	81.62	84.47	2 506
H$_2$	29.20	30.27	31.19	32.00	32.74	33.40	0

a) Construct the dependence of equilibrium composition on temperature in the 600 to 1100 K range at 1 atm pressure for the simple dehydrogenation reaction

$$\text{butane(g)} \;=\; \text{1-butene(g)} + \text{H}_2\text{(g)}.$$

b) Using data from Fig. 11, estimate the equilibrium composition of the mixture at 1000 K and 1 atm, for the case that the chemical conversion is expressed by the equations

$$\text{butane(g)} \;=\; \text{2-butene, cis(g)} + \text{H}_2\text{(g)},$$
$$\text{butane(g)} \;=\; \text{2-butene, trans(g)} + \text{H}_2\text{(g)}.$$

c) Calculate the composition of the equilibrium mixture at 1 atm pressure and 800 K, if 1-butene as well as both 2-butene isomers are formed by butane dehydrogenation.

d) What pressure is required in order that 92% conversion of butane to the reaction products should be achieved at 900 K in a dehydrogenation reaction, in which 1-butene, 2-butene, cis and 2-butene, trans are formed?

e) What amount of nitrogen must be added to achieve the same conversion as in case d), if we wish to work at 1 atm?

5. Generator gas, composed of 20 molar % CO, 15 molar % H$_2$, 5 molar % CO$_2$ and 60 molar % N$_2$ is burned at atmospheric pressure with air (21 molar % O$_2$, 79 molar % N$_2$) in a molar ratio 1 : 5 moles air. Calculate the equilibrium composition, assuming that combustion is isothermal, the mixture has ideal behaviour and the conversion process is expressed by the equations:

$$\text{H}_2\text{(g)} + 0{,}5\,\text{O}_2\text{(g)} \;=\; \text{H}_2\text{O(g)} \qquad (K_a = 150)$$
$$\text{CO(g)} + 0{,}5\,\text{O}_2\text{(g)} \;=\; \text{CO}_2\text{(g)} \qquad (K_a = 32)$$
$$0{,}5\,\text{N}_2\text{(g)} + 0{,}5\,\text{O}_2\text{(g)} \;=\; \text{NO(g)} \qquad (K_a = 0.1)$$

(Assume unit fugacity as the standard state.)

89

5 Chemical equilibrium of complex system in the ideal gas state

5.1 INTRODUCTION

The task of determining the equilibrium composition of multi-component systems is always linked to the ability of solving sets of algebraic equations. Therefore, procedures involving rational and speedy calculations were developed only recently, when numerical mathematics had already made available a number of methods of solving sets of equations, and the use of computers had enabled quick and sufficiently precise solutions, although the thermodynamic principles of the calculations were known for some time already. Thus there are at present several methods which can be used for practical purposes, permitting investigation of the chemical equilibrium of systems in the ideal gas state at constant temperature and pressure, practically unlimited in terms of complexity of the systems involved.

Several criteria may be employed to classify the methods[70,71,171]. One of these criteria is the nature of the method. In this case, methods are classified as optimation methods, gradient methods and methods demanding solution of a set of non-linear equations. Each of this group of methods differs from the rest by the use of a distinctly different mathematical apparatus. Another classification criterion may be the rate of convergence of the numerical method employed. In this case, methods are said to be of first, second, third and in general p-th order (see Appendix 1). Another possibility involves classification of the methods by the degree of generality of their application: one extreme are methods which were developed for a specific type of chemical reactions or a certain kind of processes, the other extreme includes methods which can be used quite generally irrespective of the dimensionality of the reaction system involved. As the most purposeful criterion we have selected one which corresponds to the two basic possibilities of expressing the mass balance, as stated in section 2.4.

When the mass balance is expressed by means of stoichiometric coefficients, i.e. in the form of

$$n_i = n_i^o + \sum_{r=1}^{R} v_{ri}\xi_r \quad i = 1, 2, ..., N ,\tag{5.1}$$

the method of calculating the chemical equilibrium can be characterised by the fact that we are searching for the minimum of the Gibbs function G

$$G = \sum_{i=1}^{N} n_i \mu_i = RT \sum_{i=1}^{N} n_i \left(c_i + \ln \frac{n_i}{n} \right), \tag{5.2}$$

substituting for n_i values from equations (5.1). There applies for the parameter c_i (an ideal gas system is involved)

$$c_i = \mu_i^{\circ}/RT + \ln P = G_i^{\circ}/RT + \ln P, \tag{5.3}$$

where μ_i° and G_i° is the molar free enthalpy of the pure component at the temperature of the system, pressure of 1 atm in the ideal gas state. The solution of this task may be approached from different point of view. On the one hand, a direct method may be employed, consisting of derivating the function G from equation (5.2) by ξ_r ($r = 1, 2, \ldots, R$) and solving the resulting set of R non-linear equations. This procedure, often called the method of equilibrium constants, is used often particularly for $R = 1$. Since the function G is a convex function of the variables $\xi_1, \xi_2, \ldots, \xi_R$ (see Appendix 6), different variants of the gradient method are the most successful methods. All these procedures have one condition in common, namely knowledge of the stoichiometry of the system, i.e. constuction of the matrix of stoichiometric coefficients.

When, on the other hand, the mass balance equations are expressed in terms of constitution coefficients, i.e. in the form of

$$\sum_{i=1}^{N} a_{ij} n_i = b_j \quad j = 1, 2, \ldots, M, \tag{5.4}$$

then the solution of the chemical equilibrium is equivalent to finding the minimum of the Gibbs function (5.2) on a set of points $(n_1, n_2, \ldots, n_N)$ satisfying the equation (5.4). This procedure does not necessitate a stoichiometric analysis of the system, since knowledge of the matrix of constitution coefficients will suffice for the application of the methods. It will be shown later (see section 5.4), that the problem of finding the minimum of the function (5.2) on a set of points (5.4) can be converted to solving a set of at most $(M + 1)$ non-linear equations for $(M + 1)$ unknown variables. It was only the development of methods of this kind that enabled solution of large system $(N \gg M)$, since even in these systems values of M are rarely greater than six or seven, while the value of $R = N - M$ is in no wise limited.

At a time when solving more complicated sets of algebraic equations caused considerable difficulties, the two groups of methods of truly general nature were preceded by the development of rather simpler single-purpose calculation schemes, designed for the study of chemical equilibria in specific reaction systems or groups of similar systems. Although these single-purpose procedures belong to that category of the above classification, in which stoichiometric analysis is required, yet they do not quite fall into this classification by their design and possibilities. Therefore it is better to discuss them separately as a historical introduction, presenting evidence of the manner in which chemical equilibria were treated before general solutions had been developed.

5.2 SINGLE-PURPOSE PROCEDURES

The first, practically usable procedure, which is at the same time the first attempt at solving complicated equilibria, was proposed by Damköhler and Edse[37] in connection with studies of the oxidation of propane by air at 2200 K and pressure of 40 atm. The authors assumed that the constituents H, H_2, OH, O, O_2, H_2O, NO, CO, CO_2 and N_2 would be present in equilibrium. Of the ten equations needed for the solution, they determined six by intuition, assuming the course of six independent reactions

$$CO_2 \; = \; CO + \tfrac{1}{2}O_2 \qquad K_1 = 0.0062 \tag{5.5a}$$

$$H_2O \; = \; H_2 + \tfrac{1}{2}O_2 \qquad K_2 = 0.0012 \tag{5.5b}$$

$$H_2O \; = \; \tfrac{1}{2}H_2 + OH \qquad K_3 = 0.00152 \tag{5.5c}$$

$$\tfrac{1}{2}H_2 \; = \; H \qquad K_4 = 0.0059 \tag{5.5d}$$

$$\tfrac{1}{2}O_2 \; = \; O \qquad K_5 = 0.0029 \tag{5.5e}$$

$$\tfrac{1}{2}N_2 + \tfrac{1}{2}O_2 \; = \; NO \qquad K_6 = 0.0348 \; . \tag{5.5f}$$

The form of the equilibrium relations was

$$K_r = \sum_{i=1}^{10} P_i^{v_{ri}} \quad (r = 1, 2, ..., 6), \tag{5.6}$$

where P_i is the partial pressure of the i-th constituent, v_{ri} the stoichiometric coefficient. The other four equations were then derived from the balance equations

$$n_j = \sum_{i=1}^{10} a_{ij} P_i \quad (j = 1, 2, ..., 4), \tag{5.7}$$

where n_j are numbers of gramatoms of the individual elements, i.e.:

$$n_C = P_{CO_2} + P_{CO}, \tag{5.8a}$$

$$n_H = 2P_{H_2O} + 2P_{H_2} + P_{OH} + P_H, \tag{5.8b}$$

$$n_O = 2P_{CO_2} + P_{CO} + P_{H_2O} + 2P_{O_2} + P_{OH} + P_O + P_{NO}, \tag{5.8c}$$

$$n_N = 2P_{N_2} + P_{NO}. \tag{5.8d}$$

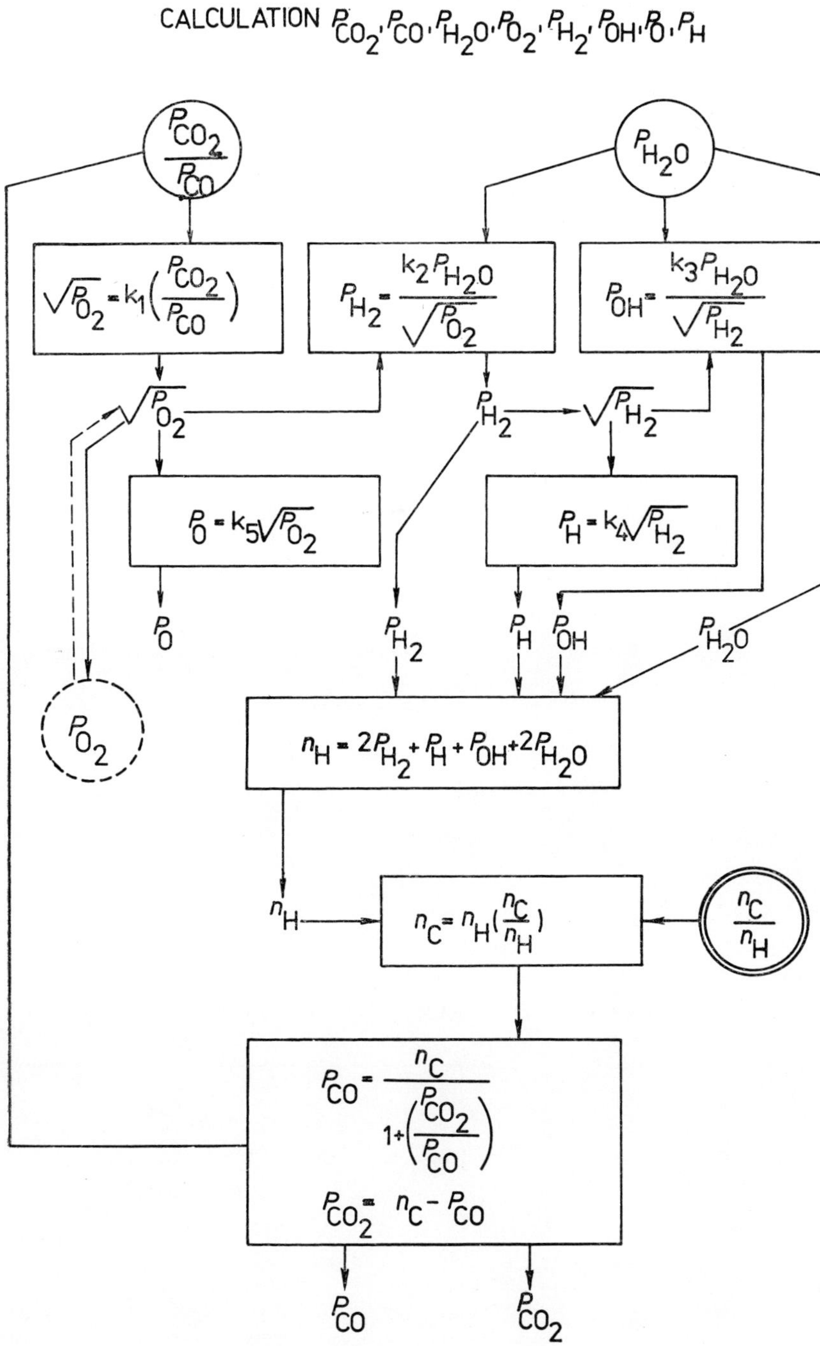

Fig. 14. Diagram of calculation of P_{CO} and P_{CO_2} according to the Damköhler — Edse procedure.

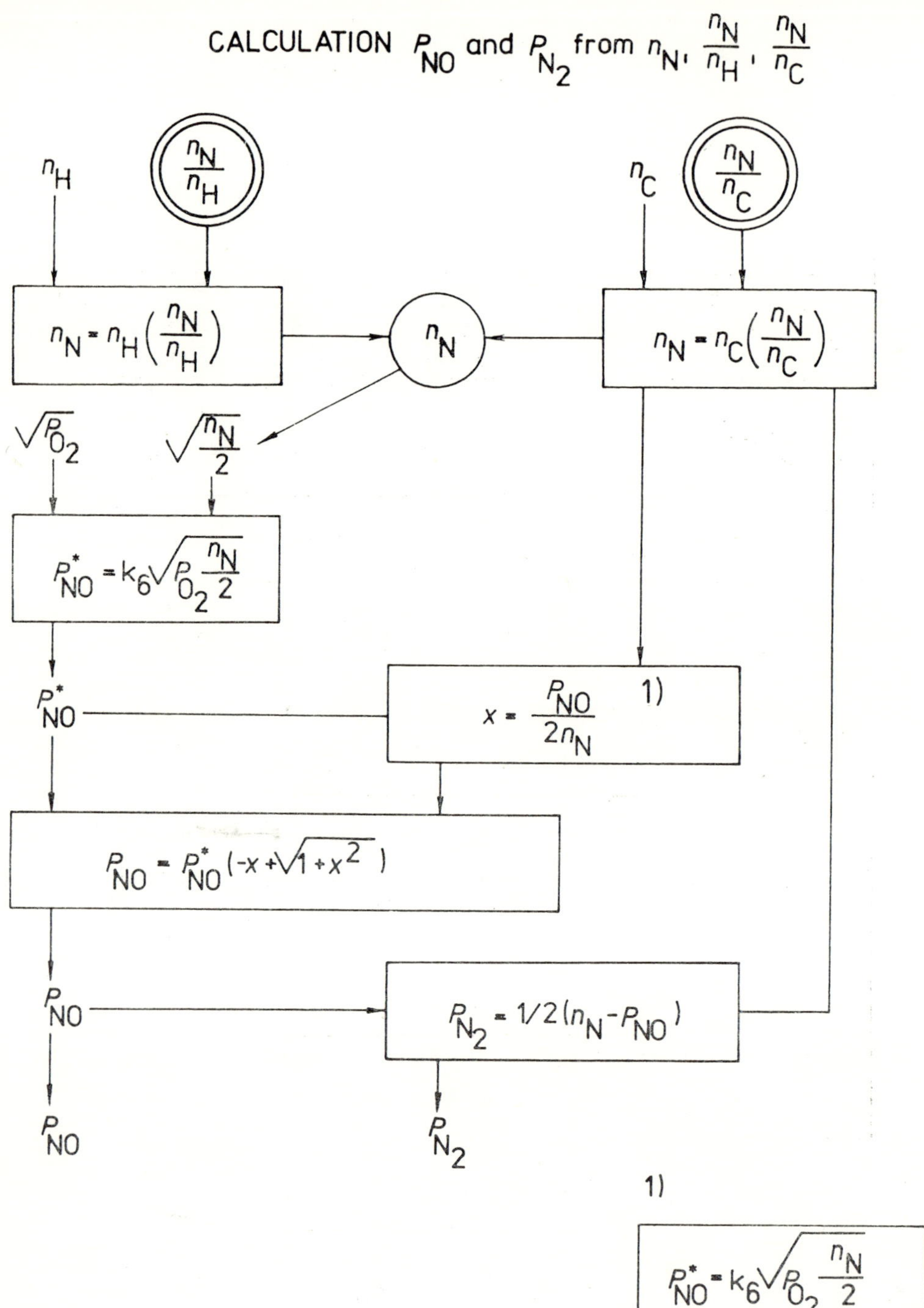

Fig. 15. Diagram of calculation of P_{NO} and P_{N_2} by the Damköhler — Edse procedure.

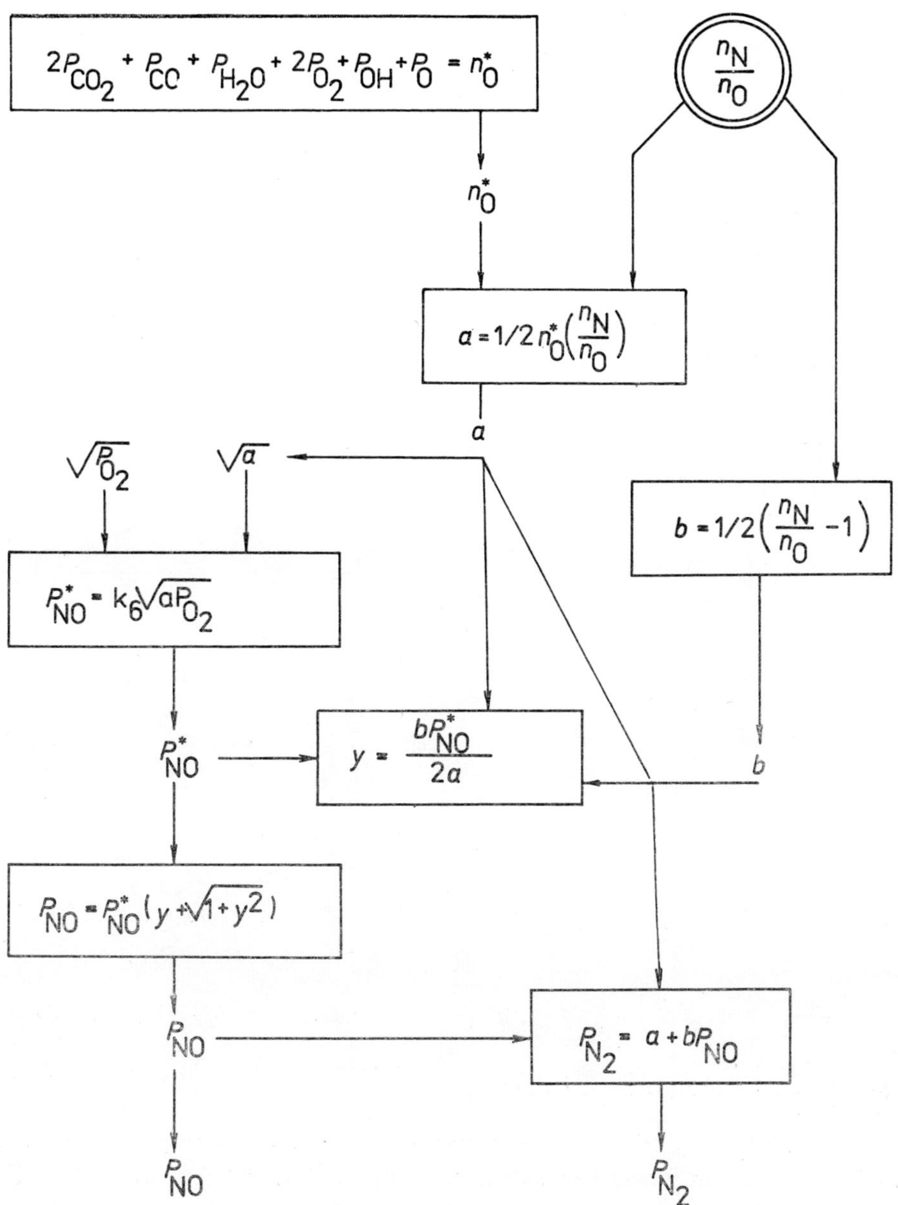

Fig. 16. Diagram of determination of partial pressures of remaining constituents by the Damköhler — Edse procedure.

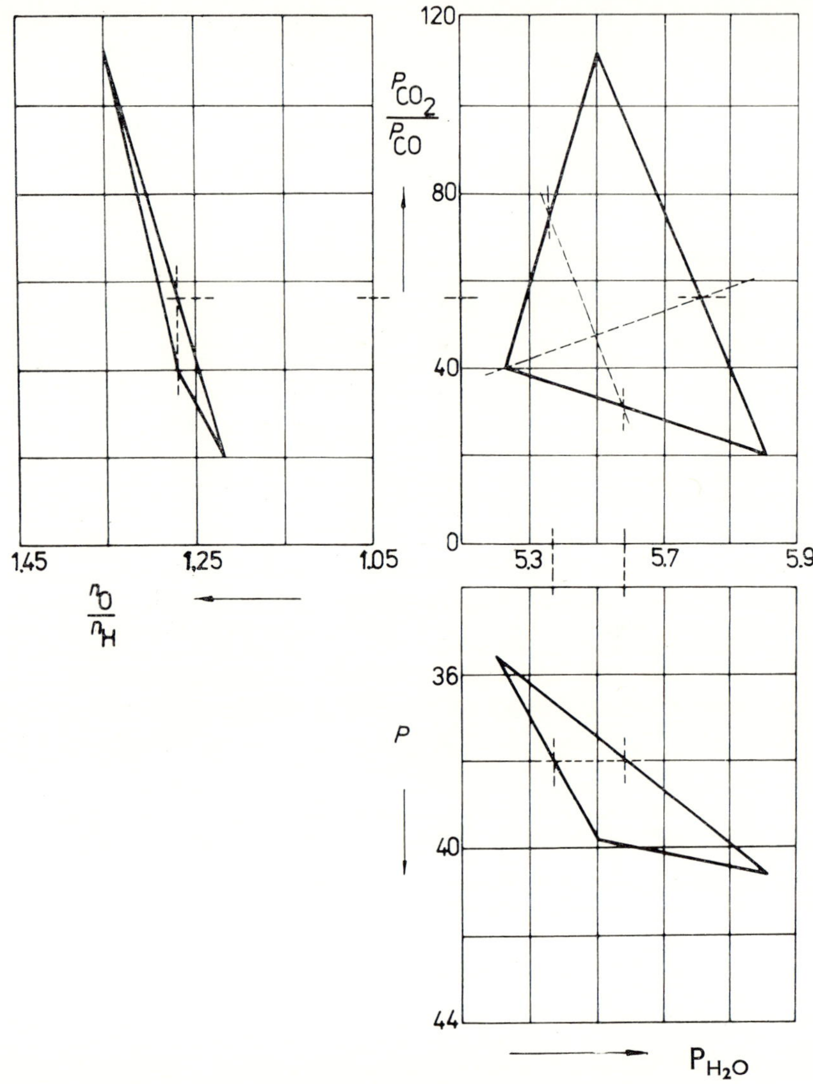

Fig. 17. Graphical determination of partial pressure values or the following approximation.

Damköhler and Edse proceeded thus:

1. They selected, by trial and error, values of the constituent P_{H_2O} and of the ratio P_{CO_2}/P_{CO}. From these, they calculated values of P_{CO} and P_{CO_2} according to the scheme shown in Fig. 14.

2. From the mass balance, they derived the ratio of n_C/n_H and any one of the ratios n_N/n_H, n_N/n_C or n_N/n_O.

3. They then calculated the complete set of values including the overall pressure according to the schemes shown in Figs. 15 and 16. They repeated this process two more times, thus obtaining three complete sets of values.

4. They now used the data obtained to construct three graphs with the coordinates

$$\frac{\dfrac{P_{CO_2}}{P_{CO}}}{P_{H_2O}}, \quad \frac{P}{P_{H_2O}} \quad \text{and} \quad \frac{\dfrac{P_{CO_2}}{P_{CO}}}{\dfrac{n_O}{n_H}} \quad \text{(see Fig. 17)}. \tag{5.9}$$

Planar sections are then taken through the graphs for the known initial conditions $P = 40$ atm and $n_O/n_H = 1.25$, from which initial values can be obtained for the calculation of the next approximation. The procedure is repeated until the required accuracy is achieved, which usually requires a series of five approximations.

Similar single-purpose procedures, the algorithm of which must be adapted or newly worked out separately for every new case, were independently derive by Kühl[97], Stein[146], Traustel[161], Peters, Kappelmacher a Voetter[122], Stein and Voetter[147], Fuchs and Glaser[56] and Kaeppeler and Baumann[86].

In an effort to generalize the calculation scheme, Villars[164] adopted a somewhat different procedure. Again purely by intuition, he divided all constituents which can occur in equilibrium, into basic and derived ones. For each derived constituent he constructed a stoichiometric equation in the form of

$$\sum_{i=1}^{10} v_{ri} A_i = 0 \quad (r = 1, 2, ..., R), \tag{5.10}$$

and he then defined for each constituent an equilibrium relation in the form of

$$K_r = \prod_{i=1}^{N} \left(n_i^o + \sum_{r=1}^{R} v_{ri}\xi_r \right)^{v_{ri}} \left(\frac{P}{n^o + \sum_{r=1}^{R} \xi_r v_r} \right)^{v_r} \quad (r = 1, 2, ..., R), \tag{5.11}$$

where n_i^o are initial numbers of moles of the individual constituents, $n^o = \sum_{i=1}^{N} n_i^o$, ξ_r is the reaction coordinate $v = \sum_{i=1}^{N} v_1$ and P is the overall pressure.

Villars mainly investigated equilibria in combustion of hydrocarbons, where a limited number of types of equilibrium relations occurs. This fact enabled him to work out the solution of each type of the equilibrium relation (5.11) as a sub-programme.

The iterative procedure consisted in determining the equilibrium composition separately for each reaction, on the basis of pre-selected values of the basic constituents. In each successive approximation, calculated values were corected by a system of balance relationships which are easy to construct, since the overall amount of each element in the system is invariable throughout the calculation.

Evidently, this procedure again cannot be generally valid. It has a certain advantage in the fact that the matrices which have to be converted are not too large, which means a saving in the magnitude of the operative memory of the computer: on the other hand, the rate of convergence of the calculation depends on the selection of the values of basic constituents.

To achieve more rapid convergence, Browne[20] suggested that the numbers of moles of individual constituents be arranged in series by decreasing magnitude. This makes sure that there will be a relative excess of the basic constituents, which means quicker convergence of the calculation, as far as the selected basic constituents satisfy the other essential conditions (see section 2.2).

Cruise[30] made use of this idea, modifying Villars' procedure by introducing into the calculation a criterion to determine, at every step, whether the value of the basic constituents is greater than the value of the derived constituents. In the reverse case the constituents are rearranged so as to satisfy this condition.

Goldwasser[58] set out from a different notion. His procedure is based on a stoichiometric description of the reactions which take place in the system and he also uses the equations (5.10), but instead of using the equilibrium relations (5.11), Goldwasser describes the system by means of a set of equations, in which he has introduced the quantity β_r, called correction factor, instead of the exact unknown variable ξ_r. The physical significance of this quantity may be visualized as a reaction coordinate in systems in which every reaction that can take place is considered quite isolated from the course of the other reactions, its algebraic form being simplified by neglection of higher-order terms. For example, there holds for the reaction

$$A_1 + A_2 = A_3 \tag{5.12}$$

that

$$K_1 = \frac{n_3}{n_1 n_2} \left(\frac{n}{P} \right). \tag{5.13}$$

Considering that

$$n_i = n_i^o + \xi v_i, \tag{5.14}$$

we obtain

$$K_1 = \frac{(n_3^o - \xi)}{(n_1^o - \xi)(n_2^o - \xi)} \left(\frac{n^o - \xi}{P} \right) \tag{5.15}$$

which may furthermore be written as

$$\xi^2(1 + PK_1) - \xi[(n_1^o - n_2^o) PK_1 + n^o - n_3^o] + PK_1 n_1^o n_2^o - n_3^o n^o = 0. \tag{5.16}$$

Instead of the exact equation (5.16) Goldwasser introduces the relationship

$$\beta = \frac{PK_1 n_1^o n_2^o - n_3^o n^o}{PK_1(n_1^o + n_2^o) + (n^o - n_3^o)}, \tag{5.17}$$

by neglecting terms of higher order (because of the low expected value of ξ) and after deciding upon the sign.

Equation (5.17) is solved, and the result is corrected by means of equation (5.14), which must be re-written in the form of

$$n_i = n_i^{\circ} + \beta v_i .\qquad(5.18)$$

The result is obtained, when all relations (5.18) have been satisfied for a given value of β.

For more complicated systems, simplified equilibrium relationships are employed, with initial concentrations replaced by immediate ones. Relations derived from the mass balance take the form of

$$n_i = n_i^{\circ} + \sum_{i=1}^{R} \beta_r v_{ri} .\qquad(5.19)$$

Let us describe an example for the sake of illustration. The system is composed of four elements C, H, O, N, and eleven constituents H_2, H_2O, CO, CO_2, O_2, N_2, NO, OH, H, O and N. The matrix of constitution coefficients is

$$
\begin{vmatrix}
0 & 2 & 0 & 0 \\
0 & 2 & 1 & 0 \\
1 & 0 & 1 & 0 \\
1 & 0 & 2 & 0 \\
0 & 0 & 2 & 0 \\
0 & 0 & 0 & 2 \\
0 & 0 & 1 & 1 \\
0 & 1 & 1 & 0 \\
0 & 1 & 0 & 0 \\
0 & 0 & 1 & 0 \\
0 & 0 & 0 & 1
\end{vmatrix}
$$

Overall chemical conversion is expressed by the following relations:

$$
\begin{aligned}
H_2 + CO_2 &= CO + H_2O & (r = 1) \\
H_2 + \tfrac{1}{2}O_2 &= H_2O & (r = 2) \\
H_2 + O &= OH + H & (r = 3) \\
H_2 + 2N &= N_2 + 2H & (r = 4) \\
H_2 + O &= H_2O & (r = 5) \\
H_2 + O_2 &= 2OH & (r = 6) \\
H_2 &= 2H & (r = 7) \\
H_2 + NO &= \tfrac{1}{2}N_2 + H_2O & (r = 8) .
\end{aligned}
\qquad(5.20)
$$

The correction factors β_r are determined by the relations:

$$\beta_1 = \frac{K_1 n_1 n_4 - n_2 n_3}{K_1(n_1 + n_4) + n_2 + n_3}$$

$$\beta_2 = \frac{2PK_2^2 n_1^2 n_5 - 2n_2^2 n}{PK^2 n_1(n_1 + n_5) + n_2(4n - n_2)}$$

$$\beta_3 = \frac{K_3 n_1 n_{10} - n_8 n_9}{K_3(n_1 + n_{10}) + n_8 + n_9}$$

$$\beta_4 = \frac{K_4 n_1 n_{11}^2 - n_6 n_9^2}{K_4 n_{11}(n_{11} + 4n_1) + n_9(n_9 + 4n_6)}$$

$$\beta_5 = \frac{PK_5 n_1 n_{10} - n_2 n}{PK_5(n_1 + n_{10}) + n - n_2}$$

$$\beta_6 = \frac{K_6 n_1 n_5 - n_8^2}{K_6(n_1 + n_5) + 4n_8}$$

$$\beta_7 = \frac{K_7 n\, n_1 - Pn_{10}^2}{K_7(n - n_1) + 4Pn_9}$$

$$\beta_8 = \frac{PK_8^2 n_1^2 n_7^2 - n_6 n_2^2 n}{2PK_8^2 n_1 n_7(n_1 + n_7) + 2n_2 n_6 n + \frac{1}{2} n_2^2(n - n_6)} . \tag{5.21}$$

The relations which express validity of the mass balance take the form of:

$$n_1 = n_1^\circ - \beta_1 - \beta_2 - \beta_3 - \beta_4 - \beta_5 - \beta_6 - \beta_7 - \beta_8$$

$$n_2 = n_2^\circ + \beta_1 + \beta_2 + \beta_5 + \beta_8$$

$$n_3 = n_3^\circ + \beta_1$$

$$n_4 = n_4^\circ - \beta_1$$

$$n_5 = n_5^\circ - \tfrac{1}{2}\beta_2 - \beta_6$$

$$n_6 = n_6^\circ + \beta_4 + \tfrac{1}{2}\beta_8$$

$$n_7 = n_7^\circ - \beta_8$$

$$n_8 = n_8^\circ + \beta_3 + 2\beta_6$$

$$n_9 = n_9^\circ + \beta_3 + 2\beta_4 + 2\beta_7$$

$$n_{10} = n_{10}^\circ - \beta_3 - \beta_5$$

$$n_{11} = n_{11}^\circ - 2\beta_4$$

$$n = n^\circ - \tfrac{1}{2}\beta_2 - \beta_5 + \beta_7 - \tfrac{1}{2}\beta_8 . \tag{5.22}$$

Evidently this procedure has several advantages. The equilibrium relationships are relatively simple, and their number — due to the manner in which they are derived — is not limited to linearly independent ones. Conversely, it is usually advantageous for the sake of rapid convergence to employ more equations than would follow from the stoichiometric balance. The method, however, has grave disadvantages too. It is a single-purpose method, i.e. it must be modified for every specific application, convergence of the calculation being uncertain. The rate of convergence is primarily influenced by the initial set of values, obtained by trial and error, in which zero values should not be included. Convergence may likewise be influenced by the fact, that the stoichiometric equations are constructed in a manner which makes sure that one of the constituents takes part in a maximum number of reactions. Even if all these prerequisites are satisfied, the rate of convergence is rather low and the number of steps required is large.

5.3 GENERAL METHODS REQUIRING STOICHIOMETRIC ANALYSIS OF THE SYSTEM

5.3.1 Principle of the methods

The purpose of this group of methods is to find the minimum of the function $Q(Q = G/RT)$

$$Q(\xi_1, \xi_2, ..., \xi_R) = \sum_{i=1}^{N} n_i \left(c_i + \ln \frac{n_i}{n} \right), \tag{5.23}$$

where n_i values are governed by relation (5.1). Derivating the right-hand side of equation (5.23) by ξ_r and taking the result as equal to zero, we obtain

$$\sum_{i=1}^{N} v_{ri} \left(c_i + \ln \frac{n_i}{n} \right) = 0 \quad r = 1, 2, ..., R . \tag{5.24}$$

Considering, that there holds

$$\sum_{i=1}^{N} v_{ri} c_i = \sum_{i=1}^{N} v_{ri} \left(\frac{\mu_i^o}{RT} + \ln P \right) = \frac{\Delta G_r^o}{RT} + v_r \ln P , \tag{5.25}$$

where $v_r = \sum_{i=1}^{N} v_{ri}$ and $\Delta G_r^o = -RT \ln (K_a)_r$, relations (5.24) may be written as

$$(\ln K_a)_r = v_r \ln P - v_r \ln n + \sum_{i=1}^{N} v_{ri} \ln n_i \quad r = 1, 2, ..., R , \tag{5.26}$$

or, in the form of

$$(K_a)_r = \left(\frac{P}{n}\right)^{v_r} \prod_{i=1}^{N} n_i^{v_{ri}} \quad r = 1, 2, ..., R .$$ (5.27)

which is the well-known form of the equilibrium constant of the r-th reaction.

On substitution of the mass balance equation into the equations (5.26), a set of R non-linear equations is obtained for R unknown variables $\xi_1, \xi_2, ..., \xi_R$, which may written schematically as

$$F_i(\xi_1, \xi_2, ..., \xi_R) = 0 \quad i = 1, 2, ..., R .$$ (5.28)

The set of equations (5.28) may be solved directly, which however is rather disadvantageous in the case of large systems. For this reason, several variants of the gradient method as well as some other types of procedures were developed, in order to obtain speedy convergency of the solution with respect to the convex nature of the Gibbs function (see Appendix 3).

5.3.2 Newton's method with reduction parameter

Newton's method is one of the most frequently used numerical methods of solving sets of non-linear equations[131]. Its advantage is very rapid convergence, since a second-order method is involved. The disadvantage is relatively great dependence of convergence on the choice of the first approximation. To a large extent, this disadvantage may be eliminated by selection of the reduction parameter.

Let the set Ω be a set of such R-membered groups of numbers $(\xi_1, \xi_2, ..., \xi_R)$, for which there holds

$$n_i = n_i^o + \sum_{r=1}^{R} v_{ri}\xi_r > 0 \quad i = 1, 2, ..., N .$$ (5.29)

Let us chose $(\xi_1^{(0)}, \xi_2^{(0)}, ..., \xi_R^{(0)}) \in \Omega$ and develop the left-hand side of the set (5.29) into a Taylor series in the point $(\xi_0^{(1)}, \xi_2^{(0)}, ..., \xi_R^{(0)})$, limiting the process to the linear terms only. This gives a set of linear equations

$$\sum_{r=1}^{R} \frac{\partial F_i}{\partial \xi_r} \Delta\xi_r = -F_i(\xi_1^{(0)}, \xi_2^{(0)}, ..., \xi_R^{(0)}) \quad i = 1, 2, ..., N ,$$ (5.30)

where $\Delta\xi_r = \xi_r - \xi_r^{(0)}$. Values of the derivation $\partial F_i/\partial \xi_r$ are calculated in the point $(\xi_1^{(0)}, \xi_2^{(0)}, ..., \xi_R^{(0)})$. The set of equations (5.30) is a set of R linear equations for R unknown variables $\Delta\xi_r$ $(r = 1, 2, ..., R)$. Such a set is easily solved e.g. by the Gauss method. Another approximation of the solution of the set (5.28) is selected according to the formula $\xi_r^{(1)} = \xi_r^{(0)} + \varepsilon \Delta\xi_r$ $(r = 1, 2, ..., R)$, where ε is the reduction parameter. The reduction parameter is selected so as to satisfy the condition $(\xi_1^{(1)}, \xi_2^{(1)}, ..., \xi_R^{(1)}) \in \Omega$. The reduction parameter can also help to make sure that the vector $\xi^{(1)}$ will not differ too much from the vector $\xi^{(0)}$. In the following procedure, the vector $\xi^{(1)}$ is used in the same way as the vector $\xi^{(0)}$ was used in the first step of the numerical process. In this manner, a series of vectors $\xi^{(2)}, \xi^{(3)}, ...,$ is obtained, which experience shows will converge in the absolute majority of cases very rapidly to the solution of the set of equations (5.28).

The weakest point of the above algorithm is finding the value of $\xi^{(0)}$. Although the set Ω contains an infinite number of vectors, yet it is very difficult and time-consuming to construct a general algorithm, by means of which at least one such vector would be found for a given initial composition n_i^o $(i = 1, 2, ..., N)$ and given matrix of stoichiometric coefficients.

For large systems (large values of R) this numerical procedure will take too much time. In such cases it is preferable to apply methods described in section (5.4), in which the overall enthalpy of the system is used to characterize chemical equilibrium.

5.3.3 Brinkley's method

Brinkley[15−17] contributed to the development of methods of measuring chemical equilibria in complex systems by being the first to formulate a general procedure, by means of which the equilibrium composition can be determined with no limitation. Together with Kandiner[88,89] he also proposed several methods of solving the sets of algebraic equations involved. Briefly, the procedure is as follows:

All prerequisites listed in section (5.1) being fulfilled, the mass balance equation (5.4) can be modified, e.g. by the Gaussian modification, into the form

$$n_i = q_i - \sum_{r=H+1}^{N} v_{ri} n_r \quad i = 1, 2, ..., H , \qquad (5.31)$$

where q_i, v_{ri} are coefficients which depend only on the constitutions coefficients and number of gratoms of individual elements in the system. In physical terms, equations (5.31) represent the balance relations for the basic constituents. For each derived constituent, stoichiometric equations can be designed to express the reaction of formation of the respective constituent from the basic constituents

$$0 = A_r - \sum_{i=1}^{H} v_{ri} A_i \quad r = H + 1, H + 2, ..., N . \qquad (5.32)$$

A total of $N - H$ equilibrium conditions, expressed in the following form, can be set up for the stoichiometric equations (5.32):

$$n_r = K_r \left(\frac{P}{n}\right)^{(v_r - 1)} \prod_{i=1}^{H} n_i^{v_{ri}} \quad r = H + 1, H + 2, ..., N , \qquad (5.33)$$

where

$$n = n_r + \sum_{i=1}^{H} n_i , \quad v_r = \sum_{i=1}^{H} v_{ri} \quad r = H + 1, H + 2, ..., N .$$

Connecting together equations (5.31) and (5.33) we obtain a set o N equations for N unknown variables $n_1, n_2, ..., n_N$, which may be written in the general form of

$$n_i = \Phi_i(n_1, n_2, ..., n_N) \quad i = 1, 2, ..., N . \tag{5.34}$$

The right-hand sides of the first H equations of the set of equations (5.34) depend only on the variables $n_{H+1}, n_{H+2}, ..., n_N$. The set (5.34) can be solved with the use of the iterative procedure. We select an initial set of values $n_1^{(1)}, n_2^{(1)}, ..., n_N^{(1)}$, substitute it into the set (5.34) and thus obtain a new set of data $n_1^{(2)}, n_2^{(2)}, ..., n_N^{(2)}$ with which the process is repeated. Newton's method can also be used. Brinkley's method involves in principle the same difficulties as stated in section 5.3.2.

a) The initial set of data $n_1^{(1)}, n_2^{(1)}, ..., n_N^{(1)}$ must be selected so as to give positive values of $n_1, n_2, ..., n_H$ on substitution into the first H equations of the set of equations (5.34). The difficulty of this problem should not be underestimated.

b) A non-linear set of N equations is being solved for N variables. For complex systems $(N > 10)$ this method therefore is unsuitable.

5.3.4   Modification of Brinkley's method

The fact, that Brinkley's method was a significant step towards a general description of chemical equilibria of multi-component systems led to the design of numerous other procedures which, in fact, were modifications of the original method. Such procedures were proposed e.g. by Huff, Gorden, Morell[79] and Zeleznik[172] and Martin and Yachter[103].

The method reported by Horn, Schüller[74] and Troltenier[75] is based on a somewhat different principle. Compared to the preceding methods, however, this one presents no obvious advantages and, therefore, will not be discussed here.

A very advantageous procedure, again in principle based on Brinkley's scheme, was suggested by Scully[142]. Construction of two sets of equations, a set of balance equations and a set derived from equilibrium, relationships is quite identical with the preceding procedures. The result are the relations (5.31), (5.32) and (5.33).

Scully's improvement mainly involves conversion of the non-linear equations which express equilibrium relationships into linear equations by converting to logarithms

$$\log n_r = (v_r - 1)\left[\log K_r + \log \frac{P}{n}\right] + \sum_{i=1}^{H} v_{ri} \log n_i \quad r = (H + 1), ..., N . \tag{5.35}$$

The calculation is based on trial-and-error substitution of values for the basic constituents, until the results agree within the limits of required precision. The equality $n_i^{(o)} = q_i$ is chosen for the initial set of data, hence we take $n_r^{(o)} = 0$ at the start $(r = H + 1, ..., N)$.

A first set of data for the derived constituents is obtained from the relationship (5.31), and substitution into the relations (5.33) leads to a new set of data for the basic constituents $n_i^{(1)}$. In the following steps, the procedures are separate. Instead of q_1, taken in the initial step as the value of the basic constituent $n_1^{(1)}$, we now chose

$$n_1^{(1)} = q_1 - \mathrm{d}\lambda , \tag{5.36}$$

where d is the reduction parameter, λ is the magnitude of the step. The calculation is done with constant values of $n_i^{(p-1)} = n_i^{(p)}$ $(i \neq 1)$.

Then, the last n_1 value is left constant and the value of n_2 is decreased according to the relation (5.36). When all n_i data have been dealt with in this manner, the step magnitude is decreased (usually tenfold) and the process is repeated. Thus, any required accuracy is achieved by constant repetition.

The calculation scheme is very simple and appears elegant at first sight. It seems, however, that it has no marked advantage over the original Kandiner-Brinkley method, since the simplicity of the calculation is lost by the need of introducing a number of logical conditions to secure convergence.

Attempting to avoid the drawbacks of the Goldwasser method (see equations (5.12) − (5.22)), Hutchinson[82] introduced a procedure in which the equilibrium conditions are formulated in general form. This eliminates the need of adapting the scheme to specific tasks. Each reaction is considered in the general form

$$\sum_{i=1}^{N} v_{ri} A_i = 0 \quad (r = 1, 2, ..., R) . \tag{5.37}$$

The number of reactions R is not limited by stoichiometric analysis of the system. After conversion to logarithms, the equilibrium conditions take the form of

$$\ln K_r = \sum_{i=1}^{N} (v_{ri} \ln n_i) - (v_r \ln n) + v_r \ln P \quad (r = 1, 2, ..., R) . \tag{5.38}$$

The relationship between equilibrium composition n_i $(i = 1, 2, ..., N)$ and the set of values, obtained by the p-th approximation, may then be defined as

$$n_i = n_i^{(p)} + \sum_{r=i}^{R} v_{ri} \beta_r^{(p)} \quad (i = 1, ..., N) , \tag{5.39}$$

where $\beta_r^{(p)}$ is a correction factor, expressing the deviation between the immediate state in the p-th step and equilibrium.

Substitution of equation (5.39) into (5.38), which is then developed into a series, and neglection of the higher-order terms leads to the first set of operative equations in the form of

$$\ln K_r = \sum_{i=1}^{N} \left(v_{ri} \ln n_i^{(p)} \right) - \left(v_r \ln n^{(p)} \right) + v_r \ln P +$$

$$+ \left[\sum_{i=1}^{N} \left(\frac{v_{ri}^2}{n_i^{(k)}} \right) - \frac{v_r^2}{n^{(k)}} \right] \beta_r^{(p)} \quad (r = 1, 2, ..., R). \tag{5.40}$$

In order to maintain validity of the equilibrium conditions, there must necessarily hold

$$\beta_r^{(p)} = \frac{\ln K_r - \ln K_r^{(p)}}{\sum_{i=1}^{N} \left(\frac{v_{ri}^2}{n_i^{(p)}} \right) - \frac{v_r^2}{n^{(p)}}}. \tag{5.41}$$

Together with the equations (5.39), now written as

$$n_i^{(p+1)} = n_i^{(p)} + \sum_{r=1}^{R} v_{ri} \beta_r^{(p)} \quad (i = 1, 2, ..., N), \tag{5.42}$$

we obtain a minimum of N equations with the variables β_r and n_i. The term $\ln K_r^{(p)}$ is calculated by substituting values of $n_i^{(p)}$ into equation (5.28). The calculation is done by means of the gradual approximations method, alternating balance and equilibrium sets of equations, until the required accuracy is obtained. The initial set of $n_i^{(1)}$ data is obtained from the known initial composition of the system $n_i^{(0)}$ and selected values of $\beta_r^{(0)}$, which however must be selected such as to give a real set of $\beta_r^{(1)}$ values. The author states, that an accuracy of 10^{-5} mole is required to calculate the chemical equilibrium of a system with 10 constituents involving 5 chemical reactions.

Although all these procedures were derived in an effort at improving Brinkley's original method, none of them succeeded in eliminating the basic inherent difficulties.

5.3.5 The NASA method

One modification of the method, which deserves more detailed discussion due to its interesting principle, was developed by the NASA staff in 1951[79]. In the 1960 to 1962 period, other NASA scientists, particularly Zeleznik and Gordon[60,80,172,173,174] investigated this method in more detail. It is based on the idea, that two of the following thermodynamic functions: temperature, pressure, volume, enthalpy, entropy etc. can be selected as independent variables to characterize the thermodynamic state of the system. This possibility of selection is a great practical advantage. The procedure is quite general and may be applied to any method of calculating the chemical equilibrium. Brinkley's method was selected, because it was one of the few methods of general character available in 1951. The principle involves amplification of the set of equations which determine the conditions of chemical equilibrium in the given system, by another equation the validity of which guarantees e.g. $\Delta H = 0$, where $\Delta H = H_1 - H_2$, H_1 being enthalpy of the system at the beginning of the reaction and H_2 enthalpy of the system in equilibrium. Addition of one more equation, i.e. addition of one more linking condition, increases the number of independent variables by one. Let us take an example. Assume that the enthalpy vs. temperature relationship is known for all compounds present in the system. H_T° values are usually tabulated for $T = 0$ K or $T = 298$ K. To determine the temperature dependence of the enthalpy of the i-th compound, we must also know the temperature dependence of the heat capacity C_P, since there holds

$$H_T^\circ = H_{T_1}^\circ + \int_{T_1}^{T} C_P^\circ(T)\,\mathrm{d}T. \tag{5.43}$$

We now require that enthalpy of the system in equilibrium be equal to a known value H. This gives a linking condition

$$H = \sum_{i=1}^{N} n_i (H_T^\circ)_i, \tag{5.44}$$

which can be ascribed e.g. to the set of equations (5.28) after substitution for n_i values from the relation (5.1). This gives a set of $R + 1$ equations for R variables $\alpha_1, \alpha_1, \ldots, \alpha_R$, where temperature (or pressure) can be selected as the last variable. In case temperature is selected as the last variable, we obviously have to know the temperature dependence of the equilibrium constants occurring in the sets of equations (5.26) or (5.28). When we select, e.g.

$$H = \sum_{i=1}^{N} n_i^{\circ}(H_T^\circ)_i, \tag{5.45}$$

i. e. we select the value of enthalpy in equilibrium equal to enthalpy at the start of the reaction, the linking condition (5.44) will have the physical significance of the relation $\Delta H = 0$. In the example under discussion, the system was characterized by the thermodynamic functions H and P, i.e. enthalpy and pressure values of the system were fixed. Entropy may be employed in quite the same way as enthalpy. In this case, the linking condition will be

$$S = \sum_{i=1}^{N} n_i (S_T^\circ)_i + \sum_{i=1}^{N} n_i \ln (n_i/n), \tag{5.46}$$

where S is a fixed selected entropy value of the equilibrium system. A choice, similar to that in the relation (5.45)

$$S = \sum_{i=1}^{N} n_i^{\circ}(S_T^\circ)_i + \sum_{i=1}^{N} n_i \ln (n_i^{\circ}/n^{\circ}), \tag{5.47}$$

guarantees validity of the condition $\Delta S = 0$ (adiabatic condition). In this example, the system was characterized by the thermodynamic functions, S, P.

A calculation of this kind requires large amounts of input data, particularly temperature relationships of equilibrium constants and heat capacities. Care must be taken in selecting the H and S values, since a case might occur in which no temperature would exist at which the enthalpy of the equilibrium system would be equal to H or, the entropy to S. It is likewise possible that though such a temperature exists, it lies outside the range of validity of the temperature relationship of equilibrium constants and heat capacities, which fact would then be the source of a considerable error.

Solutions are obtained to the above examples by calculating a modified set of the equations (5.28)

$$F_i(\xi_1, \xi_2, \ldots, \xi_R, T) = 0 \quad i = 1, 2, \ldots, R$$

$$\Psi(\xi_1, \xi_2, \ldots, \xi_R, T) = 0, \tag{5.48}$$

where the last equation in the set (5.48) represents the linking condition for enthalpy, entropy or another thermodynamic function. The following numerical procedure, which is independent of the choice of the pair of unknown variables, seems to be optimum for solving the set of equations (5.48):

a) Select $T = T_0$.

b) Solve the first R equations of the set (5.48) using the method described in the section 5.3.2.

c) Substitute calculated values of $\xi_1, \xi_2, \ldots, \xi_R$ and T_0 into the linking condition Ψ and store the value of $|\Psi|$.

d) Select $T = T_0 + \Delta T$ and again solve the first R equations in the set (5.48). Substitute again the newly calculated values of $\xi_1, \xi_2, \ldots, \xi_R$ and $T_0 + \Delta T$ into the linking condition Ψ. If the new value of $|\Psi|$ is less than the original one, select $T = T_0 + 2\,\Delta T$: conversely, if the new value of $|\Psi|$ is greater than the original one, take $T = T_0 - \Delta T$ and repeat the numerical process until the required accuracy has been achieved.

Mentz[106] and Wilkins[170] studied the problem of convergence of the NASA method. Bernhard and Hawkins[6], Tsao and Wiederhold[162], applied Newton's method to solving the set of non-linear equations (5.48), and Michels and Schneiderman[107] used the NASA method to obtain solutions for real gas systems.

5.3.6 Gradient methods

Gradient methods are based on searching for the minimum of a function Q, defined by equation (5.23) in a set Ω, which is described in section 5.3.2. One highly useful procedure based on the gradient method is Naphtali's procedure[108,109], which deserves more detailed discussion.

Naphtali set out from the fact, that

$$dG = \sum_{r=1}^{R} \Delta G_r \, d\xi_r \,, \tag{5.49}$$

where

$$\Delta G_r = \sum_{i=1}^{N} \nu_{ri}\mu_i \quad r = 1, 2, ..., R \,. \tag{5.50}$$

Naphtali selected

$$d\xi_r = -\Delta G_r \, d\lambda \quad r = 1, 2, ..., R \,, \tag{5.51}$$

λ being a scalar parameter which characterize the rate of progress in the selected direction. Substituting relation (5.51) into equation (5.49) we obtain

$$dG = -\left(\sum_{r=1}^{R} \Delta G_r^2\right) d\lambda \,. \tag{5.52}$$

For $d\lambda > 0$ there evidently holds $dG < 0$. From the material balance equations (5.1) follows

$$dn_i = \sum_{r=1}^{R} \nu_{ri} \, d\xi_r \,. \tag{5.53}$$

Substituting the relations (5.50) and (5.51) into relation (5.53) we obtain, after slight modification,

$$dn_i = -\left(\sum_{k=1}^{N} r_{ik}\mu_k\right) d\lambda \,, \tag{5.54}$$

where the coefficients r_{ik} form a symetric matrix, with

$$r_{ik} = \sum_{r=1}^{R} \nu_{ri}\nu_{kr} \quad \begin{array}{l} i = 1, 2, ..., N \\ k = 1, 2, ..., N \,. \end{array} \tag{5.55}$$

The preceding equations linked together give

$$\frac{dG}{d\lambda} = \sum_{i=1}^{N} \mu_i \frac{dn_i}{d\lambda} = -\sum_{r=1}^{R} (\Delta G_r)^2 \,. \tag{5.56}$$

The overall free enthalpy of the system is therefore a decreasing function of the parameter λ. The numerical procedure involved in the application of Naphtali's method may be divided into the following steps:

a) Chose a point $(\xi_1, \xi_2, ..., \xi_R) \in \Omega$. Again, this is a weak point in the method. It is difficult to find a sufficiently rapid algorithm capable of finding such a point for quite a general case (see also the conclusion of section 5.3.3). The reason is, that the relation $n_i > 0$ must necessarily hold for all compounds, otherwise we would come into difficulties when calculating the chemical potential $\mu_i = c_i + \ln n_i/n$ (see relation (5.2)). Denote the first approximation to the equilibrium composition $n_i^{(1)}$ $i = 1, 2, ..., N$.

b) Select the second approximation of the equilibrium composition by coverting relation (5.57) to discrete form

$$n_i^{(2)} = n_i^{(1)} + \Delta n_i = n_i^{(1)} - \left(\sum_{k=1}^{N} r_{ik}\mu_k \right) \Delta\lambda . \tag{5.57}$$

We select $\Delta\lambda > 0$ in relation (5.57) in order to guarantee a decrease of overall free enthalpy of the system. Calculate the chemical potential μ_k in the point $n_i^{(1)}$ $(i = = 1, 2, ..., N)$. It is clear that the second approximation of the equilibrium composition also satisfies the mass balance equations, and that with a "sufficiently" small $\Delta\lambda > 0$ the value of the function G decreases, i.e. a better approximation to the equilibrium composition has been achieved.

c) Repeat step b), with the difference of using $n_i^{(2)}$ values in place of $n_i^{(1)}$. This gives a new approximation $n_i^{(3)}$ $(i = 1, 2, ..., N)$ and the process is again repeated.

Another problem involved in Naphtali's method is selection of the magnitude of the step $\Delta\lambda > 0$. We must make sure beforehand, that the new approximation should consist of positive number only. This can be secured by using a sufficiently low value of $\Delta\lambda > 0$. This optimum value of $\Delta\lambda$ is most probably obtained by gradually increasing the value of $\Delta\lambda > 0$ and testing whether total free enthalpy decreases. The last successful value of $\Delta\lambda > 0$ is then used to find a new approximation of the equilibrium composition. The Schwartz-Cauchy-Bunakovsky[133] inequality can also be used to estimate the optimum value of $\Delta\lambda > 0$. Storey and van Zeggeren[148] also studied the problem of finding the optimum value of $\Delta\lambda$.

5.4 GENERAL METHODS WHICH DO NOT REQUIRE STOICHIOMETRIC ANALYSIS OF THE SYSTEM

5.4.1 Principle of the methods

The methods of this group are based on searching for minimum of the function

$$Q = \sum_{i=1}^{N} n_i(c_i + \ln n_i - \ln n) \tag{5.58}$$

in a set of points $(n_1, n_2, ..., n_N)$ satisfying the linking condition

$$\sum_{i=1}^{N} a_{ij}n_i = b_j \quad j = 1, 2, ..., M . \tag{5.59}$$

The equations (5.59) express the material balance of the system. Let us assume that the matrix of constitution coefficients A is of rank M. This can always be achieved by eliminating linearly dependent columns in matrix **A**, thus formally decreasing the number of elements in the system. It should be stated, however, that very few cases are encountered in technical practice in which this procedure would be necessary. Usually there holds $N \gg M$ and thus the probability of finding a linearly dependent column in matrix **A** is slight.

The methods may be classified in two groups. The first group of methods is based on the classical mathematical theory of Lagrangian multipliers. The second includes methods in which use is made of the theory of linear or convex programming.

Let us describe the fundamentals of the two procedures. Let us define a function

$$F(n_1, n_2, ..., n_N, \lambda_1, \lambda_2, ..., \lambda_M) = Q(n_1, n_2, ..., n_N) + \sum_{j=1}^{M} \lambda_j \left(b_j - \sum_{i=1}^{N} a_{ij} n_i \right), \qquad (5.60)$$

where the parameters $\lambda_1, \lambda_2, ..., \lambda_M$ are called Lagrangian multipliers[54]. It can be proved, that the minimum of the function is identical in its first N coordinates $n_1, n_2, ..., n_N$ with the minimum of the function Q under the linking conditions (5.59). In the point of minimum of the function F, the conditions must be satisfied

$$\frac{\partial F}{\partial n_i} = 0 \quad i = 1, 2, ..., N$$

$$\frac{\partial F}{\partial \lambda_j} = 0 \quad j = 1, 2, ..., M . \qquad (5.61)$$

The equations (5.61) are a set of $N + M$ variables $n_1, n_2, ..., n_N, \lambda_1, \lambda_2, ..., \lambda_M$. It will be shown later, that the set of equations (5.61) can be simplified to a set of only M equations. Solution of the set thus obtained then gives the solution of the problem (5.58), (5.59).

Linear programming methods can be applied by approximating the function Q by parts, using a linear function and applying the simplex method directly. Convex programming methods can also be used, since it can easily be shown that the function Q is a convex function of the number of moles. Methods of the second group are used less often, since their rate of convergence to the solution is far slower, in general, than with methods of the first group.

5.4.2 The White-Johnson-Dantzig method

The first method in which Lagrangian multiplier theory was used, was the White-Johnson-Dantzig method[38,93,168,169]. In spite of some disadvantages which shall be mentioned at the respective points of the following discussion, this method was a considerable step forward in determining chemical equilibria. For its generality and relative simplicity, it continues to be one of the most frequently employed methods of calculating equilibrium compositions.

The principle of the method involves development of the function Q from relation (5.60) into a Taylor series in the point $\boldsymbol{n}^{(1)} = \left(n_1^{(1)}, n_2^{(1)}, \ldots, n_N^{(1)}\right)$, which satisfies the mass balance equations (5.59) and moreover, there applies $n_i^{(1)} > 0$ $(i = 1, 2, \ldots, N)$.

Finding such an initial set of numbers of moles, however, is a very difficult problem in the general case. Considering the validity of the relations

$$\frac{\partial Q}{\partial n_i} = c_i + \ln n_i - \ln n$$

$$\frac{\partial^2 Q}{\partial n_i^2} = \frac{1}{n_i} - \frac{1}{n} \qquad \begin{array}{l} i = 1, 2, \ldots, N \\ k = 1, 2, \ldots, N \end{array}$$

$$\frac{\partial^2 Q}{\partial n_i\, \partial n_k} = \frac{1}{n} \qquad i = \neq k \tag{5.62}$$

the function Q can be approximated by a function Γ, obtained by neglecting third- and higher-order terms in the Taylor series

$$\Gamma = Q\left(n_1^{(1)}, n_2^{(1)}, \ldots, n_N^{(1)}\right) + \sum_{i=1}^{N} \left(c_i + \ln n_i^{(1)} - \ln n^{(1)}\right) \Delta n_i +$$

$$+ \frac{1}{2} \sum_{i=1}^{N} n_i^{(1)} \left(\frac{\Delta n_i}{n_i^{(1)}} - \frac{\Delta n}{n^{(1)}}\right)^2, \tag{5.63}$$

where

$$\Delta n_i = n_i - n_i^{(1)} \qquad i = 1, 2, \ldots, N\,.$$

$$\Delta n = n - n^{(1)}\,. \tag{5.64}$$

In the minimum point, the conditions (5.61) must be satisfied. Substitution of the approximation Γ into the relation (5.62) and implementation of the conditions (5.61) gives

$$\frac{\partial F}{\partial n_i} = c_i + \ln n_i^{(1)} - \ln n^{(1)} + \frac{n_i}{n_i^{(1)}} - \frac{n}{n^{(1)}} - \sum_{j=1}^{M} \lambda_j a_{ij} = 0 \qquad i = 1, 2, \ldots, N\,. \tag{5.65}$$

The value of n_i can be eliminated from the equations involved in relation (5.65)

$$n_i = -f_i + n y_i^{(1)} + n_i^{(1)} \sum_{k=1}^{M} \lambda_k a_{ik} \qquad i = 1, 2, \ldots, N\,, \tag{5.66}$$

where the following is valid

$$y_i^{(1)} = \frac{n_i^{(1)}}{n^{(1)}}$$

$$f_i = n_i^{(1)}\left(c_i + \ln n_i^{(1)} - \ln n^{(1)}\right). \tag{5.67}$$

The index j in relation (5.65) is converted to the index k by a purely formal process. The reason follows from the following modification. We now substitute from relation (5.66) into the mass balance equations (5.59). We obtain a set of equations

$$\sum_{k=1}^{M} r_{jk}\lambda_k + b_j u = \sum_{i=1}^{N} a_{ij}f_i \quad j = 1, 2, \ldots, M , \tag{5.68}$$

where

$$r_{jk} = r_{kj} = \sum_{i=1}^{N} a_{ij}a_{ik}n_i^{(1)}$$

$$u = \frac{n}{n^{(1)}} - 1 . \tag{5.69}$$

The set of equations (5.68) was derived with the use of the relations

$$\sum_{i=1}^{N} a_{ij}n_i^{(1)} = b_j \quad j = 1, 2, \ldots, M , \tag{5.70}$$

which are a mathematical form of the condition, that the initial set of numbers of moles (first approximation of the solution) must satisfy the mass balance equations. The relation (5.68) includes M equations for $M + 1$ variables $\lambda_1, \lambda_2, \ldots, \lambda_M, u$. The last equation is obtained by substituting equation (5.66) into the relation fot the overall number of moles $n = \sum_{i=1}^{N} n_i$. This gives the equality

$$\sum_{k=1}^{M} b_k\lambda_k = \sum_{i=1}^{N} f_i . \tag{5.71}$$

Solution of the equations (5.68) and (5.71), which are linear, gives values of $\lambda_1, \lambda_2, \ldots, \lambda_M, u$, by means of which a new approximation to the solution $n_1^{(2)}, n_2^{(2)}, \ldots, n_N^{(2)}$ is obtained from relation (5.66). This new approximation obviously satisfies the mass balance equations and the procedure can be repeated. It may happen (and this usually is the case at the beginning of the iteration process) that some of the $n_i^{(2)}$ values are negative. In this case, the reduction parameter $0 < \eta \leq 1$ must be selected such as to satisfy the condition

$$\bar{n}_i^{(2)} = n_i^{(1)} + \eta\left(n_i^{(2)} - n_i^{(1)}\right) > 0 \quad i = 1, 2, \ldots, N \tag{5.72}$$

the $\bar{n}_i^{(2)}$ values then being taken as the new approximation. With $\eta = 1$ we obtain $\bar{n}_i^{(2)} = n_i^{(2)} \quad i = 1, 2, \ldots, N$. Since the function Q is convex on a set of points

$n_1, n_2, \ldots, n_N$ which satisfy the material balance equations (see Appendix 6), the method rapidly converges to its solution. A more detailed discussion with an example will be found in the section (5.5).

Besides the already mentioned difficulties in searching for the first approximation, the method has the disadvantage of requiring rather a long time for calculation of chemical equilibrium in such systems, in which the final mixture contains a relatively small amount of one of the compounds as compared to the other compounds. This fact is due to the manner of calculating a new approximation to the equilibrium number of moles given by relation (5.66). Both these disadvantages are eliminated in the following method.

5.4.3   The method of Lagrangian multipliers

This method[165] makes use of the classical mathematical theory of Lagrangian multipliers. Differing from the White-Johnson-Dantzig method, the function F, defined by relation (5.60), is directly substituted into the conditions (5.61) instead of a quadratic approximation of the function Q from relation (5.60). In this way, we obtain the equations

$$\frac{\partial F}{\partial n_i} = c_i + \ln n_i - \ln n - \sum_{j=1}^{M} a_{ij}\lambda_j = 0 \quad i = 1, 2, \ldots, N . \tag{5.73}$$

Hence follows

$$y_i = \frac{n_i}{n} = \exp\left(\sum_{k=1}^{M} a_{ik}\lambda_k - c_i\right), \tag{5.74}$$

where the index j is replaced by the index k for purely formal reasons. Division of the left- and right-hand sides of the mass balance equations (5.59) by the overall number of moles and substitution of the relations (5.74) into the equations thus modified, leads to a set of non-linear equations

$$\sum_{i=1}^{M} a_{ij} \exp\left(\sum_{k=1}^{M} a_{ik}\lambda_k - c_i\right) - b_j t = 0 \quad j = 1, 2, \ldots, M , \tag{5.75}$$

where $t = 1/n$. The remaining equation is obtained from the condition $\sum_{i=1}^{N} y_i = 1$, i.e.

$$\sum_{i=1}^{N} \exp\left(\sum_{k=1}^{M} a_{ik}\lambda_k - c_i\right) = 1 . \tag{5.76}$$

The equations (5.75) and (5.76) form a set of $M + 1$ non-linear equations for $M + 1$ variables $\lambda_1, \lambda_2, \ldots, \lambda_M, t$. The number of equations in this set may furthermore be diminished by e.g. eliminating the variable t from the first equation and substituting into the remaining equations. This gives a new set

$$\sum_{i=1}^{M} q_{ij} \exp\left(\sum_{k=1}^{M} a_{ik}\lambda_k - c_i\right) = d_j \quad j = 1, 2, \ldots, M , \tag{5.77}$$

where

$$q_{i1} = 1 \quad d_1 = 1$$

$$q_{ij} = a_{ij} - a_{i1}b_j/b_1 \quad d_j = 0 \quad j = 2, \ldots, M . \tag{5.78}$$

The equations (5.77) now form a set of M non-linear equations for M variables $\lambda_1, \lambda_2, \ldots, \lambda_M$. The equilibrium composition is now easily determined from the calculated values of Lagrangian multipliers and the value of t by substitution into the equations (5.74). The set of non-linear equations (5.77) is best solved by means of Newton's method with reduction parameter. The choice of the first approximation $\lambda_1^{(1)}, \lambda_2^{(1)}, \ldots, \lambda_M^{(1)}$ will be discussed in the course of a more detailed description of the method in section (5.5).

One more undoubted advantage of this method is the fact, that there is no need of elaborating a first approximation of the solution to satisfy the mass balance equations. A method suggested by Powell and Sarner[129], utilizing elementary potentials, is based on largely similar ideas[24,112,168]. It will be described in more detail in section (5.4).

5.4.4   Linear and convex programming methods

Dantzig[39], who investigated linear programming problems for a long time, suggested a procedure of calculating chemical equilibria which enables the simplex method to be used. Let us mention that the simplex method allows the minimum or maximum of the linear function $Z\left(x_1, x_2, \ldots, x_N\right)$

$$Z = c_1 x_1 + c_2 x_2 + \ldots + c_N x_N , \tag{5.79}$$

where c_i $i = 1, 2, \ldots, N$ are known constants, for a set of points $x_1, x_2, \ldots, x_N$ satisfying K relationships

$$\left(\sum_{i=1}^{N} p_{ij}x_i\right)(r_j)(d_j) \quad j = 1, 2, \ldots, K , \tag{5.80}$$

where r_j is a relation symbol for which one of the symbols $\geq$, $\leq$ or $=$ can be substituted. Values of p_{ij}, d_j are known constants.

Thus, in order to be able to use the simplex method for calculating a chemical equilibrium, we must some who linearise the function Q

$$Q = \sum_{i=1}^{N} n_i(c_i + \ln n_i - \ln n) = \sum_{i=1}^{N} n_i c_i + n \sum_{i=1}^{N} y_i \ln y_i , \tag{5.81}$$

where $y_i = n_i/n$. This is achieved by linearising the function u

$$u = \sum_{i=1}^{N} y_i \ln y_i .$$ (5.82)

Since the individual terms can be separated, each can be approximated separately. The form of the function $u = y \ln y$ is shown by the curve in Fig. 18. The curve can

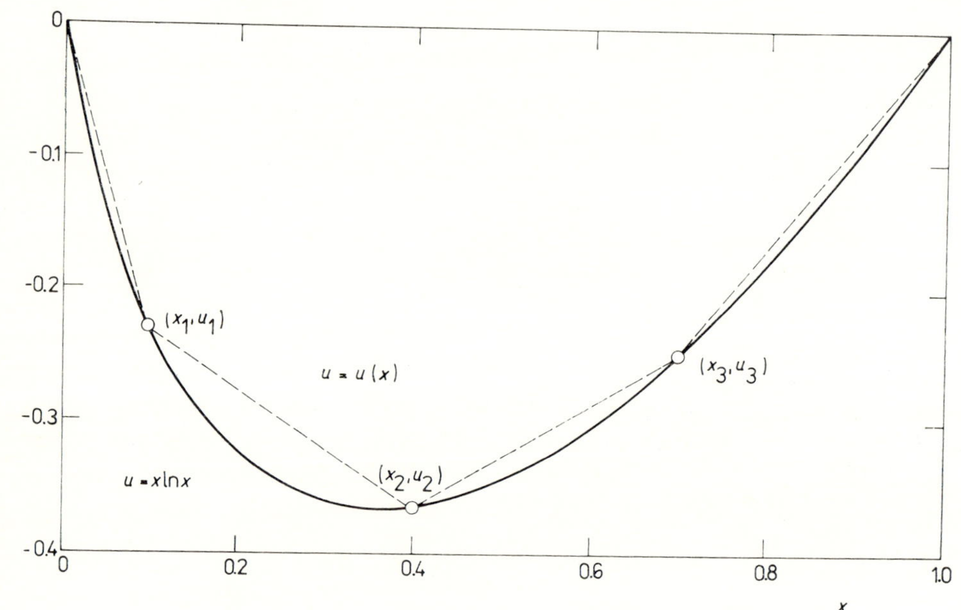

Fig. 18. Approximation of the function $u = y \ln y$ by the line $u = u(x)$.

be replaced by a linear function $u = u(y)$, composed of straight lines which touch the original curve in K points with the coordinates (y_l, u_l), $l = 1, 2, ..., K)$.

Under these conditions the calculation is equivalent to finding values of $n_i \geqq 0$, $n_{il} \geqq 0$ and the minimum z by satisfying the relationships

$$\sum_{i=1}^{N} a_{ij} n_i = b_j \qquad j = 1, 2, ..., M ,$$ (5.83)

$$\sum_{i=1}^{N} n_i - n = 0 ,$$ (5.84)

$$\sum_{l=1}^{K} n_{il} - n = 0 \qquad i = 1, 2, ..., N ,$$ (5.85)

$$\sum_{l=1}^{K} y_i n_{il} - y_i = 0 \qquad i = 1, 2, ..., N ,$$ (5.86)

$$\sum_{i=1}^{N} c_i n_i + \sum_{i=1}^{N} \sum_{l=1}^{K} u_l y_{il} = z .$$ (5.87)

116

Thus the problem has been converted to one of linear programming, in solving which a simplex code is used with advantage. To economize the operative store of the computer and speed up the calculation, the following strategy is recommended: In the first approximation, calculate values of the function $u_i = y_i \ln y_i$ for all values of i and three values of y_i; 0.0; 0.5 and 1.0. In the following approximations, narrow down the interval to one half each time, which represents the calculation of another value of u_i. Halving of the interval with each new approximation continues, until the required accuracy has been achieved.

A different procedure can also be applied when linearising the function Q. There evidently holds, that

$$Q = \sum_{i=1}^{N} n_i(c_i + \ln n_i - \ln n) = \sum_{i=1}^{N} n_i c_i + \sum_{i=1}^{N} n_i \ln n_i - n \ln n . \qquad (5.88)$$

From relation (5.59) follows a limitation of the number of moles of each compound in equilibrium

$$n_i \leqq \min_{j, a_{ij} \neq 0} (b_j/a_{ij}) . \qquad (5.89)$$

The resulting equilibrium composition depends (with the exception of the multiple) solely on the ratio between the numbers of moles of the input constituents. Therefore, such absolute values of the numbers of moles of input constituents can be selected (while maintaining the required ratio), that the number of gramatoms on the j-th element b_j will be low enough to give

$$0 < n_i < n < 1 . \qquad (5.90)$$

Controlling the validity of the condition (5.88) in this manner, the linearisation procedure described can be done directly with the variables $n_1, n_2, \ldots, n_N$: this avoids difficulties with the overall number of moles n which is involved in the relation (5.81).

When making use of convex programming which, however, is more difficult in general, there is no need of linearising the convex function Q. Application of optimation methods to the solution of chemical equilibrium problems was studied by Anthony and Himmelblau[2]. The principle of these methods may be formulated thus: consider a fully determined set of non-linear equations

$$f_i(n_1, n_2, \ldots, n_N) = 0 \quad i = 1, 2, \ldots, N \qquad (5.91)$$

a solution can be obtained when such values of n_i ($i = 1, 2, \ldots, N$) will be determined as to achieve a total minimum of the function F

$$F = \sum_{i=1}^{N} f_i^2 . \qquad (5.92)$$

At the same time, however, the mass balance equations must be satisfied, and there must apply that

$$n_i > 0 \quad i = 1, 2, \ldots, N \,. \tag{5.93}$$

The so-called direct search method, described by Hooke and Jeeves[73], is used for the minimum of the function F. The above mentioned authors Anthony and Himmelblau also studied different possibilities of selecting equilibrium criteria and thus also of choosing the function $f_1 \; i = 1, 2, \ldots, N$ or F. One of the existing possibilities is to choose the function F equal to the overall enthalpy of the system. However, the state of equilibrium may also be characterized by R linearly independent stoichiometric equations, satisfying the relations (see (3.29))

$$f_r = \left(\Delta G^\circ\right)_r - \sum_{i=1}^{N} v_{ri}\mu_i^\circ = 0 \,. \tag{5.94}$$

It is clear on closer investigation, that the procedure involved is rather similar to a method developed by Scully[142], which belongs to the group of methods which require a stoichiometric description of the system.

The time consumer in calculating chemical equilibria by one of the linear or non-linear programming methods is usually considerably longer than the time needed for methods based on the theory of Lagrangian multipliers[165]. For this reason, linear and non-linear programming are rather seldom employed in calculating chemical equilibria.

5.5   EVALUATION

There follows from the preceding sections, that a number of procedures are nowadays available for calculating the chemical equilibria of ideal gas systems. Most of the procedures can also be employed for very complicated systems, the number of elements and compounds being practically unlimited.

The individual, generally applicable methods may be compared mutually with respect to the degree of complication of the set of non-linear equations involved. Up to recent time, the level of data processing equipment required was another criterion. Now, however, when a sufficiently rapid computer is available to nearly every research institution, this circumstance is becoming less important.

Calculation of chemical equilibria by means of equilibrium constant, i.e. with the aid of methods requiring stoichiometric analysis of the system, a set of R non-linear equations must be solved, where $R = N - H \geqq N - M$. In the second case, when studying equilibria by means of minimalization of overall free enthalpy, i.e. using methods which do not require stoichiometric analysis of the system, a set of not more than $(H + 1)$ non-linear equations will be involved, where $H \leq M$. It is clear therefore, that procedures with equilibrium constants have a certain advantage in the calculation of chemical equilibria of more simple systems, since their approach is more easily understood by chemists, and they can be calculated with the aid of the simplest data processing equipment.

Nowadays chemical equilibria of highly complicated systems must be solved more often, the number of compounds being of the order of tens. Since the number of elements M usually is not greater than 5 or 6, the value of R is likewise of the order of tens. Treatment of sets of non-linear equations of this size (irrespective of the method employed) is highly demanding in terms of numerical effort as well as of time. Therefore, the group of methods for which no stoichiometric analysis is needed is being increasingly used. The exception are systems of isomers, where calculation by means of equilibrium constants can be simpler. Methods based on linear or non-linear programming can only be recommended in those cases, where a computer of sufficient speed and store capacity is available.

When a calculating machine only is available, the White-Johnson-Dantzig method is recommended, as it is based on two simple procedures and converges rather quickly. Obviously this method is likewise suitable for solving complicated systems.

5.6   DESCRIPTION OF SELECTED METHODS

The preceding section presented a brief survey of methods based on two main procedures, by means of which virtually every case of chemical equilibrium in the ideal gas state can be solved irrespective of its degree of complexity. Considering the great number of methods suggested, which often are very similar to each other, we cannot discuss each method individually. It appears useful, however, to describe in detail several selected procedures which fully cover the range of problems involved and which may at the same time be considered to be the best.

5.6.1   Newton's method with reduction parameter

Section 5.3.2 described the principle of application of Newton's method with reduction parameter for solving systems of non-linear equations

$$F_r(\xi_1, \xi_2, ..., \xi_R) = 0 \quad r = 1, 2, ..., R , \tag{5.95}$$

on the set of points $(\xi_1, \xi_2, ..., \xi_R)$, for which is satisfied

$$(\xi_1, \xi_2, ..., \xi_R) \in \Omega , \tag{5.96}$$

which originated as the general way of writing the equilibrium conditions for R

linearly independent reactions

$$\left(K_a\right)_r = \left(\frac{P}{n}\right)^{v_r} \sum_{i=1}^{N} n_i^{v_r} \quad r = 1, 2, ..., R \tag{5.97}$$

or,

$$\ln\left(K_a\right)_r - v_r \ln P + v_r \ln n - \sum_{i=1}^{N} v_{ri} \ln n_i = 0 \quad r = 1, 2, ..., R, \tag{5.98}$$

where the number of moles n_i $i = 1, 2, ..., N$ must satisfy the mass balance equations

$$n_i = n_i^o + \sum_{i=1}^{R} v_{ri}\xi_r \quad i = 1, 2, ..., N. \tag{5.99}$$

When using this method, we must overcome two basic difficulties, which however are greatly similar for all methods of solving the problem of chemical equilibrium:

a) construction of the first approximation of the solution

$$n_i^{(1)} > 0 \quad i = 1, 2, ..., N,$$

b) securing convergence of the numerical process.

The two problems can be solved easily and completely for the case of $R = 1$, i.e. the case when a single reaction is being considered. In this case the relationships (5.98) and (5.99) can be rewritten in the form of

$$\sum_{i=1}^{N} v_i \ln n_i - v \ln n + v \ln P - \ln K = 0 \tag{5.100}$$

and

$$n_i = n_i^o + v_i\xi \quad i = 1, 2, ..., N, \tag{5.101}$$

where the index $r = 1$ is left out. There evidently applies that

$$0 < \xi < \min_{i \in w} \frac{n_i^o}{|v_i|}, \tag{5.102}$$

where W is the symbol for the set of initial compounds. The relation (5.102) obviously applies under the assumption that at least one product is not contained in the initial mixture. In the reverse case the value of the reaction coordinate might even achieve negative values. The set Ω, which in the general case is defined as the set of all R-membered groups $(\xi_1, \xi_2, ..., \xi_R)$ for which the expression $n_i > 0$ $i = 1, 2, ..., N$ hold after substitution into relation (5.99), contains in the case of $R = 1$ precisely all such points ξ, which satisfy the relationship (5.100) may be written in the general form as

$$F(\xi) = 0, \tag{5.103}$$

where $\xi \in \Omega$ applies. Taking the derivative of the left-hand side of the equation (5.103) we obtain the relation

$$\frac{dF}{d\xi} = \sum_{i=1}^{N} \frac{v_i^2}{n_i} - \frac{v^2}{n} . \tag{5.104}$$

It can easily be proved (see Appendix 3), that the right-hand side of the equation (5.104) is always positive for any value of $n_i > 0$ $i = 1, 2, ..., N$. From the validity of the relations

$$\lim_{\xi \to 0+} F(\xi) = -\infty$$

$$\lim_{\xi \to \xi^*-} F(\xi) = +\infty , \tag{5.105}$$

where ξ^* denotes the expression on the right-hand side of the relation (5.102) follows, that the function $F(\xi)$ is a rising function in its definition range Ω, having just one zero point.

Construction of the first approximation of the solution of the equation (5.100), i.e finding arbitrary values of the reaction coordinate $\xi^{(1)}$ such as to satisfy the condition $\xi^{(1)} \in \Omega$ therefore is not difficult. When applying Newton's method with reduction parameter we determined the $(p + 1)$-th approximation, knowing already the p-th approximation, according to the expression

$$\xi^{(p+1)} = \xi^{(p)} - \lambda_p \frac{F(\xi^{(p)})}{F'(\xi^{(p)})} \quad p = 1, 2, ... , \tag{5.106}$$

where $F' = dF/d\xi$ and $\lambda_p \in (0, 1\rangle$ for $p = 1, 2, ...$ Selecting $\lambda_p = 1$ $(p = 1, 2, ...)$ we obtain Newton's classical method. The reduction parameter $\lambda_p \in (0, 1\rangle$ $p = 1, 2, ...$ is selected at every p-th step such as to satisfy $\xi^{(p+1)} \in \Omega$: considering the properties of the function $F(\xi)$ we thus secure convergence of the numerical process. Starting from a given iteration step the reduction parameter may be chosen as equal to one, so that the high rate of convergence of Newton's classical method in the vicinity of the solution is likewise retained in Newton's method with reduction parameter (2nd order method). The procedure described is best illustrated by means of an example.

Example 19

The equilibrium constant of ammonia synthesis

$$\tfrac{3}{2} H_2(g) + \tfrac{1}{2} N_2(g) = HN_3(g) ,$$

is $K_a = 0.30076$ at 500 K. The initial mixture contains 1 mole hydrogen and 1 mole nitrogen. Determine the equilibrium composition at 5 atm, disregarding the real behaviour of the mixture.

The mass balance equations (5.101) will in this case be

$$n_1 = n_{H_2} = 1 - 1.5\xi$$

$$n_2 = n_{H_2} = 1 - 0.5\xi$$

$$n_3 = n_{NH_3} = \xi .$$

The equation (5.100) takes the specific form of

$$F(\xi) \equiv -1.5 \ln\left(1 - 1.5\xi\right) - 0.5 \ln\left(1 - 0.5\xi\right) + \ln \xi +$$
$$+ \ln\left(2 - \xi\right) - \ln 5 - \ln 0.30076 = 0$$

and similarly, equation (5.104) will be

$$\frac{dF}{d\xi} = \frac{9}{4(1 - 1.5\xi)} + \frac{1}{4(1 - 0.5\xi)} + \frac{1}{\xi} - \frac{1}{2 - \xi} .$$

From relation (5.102) follows that the set Ω is formed by points ξ, for which the expression $\xi \in (0;0.6\overline{6})$ applies. The calculation was carried out for three different values of the first approximation $\xi^{(1)} \in (0;0.6\overline{6})$ and the results are given in the following table. The tabulated data show, that convergence is very steep. It is an interesting trait of this example, that there was no need of using a reduction parameter of less than one.

p	$\xi^{(p)}$	$F(\xi^{(p)})$	$F'(\xi^{(p)})$	λ_p
1	0.01	-4.299	102.0	1
2	0.05214	-2.560	21.36	1
3	0.17196	-1.073	8.574	1
4	0.29709	-0.124	7.131	1
5	0.31448	0.00004	7.142	1
6	0.31448	0.0000005	7.142	1

p	$\xi^{(p)}$	$F(\xi^{(p)})$	$F'(\xi^{(p)})$	λ_p
1	0.2	-0.842	7.937	1
2	0.30609	-0.059	7.132	1
3	0.31448	$+0.00004$	7.142	1
4	0.31448	$+0.0000005$	7.142	1

p	$\xi^{(p)}$	$F(\xi^{(p)})$	$F'(\xi^{(p)})$	λ_p
1	0.66	6.577	226.1	1
2	0.63092	4.024	43.18	1
3	0.52773	1.973	13.15	1
4	0.38773	0.537	7.647	1
5	0.31751	0.022	7.148	1
6	0.31448	0.000009	7.142	1

The general case $(R > 1)$ cannot be solved as simply as in the preceding case of a simple chemical reaction. Let us first turn to the first difficulty, that of constructing the first approximation of the solution. The norm of the vector

$$\xi = (\xi_1, \xi_2, \ldots, \xi_R), \quad \text{i.e.} \quad \|\xi\| = \sqrt{(\xi_1^2 + \xi_2^2 + \ldots + \xi_R^2)},$$

is a measure of the distance of the equilibrium mixture from the initial composition. For sufficiently low values of equilibrium constants $(K_a)_r$ $r = 1, 2, \ldots, R$ the value of $\|\xi\|$ is likewise low. Thus, there exists for an arbitrary value of $\varepsilon > 0$ a value $\xi \in \Omega$ such, that there holds $\|\xi\| < \varepsilon$. Let us assume that a random number generator is available for values of $x \in \langle 0, 1 \rangle$, which nowadays is true of most computers. Substituting $\zeta = a + (b - a)\, x$ we may now generate random numbers in an arbitrary finite interval $\langle a, b \rangle$. Evidently, for any selected $a > 0$ there exists a value $\xi \in \Omega$ such as to satisfy the condition $\xi_r \in (-a, a)$ $r = 1, 2, \ldots, R$. Let us chose e.g. $a = 1$ and let us use the first R random numbers of the interval $\langle -1, 1 \rangle$ to construct the vector $\xi^{(1)}$. If there holds that $\xi^{(1)} \in \Omega$ the above vector will be the first approximation of the solution. In the reverse case, we shall take another group of R random numbers from the interval $\langle -1, 1 \rangle$ to construct the vector $\xi^{(2)}$, testing again for the condition $\xi^{(2)} \in \Omega$. If even the e.g. hundredth attempt will be unsuccessful, we may apply the above algorithm to the interval $(-\frac{1}{2}, \frac{1}{2})$ in order to increase the probability of finding the first approximation of $\xi \in \Omega$, etc. According to the laws of probability this process must lead to success. Should random number generation be time-consuming, the following procedure may also be used: Use the first R random numbers to construct the vector $\xi^{(1)}$. Use the second component of the vector $\xi^{(1)}$ as the first component of the vector $\xi^{(2)}$, the third component of the vector $\xi^{(1)}$ as the second component of the vector $\xi^{(2)}$, etc. Take another random number for the R-th component of the vector $\xi^{(2)}$. In this way we make sure that only one random number is generated with every following vector ξ. It is difficult to estimate the time required to find a solution by means of methods based on sequences of random numbers (Monte Carlo method), but experience shows that this numerical process is very advantageous in terms of time.

By obtaining the first approximation $\xi^{(1)} \in \Omega$ the necessary conditions have been created for application proper of Newton's method with reduction parameter. It is particularly important, that selection of the reduction parameter $\lambda \in (0, 1\rangle$ secures in every iteration step the condition $\xi^{(p)} \in \Omega$ $i = 2, 3, \ldots$ Moreover, it is adantageous to chose the parameter λ such that every new approximation of the vector ξ differs from the component of the original approximation by less than d%, the value of d usually being chosen equal to five or ten. This approach is particularly suitable for the first steps of the iteration process, where values of $\Delta\xi_r$ $r = 1, 2, \ldots, R$ have the correct sign and mutual ratio, but their absolute values are too great. Starting from a certain iteration step, $\lambda = 1$ may be selected, by means of which the properties of a second-order method are retained.

Notes on constructing the programme

The block diagram (Fig. 19) shows the main steps of the numerical algorithm of solving chemical equilibrium by Newton's method with reduction parameter. The algorithm may be divided into the following blocks:

a) Determination of R linearly independent reactions. The reactions are either known beforehand or, they can be obtained from the matrix of constitution coefficients. The latter method has been described in section 2.3.

b) Calculation of equilibrium constants. Here, two cases may again set in. Either the equilibrium constant of the reaction involved is known beforehand (the equilibrium constant vs. temperature relationship is tabulated for many reactions), or it can be determined from tabulated thermochemical data of the respective compound $-(G_T^o - H_{T_1}^o)/T$, $H_{T_1}^o$, T_1 usually being either 0 K or 298.15 K. The following applies to the equilibrium constant of the r-th reaction

$$-RT\ln(K_a)_r = \Delta G_r^o = \sum_{i=1}^{N} v_{ri}(G_T^o)_i \,.$$

c) Determination of the first approximation $(\xi_1^{(1)}, \xi_2^{(1)}, ..., \xi_R^{(1)})$. This problem was discussed in the final part of the preceding section.

d) Solution of a system of non-linear equations by Newton's method with reduction parameter. It is advantageous to calculate in every iteration step the expression D

$$D = \sum_{r=1}^{R} \left| \Delta\xi_r^{(p)}/\xi_r^{(p)} \right| \,,$$

where $\Delta\xi_r^{(p)}$ is the increment of the value $\xi_r^{(p)}$ calculated by Newton's method for the r-th reaction coordinate in the p-th iteration step. The calculation is stopped as soon as the relation $D < \varepsilon$ becomes valid, the value of ε being usually chosen equal to 0.0001.

e) Print-out of results.

5.6.2   Brinkley's method

As was shown in the brief discussion in section 5.3.3, the method starts out from two basic ideas:

1. The balance of chemical elements in a closed system is independent of chemical conversions.

2. The state of equilibrium does not depend on the manner in which the system arrived at equilibrium.

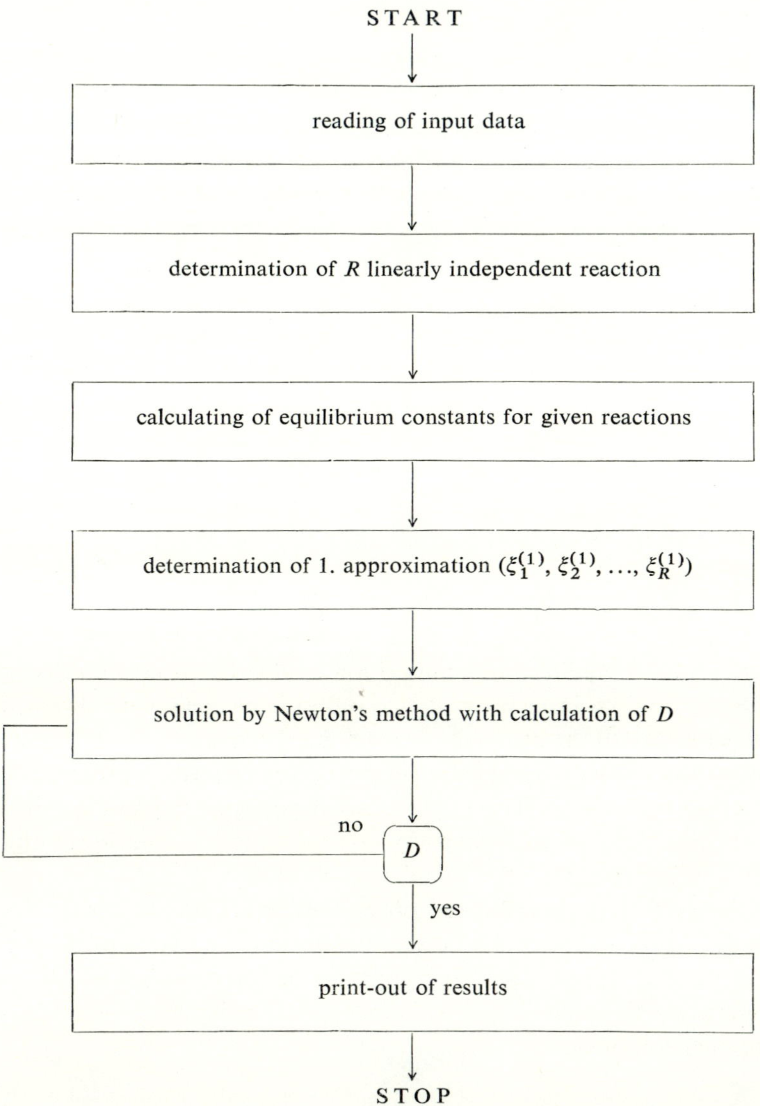

Fig. 19. Block diagram of Newton's method with reduction parameter.

Using these principles it is possible to construct as many independent equations as there are constituents present in the system. In these equations, temperature and pressure play the role of parameters, numbers of moles or molar fractions are the unknown variables. Thus, considering again a closed single-phase system of ideal gases, formed by N constituents and M elements or basic particle species, the required number N of algebraic equations will include two types of relationship: a number H of balance equations and equilibrium relationships forming the remaining number.

It is essential for the further procedure that the matrix of constitution coefficients be rearranged and modified so as to contain only H linearly independent columns: the first H rows, representing a qualitative description of the basic constituents, must likewise be linearly independent. An essential condition of the selection of basic constituents is the fact, that they must include all elements. It is then possible to define the two sets of algebraic relationships: balance relationships for every basic constituent, using the equations (5.32), and equilibrium relationships (5.34) for individual chemical conversions, expressed as linear combinations of row vectors of the basic constituents by the equations (5.33).

Solution of equations

Calculation of the equilibrium composition is now based on iterative solution of the two sets of equations in such a way, that results from one set are always substituted into the other set, until the required accuracy has been achieved. This process, however, involves the difficulty of the convergence of successive approximations, which must be secured beforehand. This convergence depends on the absolute ratio of equilibrium concentrations of basic and derived constituents. The lower this ratio is, the slower is convergence of the calculation. Brinkley[15−17] was aware of this difficulty, and therefore proposed three procedures applicable to cases with different numbers of moles of the individual constituents. Although none of these procedures are quite general in application, they ought to be mentioned briefly at least, since they are particularly applicable to solution of less complicated systems by means of conventional equipment.

The simple iteration method

In cases where we can predict, by intuition or on the grounds of experimental data, which constituents will prevail in the state of equilibrium, we may apply the simple iteration method with advantage[140]. The following usually is a sufficient condition

$$K_r \left(\frac{P}{n}\right)^{v_r} \ll 1 . \tag{5.107}$$

The method involves selection of a set of initial data n_i and calculation of the corresponding values of n_r by means of the equilibrium relationships (5.22). If the values which n_i can attain are not known, it is advisable to chose $n_i = q_i$. New values of n_i are then calculated from the balance relationships (5.32) on the basis of the n_r values, to be used for the calculation of new n_r values in the next approximation. In order that the step in the next calculation should not be too large, it is advantageous to calculate values of the numbers of moles of individual basic constituents for the next approximation either as the weighted mean of two preceding values from the relations

$$n_i^{(p+1)} = 0.4n_i^{(p-1)} + 0.6n_i^{(p)} , \tag{5.108a}$$

or

$$(n_i)_5 = \frac{(n_i)_2 (n_i)_3 - (n_i)_1 (n_i)_4}{(n_i)_2 + (n_i)_3 - (n_i)_1 - (n_i)_4} , \tag{5.108b}$$

where

$(n_i)_1$ ... value selected in the 1. step

$(n_i)_2$ ... value calculated in the 1. step

$(n_i)_3$ ... value selected in the 2. approximation, e.g. with the use of (5.108a)

$(n_i)_4$ ... value calculated in the 2. approximation

$(n_i)_5$ ... initial value for the following step.

The following set of basic constituent values can also be selected graphically. Calculated values of numbers of moles of the basic constituents are first plotted vs. assumed values for a number of steps. The initial data for the next step is then given by the point of intersection. The procedure is easily understood from the following example.

Example 20

Let us, for the sake of comparison, take the same system for which the Damköhler-Edse method[37] was originally suggested, i.e. oxidation of propane by a stoichiometric amount of air at 2200 K and 40 atm, according to the reaction scheme

$$C_3H_8(g) + 5 O_2(g) + 20 N_2(g) \rightarrow \text{products} .$$

For the sake of simplicity we take the $O_2 : N_2$ ratio of air to be $1 : 4$ moles. The following constituents are considered in equilibrium: CO_2, H_2O, N_2, CO, H_2, H, O_2, O, OH and NO. The matrix of constitution coefficients takes the form of:

	j	C	O	H	N
i		1	2	3	4
CO_2	1	1	2	0	0
H_2O	2	0	1	2	0
N_2	3	0	0	0	2
CO	4	1	1	0	0
H_2	5	0	0	2	0
H	6	0	0	1	0
O_2	7	0	2	0	0
O	8	0	1	0	0
OH	9	0	1	1	0
NO	10	0	1	0	1

Rank of the matrix is $H = 4$, thus the system contains four independent constituents. Considering the required condition that the independent constituents should be represented in the highest concentration, it is advisable to select CO_2, H_2O and N_2. The fourth constituent selected is CO. A determinant formed from the constitution coefficients of these four constituents is $D = 4$, providing that the constituents are linearly independent. The remaining six constituents shall be considered to be derived: using the procedure described in Example 5, we now construct the matrix of stoichiometric coefficients for six linerlay independent stoichiometric equations in the form of synthesis reactions:

	i	CO_2	H_2O	N_2	CO	H_2	H	O_2	O	OH	NO
r		1	2	3	4	5	6	7	8	9	10
5		1	-1	0	-1	1	0	0	0	0	0
6		$\frac{1}{2}$	$-\frac{1}{2}$	0	$-\frac{1}{2}$	0	1	0	0	0	0
7		-2	0	0	2	0	0	1	0	0	0
8		-1	0	0	1	0	0	0	1	0	0
9		$-\frac{1}{2}$	$-\frac{1}{2}$	0	$\frac{1}{2}$	0	0	0	0	1	0
10		-1	0	$-\frac{1}{2}$	1	0	0	0	0	0	1

We now construct balance relationships for the basic constituents in the form of:

$$n_1 = q_1 + n_5 + n_6 + n_7 - n_8 - 2n_9 - n_{10}$$

$$n_2 = q_2 - n_5 - n_6 - n_7$$

$$n_3 = q_3 - \tfrac{1}{2}n_{10}$$

$$n_4 = q_4 - n_5 - n_6 + n_7 + n_8 + 2n_9 + n_{10}\,.$$

The constants q_j are calculated from the relation (5.4), i.e.

$$q_1 + q_4 = 3$$
$$2q_1 + q_2 + q_4 = 10$$
$$2q_2 = 8$$
$$2q_3 = 40 \, ,$$

where $q_1 = 3$, $q_2 = 4$, $q_3 = 20$, $q_4 = 0$.

Therefore the final form of the balance relationships will be:

$$n_1 = 3 + n_5 + n_6 + n_7 - n_8 - 2n_9 - 2n_{10}$$
$$n_2 = 4 - n_5 - n_6 - n_7$$
$$n_3 = 20 - \tfrac{1}{2}n_{10}$$
$$n_4 = -n_5 - n_6 + n_7 + n_8 + 2n_9 + n_{10} \, .$$

The set of equilibrium relationships will take the form of:

$$n_5 = K_5(n_1)^{-1} n_2 n_4 \qquad\qquad K_5 = 3.846 \times 10^{-5}$$

$$n_6 = K_6 \left(\frac{P}{n}\right)^{-1/2} n_1^{-1/2} n_2^{1/2} n_4^{1/2} \qquad\qquad K_6 = 1.93 \ \times 10^{-1}$$

$$n_7 = K_7 \left(\frac{P}{n}\right)^{-1} n_1^2 n_4^{-2} \qquad\qquad K_7 = 1.799 \times 10^{-5}$$

$$n_8 = K_8 \left(\frac{P}{n}\right)^{-1} n_1 n_4^{-1} \qquad\qquad K_8 = 3.448 \times 10^{-3}$$

$$n_9 = K_9 \left(\frac{P}{n}\right)^{-1/2} n_1^{1/2} n_2^{1/2} n_4^{-1/2} \qquad\qquad K_9 = 2.155 \times 10^{-4}$$

$$n_{10} = K_{10} \left(\frac{P}{n}\right)^{-1/2} n_1 n_3^{1/2} n_4^{-1} \qquad\qquad K_{10} = 2.597 \times 10^{-3} \, .$$

With the last relation, $n = n_i$, the list of equations to be solved is complete. Let us now chose four basic values for the primary constituents, e.g.

$$n_1 = 2.9$$
$$n_2 = 4.0$$
$$n_3 = 20.0$$
$$n_4 = 0.1 \, .$$

Table 2. Iterative calculation of equilibrium composition of oxidation of propane with air

		Step							
		1	2	3	4	5	6	7	8
Chosen values	n	27.000	27.0709	27.0724	27.0591	27.0617	27.0614	27.0620	27.0621
	n_{CO_2}	2.900	2.9269	2.9023	2.9123	2.9125	2.9145	2.9152	2.9156
	n_{H_2O}	4.000	3.9746	3.9674	3.9633	3.9620	3.9613	3.9611	3.9610
	n_{N_2}	20.000	19.9931	19.9877	19.9870	19.9864	19.9864	19.9863	19.9863
	n_{CO}	0.100	0.0731	0.0854	0.0828	0.0842	0.0842	0.0843	0.0842
Calcu-lated values	n_{H_2}	0.02668	0.01920	0.02258	0.02179	0.02215	0.02213	0.02215	—
	n_H	0.00079	0.00067	0.00073	0.00072	0.00072	0.00072	0.00072	—
	n_{OH}	0.03052	0.03579	0.03294	0.03349	0.03320	0.03321	0.03320	—
	n_O	0.00035	0.00049	0.00041	0.00043	0.00042	0.00042	0.00042	—
	n_{NO}	0.02297	0.03174	0.02694	0.02788	0.02741	0.02743	0.02741	—
	n_{O_2}	0.02185	0.04173	0.03007	0.03221	0.03113	0.03118	0.03114	—
	n_{CO_2}	2.9448	2.8860	2.9190	2.9127	2.9158	2.9157	2.9158	—
	n_{H_2O}	3.9577	3.9526	3,9606	3.9609	3,9611	3.9609	3.9609	—
	n_{N_2}	19.9885	19.9841	19.9865	19.9861	19.9863	19.9863	19.9863	—
	n_{CO}	0.0552	0.1140	0.0810	0.0873	0.0842	0.0843	0.0842	—

These are substituted into the set of equilibrium relationships and the numbers of moles of derived constituents are calculated. From these $n_5 - n_{10}$ values of the basic constituents are again obtained, to be used in calculating a new set of $n_5 - n_{10}$ data. In the next iteration step values were calculated with the use of relation (5.108b). In this way the calculation continues until the required accuracy has been achieved. Numbers of moles of constituents in the individual steps are listed in Table 2.

The modified iteration method

The modified iteration method can be employed in cases, where the equilibrium concentration value of one or more derived constituents is comparable with the lowest concentration of a basic constituent.

We may assume in the case discussed in Example 20, that the hydrogen concentration (constituent 5) in the equilibrium mixture will be comparable to the CO (n_4) concentration. Let us chose

$$\alpha = n_1 - n_5 .$$

$$\beta = n_2 + n_5 ,$$

$$\gamma = n_4 - n_5 . \tag{5.109}$$

Introducing this substitution into the equilibrium relation for n_5, we obtain after modification of the second-power equation,

$$n_5^2(1 - K_5) + n_5[\alpha + K_5(\beta + \gamma)] - K_5\beta\gamma = 0. \tag{5.110}$$

For the following calculation we may chose for the initial values $\alpha = q_1$, $\beta = q_2$, $\gamma = q_4$, obtaining hereby a value for n_5 and from it, employing relations (5.109), values of n_1, n_2 and n_4. Taking the value of q_3 for that of n_3, we may calculate $n_6 - n_{10}$ from the equilibrium relationships (5.33). From the balance relationships we than determine values of n_1, n_2, n_3 and n_4, modified values of α, β, γ and a new value for n_5. The calculation proceeds until the required accuracy is achieved.

Newton's method with reduction parameter

Newton's method with reduction parameter may be employed for any set of non-linear equations, i.e. also for the set of non-linear equations obtained by Brinkley's method.

The starting point again are two sets of equations, equilibrium equations and balance equations, determined by the relations (5.33) and (5.31), which however must be modified slightly for this purpose. Dividing the left- and right-hand sides of equation (5.33) by the total number of moles n we obtain the set of equations

$$y_r = K_r P^{(v_r - 1)} \prod_{i=1}^{H} y_i^{v_{ri}} \quad r = H + 1, \ldots, N. \tag{5.111}$$

The set of balance equations is obtained from equation (5.4) by conversion to the form

$$b_j = \sum_{i=1}^{N} a_{ij} n_i^o \quad j = 1, 2, \ldots, H, \tag{5.112}$$

where n_i^o is the initial number of moles of the i-th initial constituent. Using the symbol q_i to denote the number of moles of basic constituents which would provide the same balance,

$$\sum_{i=1}^{H} q_i a_{ij} = b_j \quad j = 1, 2, \ldots, H, \tag{5.113}$$

we solve the set (5.113) to obtain values of q_i, which we now normalize to $\bar{q}_i$ values so as to obtain $\sum_{i=1}^{H} \bar{q}_i = 1$. Introducing again molar fractions y_i instead of number of moles of basic constituents and using n_s to denote the overall number of moles in equilibrium with respect to values of $\bar{q}_i$, we may finally convert equation (5.31) to the form

$$y_i + \sum_{r=H+1}^{N} v_{ri} y_r = \frac{\bar{q}_i}{n_s} \quad i = 1, 2, \ldots, H. \tag{5.114}$$

Summation leads to

$$1 + \sum_{r=H+1}^{N} (v_r - 1)\, y_r = \frac{1}{n_s}, \qquad (5.115)$$

where

$$v_r = \sum_{i=1}^{H} v_{ri}.$$

When the expressions (5.114) and (5.115) are joined together, the balance equations convert to

$$y_i = \bar{q}_i - \sum_{r=H+1}^{N} \left[v_{ri} - \bar{q}_i(v_r - 1) \right] y_r, \quad i = 1, 2, ..., H. \qquad (5.116)$$

The set of non-linear equations, formed by joining together the relations (5.111) and (5.116) is then solved by means of the procedure described in section 5.5.1. Obviously the difficulty of determining the initial step again comes up at the start. Besides procedures which were recommended in the previous sections, the following also appears to be very useful:

When no details of the system are known, we may chose $y_i^{(0)} = 1/N, i = 1, 2, ..., H$. This procedure may fail, however, if the equilibrium concentration of one of the basic constituents is close to zero. The system then becomes greatly unstable, as may be judged from the form of equilibrium relationships (5.111). In these cases different basic constituents must be selected and the calculation must be repeated.

Another possible approach is based on the following thoughts: In ideal systems, the overall composition of the system is a function of the basic constituents only. We therefore select several mutual ratios of molar fractions of basic constituents, as well as several different proportions of basic constituents in the system. For all possible combinations we determine the value of

$$\Theta = \sum_{i=1}^{N} f_1^2,$$

where

$$f_i(y_1, y_2, ..., y_N) = 0 \quad i = 1, 2, ..., N,$$

is the general form of the set of non-linear equations. The calculation will then be started by that combination, which corresponds to the lowest value of Θ.

The Newton-Raphson method

Very often, Newton's method is applied to chemical equilibrium calculations according to Brinkley's procedure in the following manner. Let us substitute for the value of n_r $(r = H + 1, ..., N)$ into the balance equations (5.31) from the equilibrium relations (5.33). This leads to a set of equations

$$F_i \equiv q_i - n_i - \sum_{r=H+1}^{N} v_{ri} K_r \left(\frac{P}{n}\right)^{v_r-1} \prod_{j=1}^{H} n_j^{v_{rj}} = 0 \quad i = 1, 2, \ldots, H$$

$$n = \sum_{r=1}^{H} n_i + \sum_{r=H+1}^{N} K_r \left(\frac{P}{n}\right)^{v_r-1} \prod_{j=1}^{H} n_j^{v_{rj}} . \tag{5.117}$$

Let us select the first approximation of $n_i^{(1)}$ $i = 1, 2, \ldots, H$ and develop the left-hand side of the relation (5.117) into a Taylor series, neglecting second- and higher-order terms. This results in a set of linear equations

$$\sum_{i=1}^{H} q_{ik}\Pi_i + d_k w = F_k \quad k = 1, 2, \ldots, H$$

$$\sum_{i=1}^{H} d_i\Pi_i = W , \tag{5.118}$$

where

$$q_{ik} = n_i + \sum_{r=H+1}^{N} v_{ri} v_{kr} n_r \quad i, k = 1, 2, \ldots, H$$

$$d_i = \frac{1}{n} \left[n_i + \sum_{r=H+1}^{N} v_{ri} n_r \right]$$

$$W = 1 - \frac{1}{n} \left[\sum_{i=1}^{H} n_i + \sum_{r=H+1}^{N} n_r \right] , \tag{5.119}$$

for the unknown variables Π_i $i = 1, 2, \ldots, H$ and w. The second approximation $n_i^{(2)}$ $i = 1, 2, \ldots, H$ is then obtained from

$$n_i^{(2)} = \frac{n_i^{(1)} \cdot n^{(2)}}{n^{(1)}} (1 + \lambda \Pi_i) \quad i = 1, 2, \ldots, H$$

$$n^{(2)} = n^{(1)} + \lambda w , \tag{5.120}$$

where λ is the reduction parameter. This procedure is used to solve the following example.

Example 21

One of the industrial processes used to synthesize methylamines is the reaction of methanol with ammonia on dehydrating catalysts at elevated pressure and temperatures of 550 to 800 K. Calculate the equilibrium composition at 700 K, 40 atm and a ratio of initial components equal to 4.5 moles ammonia per 1 mole methanol.

Analysis of the reaction mixture shows, that besides three methylamines and the non-reacted initial components we must also consider hydrogen, water, dimethylether and formaldehyde. The system thus determined therefore contains nine constituents, for which both fundamental matrixes can be constructed: the matrix of constitution coefficients in the form of

		$(j=1)$ C	$(j=2)$ H	$(j=3)$ O	$(j=4)$ N
CH_3OH	$(i=1)$	1	4	1	0
NH_3	$(i=2)$	0	3	0	1
H_2O	$(i=3)$	0	2	1	0
H_2	$(i=4)$	0	2	0	0
CH_3NH_2	$(i=5)$	1	5	0	1
$(CH_3)_2NH$	$(i=6)$	2	7	0	1
$(CH_3)_3N$	$(i=7)$	3	9	0	1
$(CH_3)_2O$	$(i=8)$	2	6	1	0
CH_2O	$(i=9)$	1	2	1	0

with a rank of $H = 4$ and a derived matrix of stoichiometric coefficients in the form of

$$\begin{vmatrix} -1 & -1 & 1 & 0 & 1 & 0 & 0 & 0 & 0 \\ -2 & -1 & 2 & 0 & 0 & 1 & 0 & 0 & 0 \\ -3 & -1 & 3 & 0 & 0 & 0 & 1 & 0 & 0 \\ -2 & 0 & 1 & 0 & 0 & 0 & 0 & 1 & 0 \\ -1 & 0 & 0 & 1 & 0 & 0 & 0 & 0 & 1 \end{vmatrix},$$

which determined five linearly independent chemical reactions

$$
\begin{aligned}
CH_3OH + NH_3 &= CH_3NH_2 + H_2O & (r=1) \\
2\,CH_3OH + NH_3 &= (CH_3)_2NH + 2\,H_2O & (r=2) \\
3\,CH_3OH + NH_3 &= (CH_3)_3N + 3\,H_2O & (r=3) \\
2\,CH_3OH &= (CH_3)_2O + H_2O & (r=4) \\
CH_3OH &= CH_2O + H_2 & (r=5).
\end{aligned}
$$

Clearly, the basic constituents should be methanol, ammonia and water: hydrogen was chosen as the fourth. This selection is justified, since the constituents selected form a non-zero determinant of consitution coefficients.

The equilibrium relationships will be

$$
\begin{aligned}
n_5 &= K_5 n_1 n_2 n_3^{-1} \\
n_6 &= K_6 n_1^2 n_2 n_3^{-2} \\
n_7 &= K_7 n_1^3 n_2 n_3^{-3} \\
n_8 &= K_8 n_1^2 n_3^{-1} \\
n_9 &= K_9 \left(\frac{P}{n}\right)^{-1} n_1 n_4^{-1}
\end{aligned}
$$

and the balance relationships are the following

$$n_1 = q_1 - n_5 - 2n_6 - 3n_7 - 2n_8 - n_9$$

$$n_2 = q_2 - n_5 - n_6 - n_7$$

$$n_3 = q_3 + n_5 + n_6 + n_7 + n_8$$

$$n_4 = q_4 + n_9 \,,$$

where $q_1 = 1$, $q_2 = 4.5$, $q_3 = q_4 = 0$.

The thermochemical data of the individual constituents are[150]:

i	1	2	3	4	5	6	7	8	9
$(\log K_f)_{i700K}$	7.756	-2.044	15.585	0	-8.521	-14.562	-20.635	0.606	7.383

and hence, calculated values of equilibrium constants for the individual reactions are:

r	$\log K_a$	K_a
(1)	1.352	22.4
(2)	3.140	1.38×10^3
(3)	4.896	7.87×10^4
(4)	0.679	4.77
(5)	-0.394	2.47×10^{-1}

Solution obtained in the individual steps are given in the table. It is evident, that in this case Newton's method brought a solution in a few steps, although the numbers of moles of two basic constituents (methanol and hydrogen) in the equilibrium mixture are considerably lower than the number of moles of most derived constituents. It is assumed, that one of the causes may be a relatively very favourable selection of the initial set of mole numbers.

The results agree well with calculated and experimental data published by Šeha and Habada[154] (slight differences in the equilibrium composition must be ascribed to different thermochemical values employed).

Constituent	Number of moles of individual constituents in the course of the calculation step							
	(0)	(1)	(2)	(3)	(4)	(5)	(6)	(7)
CH_3OH $(i = 1)$	0.1	0.010	0.006	0.002	0.001	0.0009	0.00084	0.00085
NH_3 $(i = 2)$	4.2	4.044	3,990	3.952	3.924	3.9074	3.90012	3.90057
H_2O $(i = 3)$	0.7	0.936	0.953	0.972	0.985	0.9924	0.99262	0.99266
H_2 $(i = 4)$	0.1	0.044	0.036	0.024	0.013	0.0064	0.00646	0.00646
CH_3NH_2 $(i = 5)$	0.1	0.161	0.225	0.265	0.301	0.3215	0.33152	0.33155
$(CH_3)_2NH$ $(i = 6)$	0.1	0.121	0.133	0.145	0.142	0.1426	0.14256	0.14257
$(CH_3)_3NH$ $(i = 7)$	0.1	0.174	0.152	0.138	0.133	0.1285	0.12530	0.12531
$(CH_3)_2O$ $(i = 8)$	0.1	0.010	0.005	0.002	0.001	0.0003	0.00008	0.00003
CH_2O $(i = 9)$	0.1	0.044	0.036	0.024	0.013	0.0064	0.00646	0.00646
$\Sigma n_i \equiv n$	5.6	5.544	5.536	5.524	5.513	5.5064	5.50596	5.50646

The Marquardt method

Marquardt[102] suggested a method for solving sets of non-linear equations which is in principle a combination of Newton's method and the gradient method. Marquardt tried to utilize the advantages of the two procedures, the steep convergence of Newton's method and the slight influence of choice of the initial set of data on convergence, which is an advantage of the gradient method. Designing an algorithm working in between the two methods mentioned, Marquardt believed that he would likewise eliminate the disadvantages of these methods.

We have found on using Marquardt's method for solving a complicated chemical equilibrium, however, that it offers no substantial improvement. Therefore we refrain from a detailed description of the method.

Other methods

A pressing need of solving sets of non-linear equations in other fields of science has recently led to the development of a number of new procedures[7,19,120,123]. Some of the authors describe the procedure (Barrar-Loeb, Pereyra), others present the algorithm (Brown, Pankiewicz). While obviously incomparable values of individual unknown variables (in the case of chemical equilibrium, numbers of moles of the constituents) always are a great difficulty, yet we may believe that it is only a matter of time until a reliable universal procedure will have been developed. Meanwhile, since data processing equipment can be used, this difficulty may be avoided by rearranging the constituents in such a way as to arrive at basic constituent values at least comparable to values of derived constituents.

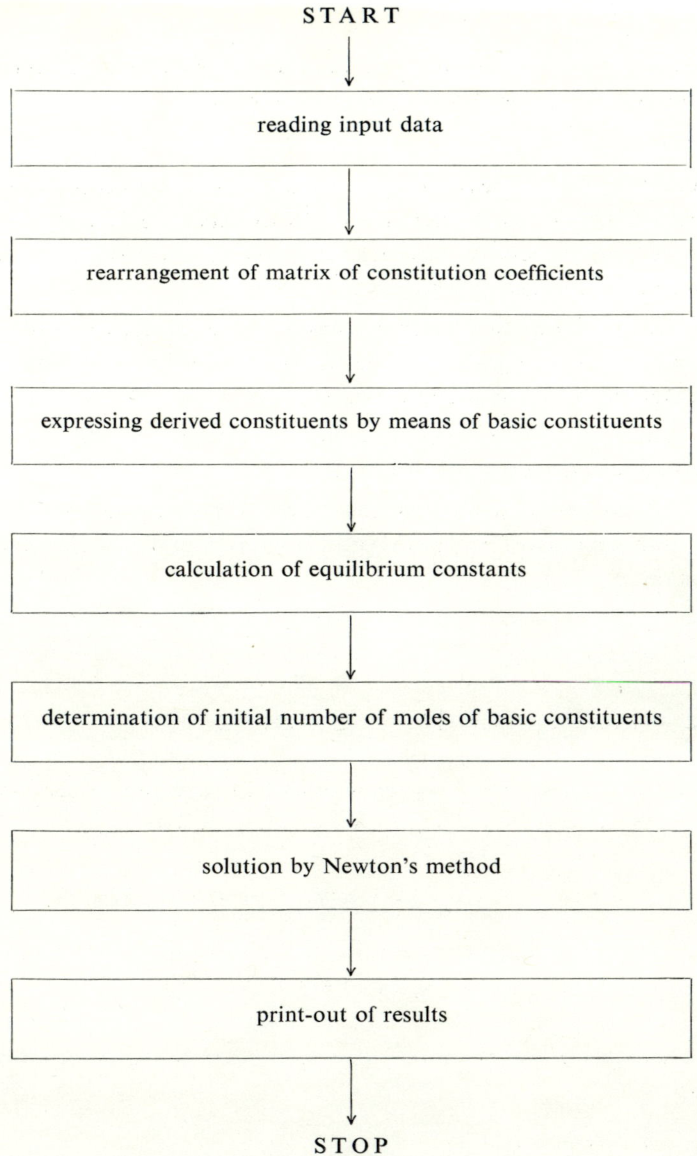

Fig. 20. Block diagram of Brinkley's method.

Notes on construction of programmes

The procedure of solving chemical equilibria by Brinkley's method is shown in the block diagram in Fig. 20. The numerical algorithm may be divided into the following blocks:

a) Determination of the rank H of the matrix of constitution coefficients and rearrangement of this matrix in such a way, as to make the first H rows satisfy the demand of linear independence. Thus the basic constituents are determined, for which a set of balance equation can be obtained.

b) Determination of the equilibrium constants of the respective chemical reactions. This can be achieved by a procedure which was described in the section on application of Newton's method with a reduction parameter.

c) Calculation of the initial set of numbers of moles of the basic constituents by one of the variants described in section 5.6.1.

d) Solution of the set of non-linear equations by one of the numerical methods. According to our experience, Newton's method with reduction parameter is the most advantageous one.

e) Print-out of results.

5.6.3   The White-Johnson-Dantzig method

The principle of this method has been described in detail in section 5.4.2. The iterative procedure involves solution of a set of linear equations

$$\sum_{k=1}^{M} r_{jk}\lambda_k + b_j u = \sum_{i=1}^{N} a_{ij}f_i \quad j = 1, 2, ..., M$$

$$\sum_{k=1}^{M} b_k\lambda_k = \sum_{i=1}^{N} f_i \tag{5.121}$$

The new approximative values of $n_i^{(2)}$ $i = 1, 2, ..., N$ being determined from the relationships

$$n^{(2)} = n^{(1)}(1 + u)$$

$$n_i^{(2)} = -f_i^{(1)} + n^{(2)}y_i^{(1)} + n_i^{(1)} \sum_{k=1}^{M} \lambda_k a_{ik} \quad i = 1, 2, ..., N , \tag{5.122}$$

where

$$f_i^{(1)} = n_i^{(1)}(c_i + \ln n_i^{(1)} - \ln n^{(1)}) .$$

The process is repeated with this new approximation of the equilibrium composition, until the required accuracy is achieved. The method is illustrated in the following example.

138

Example 22

For comparison, let us chose the same system as in Example 11, i.e.

$$CH_4(g) + H_2O(g) \rightarrow CH_4(g), H_2O(g), H_2(g), CO(g), CO_2(g)$$

at 900 K, 1 atm pressure and an initial mixture of 1 mole CH_4 : 1 mole H_2O.

Thermochemical data

Constituent		$-(G^\circ - H_0^\circ/T)$ (cal K^{-1} mole^{-1}; 900 K)	ΔH_0° (cal mole^{-1})
$i = 1$	CH_4	46.470	$-15\ 987$
$i = 2$	H_2O	46.120	$-57\ 104$
$i = 3$	H_2	32.004	0
$i = 4$	CO	48.097	$-27\ 202$
$i = 5$	CO_2	43.047	$-93\ 969$

1. Select a trial set of data according to (5.70), i.e.

$$n_1^{(1)} = 0.50\ ; \quad n_2^{(1)} = 0.25\ ; \quad n_3^{(1)} = 1.75\ ; \quad n_4^{(1)} = 0.25\ ;$$

$$n_5^{(1)} = 0.25\ ; \quad n^{(1)} = 3.00\ .$$

Calculate the respective molar fractions and values of

$$f_1 = n_i^{(1)} \left[\left(\frac{G^\circ}{RT_i} \right) + \ln P + \ln y_i \right].$$

$$f_1 = -17.057\ 16$$

$$f_2 = -14.405\ 14$$

$$f_3 = -29.126\ 27$$

$$f_4 = -10.474\ 17$$

$$f_5 = -20.429\ 53\ .$$

2. Calculate the expressions $\sum\limits_{i=1}^{5} a_{ij} f_i\ (j = 1, 2, 3)$ and $\sum\limits_{i=1}^{5} f_i$ as the right-hand sides of the set (5.120).

$$\sum_{i=1}^{5} a_{i1} f_i = -\ 47.960\ 86$$

$$\sum_{i=1}^{5} a_{i2} f_i = -155.291\ 56$$

$$\sum_{i=1}^{5} a_{i3} f_i = - 65.738\ 37$$

$$\Gamma \equiv \sum_{i=1}^{5} f_i = - 91.492\ 27\ .$$

3. Calculate the matrix of coefficients r_{jk} according to relation (5.69)

	$j = 1$	$j = 2$	$j = 3$
$k = 1$	1	2	0.75
$k = 2$	2	16	0.5
$k = 3$	0.75	0.5	1.5

Determine b_j ($b_1 = 1$; $b_2 = 6$; $b_3 = 1$) and construct and solve the set.

The solution leads to:

$$\begin{aligned}
\lambda_1 + \ \ 2\lambda_2 + 0.75\lambda_3 + \ \ u &= -47.960\ 86\,, &\qquad \lambda_1 &= - \ 0.806\ 967 \\
2\lambda_1 + 16\lambda_2 + 0.5\ \lambda_3 + 6u &= -15.529\ 15\,, &\qquad \lambda_2 &= - \ 8.339\ 12c \\
0.75\lambda_1 + 0.5\lambda_2 + 1.5\ \lambda_3 + \ \ u &= -65.738\ 37\,, &\qquad \lambda_3 &= -40.650\ 580 \\
\lambda_1 + \ \ \lambda_2 + \ \ \lambda_3 + \ \ &= -91.492\ 27\,. &\qquad u &= \ \ \ 0.012\ 2825\,.
\end{aligned}$$

4. Determine a new overall number of moles in the system $n^{(2)} = (u + 1)\, u^{(1)}$; $n^{(2)} = 3.036847$ from the relations (5.121) and calculate new, corrected values for the number of moles of individual constituents, using these to repeat the calculation (see pages 142, 143).

The above procedure requires that the initial set of numbers of moles should satisfy the mass balance equations

$$\sum_{i=1}^{N} a_{ij} n_i = b_j \quad j = 1, 2, \ldots, M$$

$$n_i > 0 \qquad i = 1, 2, \ldots, N\,. \tag{5.123}$$

The first approximation can be obtained by a process similar to that used in Newton's method with reduction parameter. The equation (5.123) can be modified by Gaussian elimination to give

$$n_i = q_i + \sum_{j=H+1}^{R} d_{ij} n_j \quad i = 1, 2, \ldots, H\,, \tag{5.124}$$

where the coefficients q_i, d_{ij} depend only on the value of constitution coefficients (the value of H defines the rank of the matrix of constitution coefficients). From equations (5.124) follows, that the value of n_i lies in the interval

$$0 < n_i < \alpha_i = \min_{j,\,a_{ij} \neq 0} (b_j / a_{ij})\,. \tag{5.125}$$

Values of $n_i \in (0, \alpha_i)$ can be generated by means of the random number generator for substitution into the righ-hand side of equation (5.124). As soon as the case $n_i > 0$ $i = 1, 2, \ldots, H$ sets in we have arrived at the first approximation of the equilibrium composition.

Before the White-Johnson-Dantzig method can be applied directly, the rank of the matrix of constitution coefficients must be determined. Let us use the example of methanol synthesis

$$CO + H_2 \rightarrow CO, H_2, CH_3OH, CH_2O. \tag{5.126}$$

The matrix of constitution coefficients is

$$\begin{vmatrix} 1 & 0 & 1 \\ 0 & 2 & 0 \\ 1 & 4 & 1 \\ 1 & 2 & 1 \end{vmatrix} \tag{5.127}$$

and therefore the mass balance equations (5.123) may be written as

$$n_1 + \quad\quad n_3 + n_4 = \alpha$$

$$2n_2 + 4n_3 + 2n_4 = 2\beta$$

$$n_1 + \quad\quad n_3 + n_4 = \alpha, \tag{5.128}$$

where α and β are initial number of moles of CO and H_2, resp. Clearly the first and third columns of the matrix (5.127) and thus also the first and third equation in the relation (5.128) become identical. Therefore one of the equations, e.g. the third one, must be left out, formally decreasing the number of elements in the system. This step is essential as otherwise the set of linear equations (5.120) would have a singular matrix, i.e. a determinant equal to zero.

The case where the overall number of moles does not change in the course of a reaction, i.e. $u = 0$ presents another possibility of decreasing the number of equations in the set (5.120). Differing from the preceding, however, this case need not be determined beforehand. It would be an advantage, however, if chemical equilibria were being calculated with the use of a desk-top calculating machine. In this case a decreased number of equations in the set (5.120) would mean a considerable saving in time.

Thus it is advantageous to classify gaseous systems in four categories:

1. Systems with $H = M, u \neq 0$ holding outside the equilibrium state.

2. Systems with $H = M, u = 0$ holding outside the equilibrium state.

3. Systems with $H < M, u \neq 0$ holding outside the equilibrium state.

4. Systems with $H < M, u = 0$ holding outside the equilibrium state.

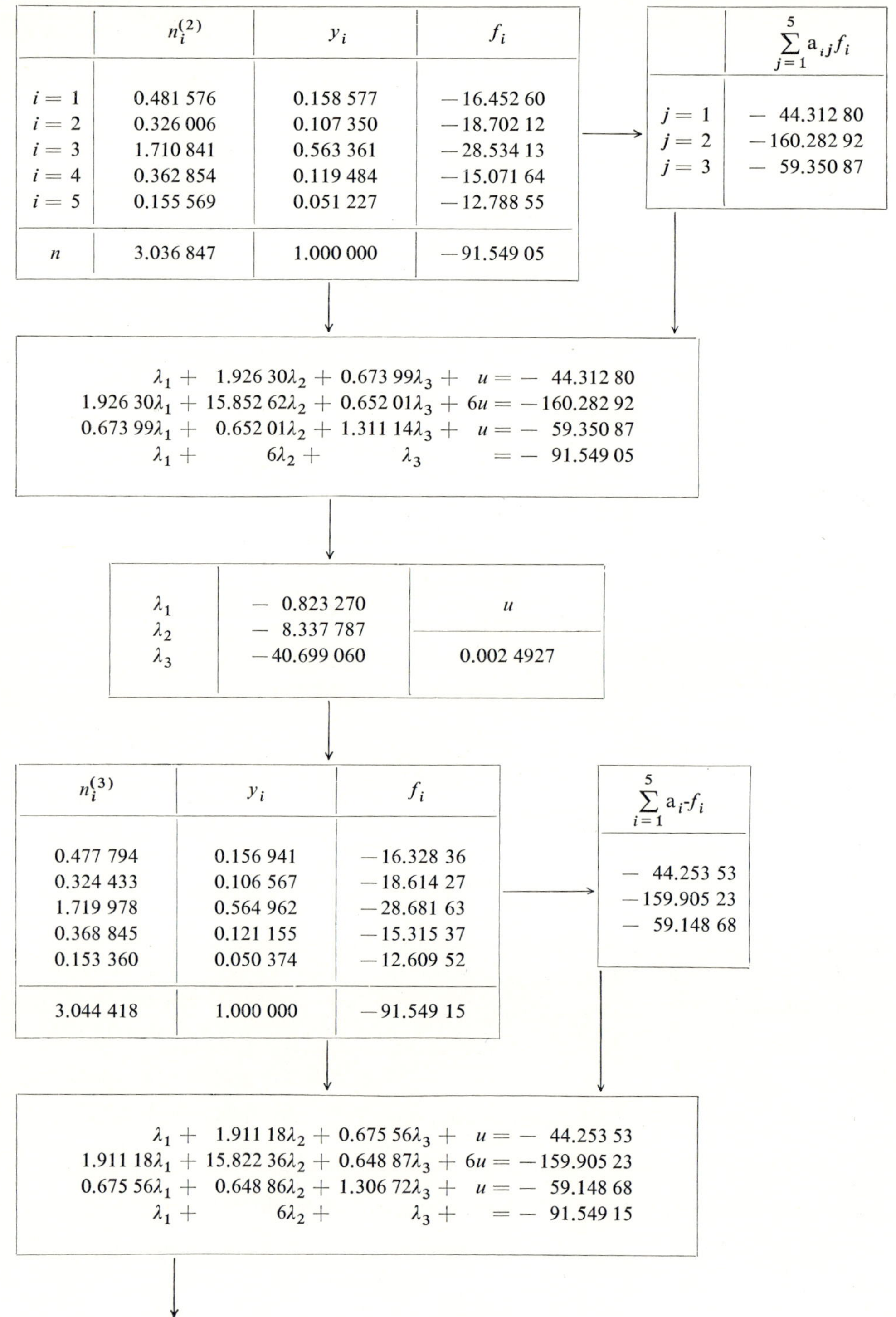

$$\lambda_1 + 1.926\,30\lambda_2 + 0.673\,99\lambda_3 + u = -44.312\,80$$
$$1.926\,30\lambda_1 + 15.852\,62\lambda_2 + 0.652\,01\lambda_3 + 6u = -160.282\,92$$
$$0.673\,99\lambda_1 + 0.652\,01\lambda_2 + 1.311\,14\lambda_3 + u = -59.350\,87$$
$$\lambda_1 + 6\lambda_2 + \lambda_3 = -91.549\,05$$

$$\lambda_1 + 1.911\,18\lambda_2 + 0.675\,56\lambda_3 + u = -44.253\,53$$
$$1.911\,18\lambda_1 + 15.822\,36\lambda_2 + 0.648\,87\lambda_3 + 6u = -159.905\,23$$
$$0.675\,56\lambda_1 + 0.648\,86\lambda_2 + 1.306\,72\lambda_3 + u = -59.148\,68$$
$$\lambda_1 + 6\lambda_2 + \lambda_3 + = -91.549\,15$$

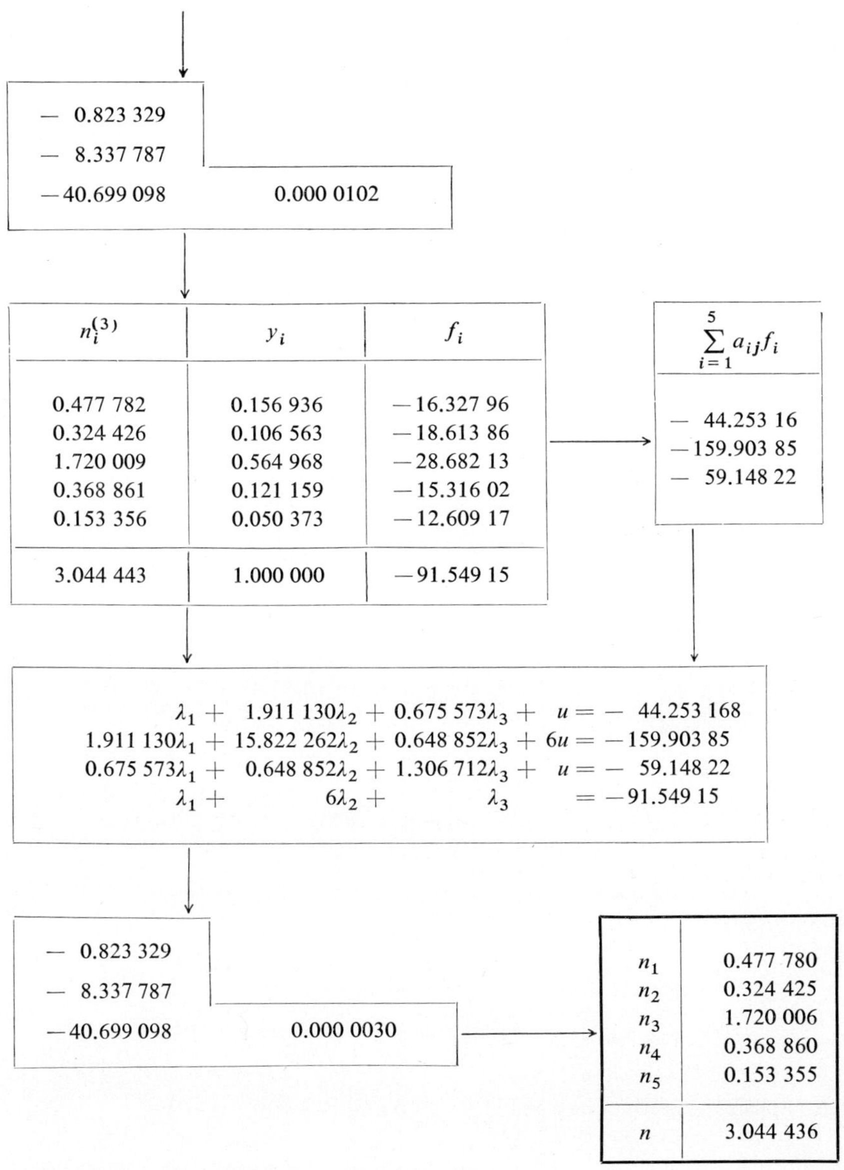
— 0.823 329
— 8.337 787
—40.699 098   0.000 0102

$n_i^{(3)}$ y_i f_i $\sum_{i=1}^{5} a_{ij} f_i$

0.477 782   0.156 936   —16.327 96   — 44.253 16
0.324 426   0.106 563   —18.613 86   —159.903 85
1.720 009   0.564 968   —28.682 13   — 59.148 22
0.368 861   0.121 159   —15.316 02
0.153 356   0.050 373   —12.609 17

3.044 443   1.000 000   —91.549 15

$\lambda_1 + 1.911\,130\lambda_2 + 0.675\,573\lambda_3 + u = -44.253\,168$
$1.911\,130\lambda_1 + 15.822\,262\lambda_2 + 0.648\,852\lambda_3 + 6u = -159.903\,85$
$0.675\,573\lambda_1 + 0.648\,852\lambda_2 + 1.306\,712\lambda_3 + u = -59.148\,22$
$\lambda_1 + 6\lambda_2 + \lambda_3 = -91.549\,15$

— 0.823 329
— 8.337 787
—40.699 098   0.000 0030

n_1 0.477 780
n_2 0.324 425
n_3 1.720 006
n_4 0.368 860
n_5 0.153 355

n 3.044 436

The first category includes systems in which the number of independent balance relationships is equal to the number of elements or basic species and where, moreover, conversion of the initial to the equilibrium mixture involves a change in the overall number of moles. The two conditions are usually satisfied when the initial mixture includes at least two constituents and the equilibrium mixture is rather complex. In such cases, there is usually at least one reaction among the many possible ones, the sum of stoichiometric coefficients of which is non-zero. For a system thus defined, the set (5.121) may be used unchanged.

It is clear in the case of systems of the second category, that the set of linear equations includes one unknown variable less, so that the number of equations solved is $M = H$. The following system for example belongs to this category:

$$CO_2(g) + H_2(g) + H_2S(g) \rightarrow$$
$$\rightarrow CO_2(g), H_2(g), H_2S(g), CO(g), H_2O(g), COS(g) .$$

The matrix of constitution coefficients will take the form of

$$\begin{vmatrix} 1 & 0 & 2 & 0 \\ 0 & 2 & 0 & 0 \\ 0 & 2 & 0 & 1 \\ 1 & 0 & 1 & 1 \\ 1 & 0 & 1 & 0 \\ 0 & 2 & 1 & 0 \end{vmatrix} . \tag{5.129}$$

The rank of the matrix is $H = 4$, i.e. two linearly independent reactions will take place in the system.

Taking the first four constituents to be the basic ones, we obtain the following matrix of stoichiometric coefficients as described in Example 5

$$CO_2 + H_2 = CO + H_2O ,$$
$$H_2S + CO = H_2 + COS . \tag{5.130}$$

Evidently, there applies to both equations $\sum\limits_{i=1}^{N} v_{ri} = 0$ and therefore also $\sum\limits_{r=1}^{R} \left| \sum\limits_{i=1}^{N} v_{ri} \right| = 0$.

The overall number of moles in the system is stable. Therefore there is no need of determining the quantity u, which is always equal to zero. Thus the set (5.121) will consist of only four equations, namely ($r_{12} = r_{21} = 0$, values of $n_i^{(1)}$ are not presented)

$$8\lambda_1 + \lambda_3 + 1{,}5\lambda_4 = \sum_{i=1}^{6} a_{i1} f_i$$

$$\lambda_2 + 1{,}5\lambda_3 + 0{,}25\lambda_4 = \sum_{i=1}^{6} a_{i2} f_i$$

$$\lambda_1 + 1{,}5\lambda_2 + 3\lambda_3 + 0{,}25\lambda_4 = \sum_{i=1}^{6} a_{i3} f_i$$

$$1{,}5\lambda_1 + 0{,}25\lambda_2 + 0{,}25\lambda_3 + \lambda_4 = \sum_{i=1}^{6} a_{i4} f_i \tag{5.131}$$

and the relationship (5.122) converts to

$$n_i^{(2)} = -f_i^{(1)} + \left(1 + \sum_{j=1}^{4} \lambda_j a_{ij}\right) n_i^{(1)} \quad i = 1, 2, \ldots, 6. \tag{5.132}$$

The end of the calculation is determined not by the condition $|u| < \varepsilon$, but by $\left|(\Gamma)^{(p+1)} - (\Gamma)^p\right| < \varepsilon$.

The third category includes systems in which one of the columns in the matrix of constitution coefficients is a linear combination of the other columns. In such cases a corresponding number of linearly dependent columns is left out: with respect to the following calculation it is advantageous to keep as far as possible those with a maximum number of zeros. This greatly simplifies the process of solving the set (5.121).

For illustration, let us take the above mentioned methanol synthesis according to the scheme

$$CO + H_2 \rightarrow CO, H_2, CH_3OH, CH_2O. \tag{5.133}$$

Evidently, $H = 2$, but $M = 3$. Since there are only two basic constituents, two balance relationships will suffice to describe the system, so that one column of the matrix of constitution coefficients must be deleted. The resulting set of linear equations according to (5.130) will take the general form of

$$r_{11}\lambda_1 + r_{12}\lambda_2 + ub_1 = \sum_{i=1}^{4} a_{i1} f_i,$$

$$r_{21}\lambda_1 + r_{22}\lambda_2 + ub_2 = \sum_{i=1}^{4} a_{i2} f_i,$$

$$b_1\lambda_1 + b_2\lambda_2 = \sum_{i=1}^{4} f_i, \tag{5.134}$$

where $j = 1$ will be either carbon or oxygen, $j = 2$ will be hydrogen.

The fourth category finally includes systems of maximum symetry with $H < M$ and $\sum_{r=1}^{R} \left|\sum_{i=1}^{N} v_{ri}\right| = 0$. In these cases the set of linear equations is reduced as in the second category, i.e. by eliminating the respective number of columns in the matrix of stoichiometric coefficients, as well as due to the unchanging number of moles in the system.

Let us employ the example of 1-hexene isomerisation. The reaction mixture is formed by a total of 17 isomeric hexenes. Clearly the matrix of constitution coefficients will consist of 17 equal rows (6 12). The rank of this matrix will be $H = 1$; 16 linearly independent reactions are possible, i.e. skeleton isomerisation, double-bond shift along the chain and cis-trans conversion. Basing the calculation on 1 mole of 1-hexene, the set (5.121) will be reduced to a single equation of the type

$$r_{11}\lambda_1 = \sum_{i=1}^{17} a_{i1}f_i \tag{5.135}$$

and the relation (5.122) is reduced to 17 relations of the type

$$n_i^{(2)} = -f_i^{(1)} + n_i^{(1)}(1 + a_{i1}\lambda_1) \quad i = 1, 2, \ldots, 17 . \tag{5.136}$$

Notes on construction of the programme.

The block diagram in Fig. 21 shows the process of constructing a programme for calculation of chemical equilibria by the White-Johnson-Dantzig method. The procedure may be divided into the following main blocks:

a) Reading input data. The basic input data are temperature and pressure of the system, number of compounds and elements, number of input compounds and numbers of moles of input compounds and the matrix of constitution coefficients. An important input are thermochemical data for each compound, serving for determination of the value of $c_i i = 1, 2, \ldots, N$.

b) Determination of the rank of the matrix of constitution coefficients. In the case of $H < M$ a number $M - H$ of columns in the matrix of constitution coefficients must be deleted (obviously so as not to alter the rank of the matrix).

c) Obtaining the first approximation of equilibrium composition. The procedure for determining the first approximation was already described in this section.

d) The iteration process proper, based on equations (5.121) and (5.122). It is essential to avoid negative values of mole numbers (see relation (5.72)).

e) End of the iteration process and print-out of results. A logical condition, establishing the end of the calculation, may either be numerical value of the quantity u or, the change of overall free enthalpy of the system in two successive approximations, since both these values should be equal to zero when equilibrium is reached. Practical experience, however, shows that it is advantageous to include both conditions in the programme. Indication by means of the change of overall free enthalpy is quite general, but it may differ in accuracy depending on whether the form of the overall free enthalpy vs. composition relationship has a sharp or a shallow minimum in the process of conversion from the initial to the equilibrium state. On the other hand, checking the calculation by means of changes in the value of u is very useful because of the direct relationship to the values of individual numbers of moles. This, however, is not quite a general phenomenon as we have shown in section 5.2.2.3, as it will

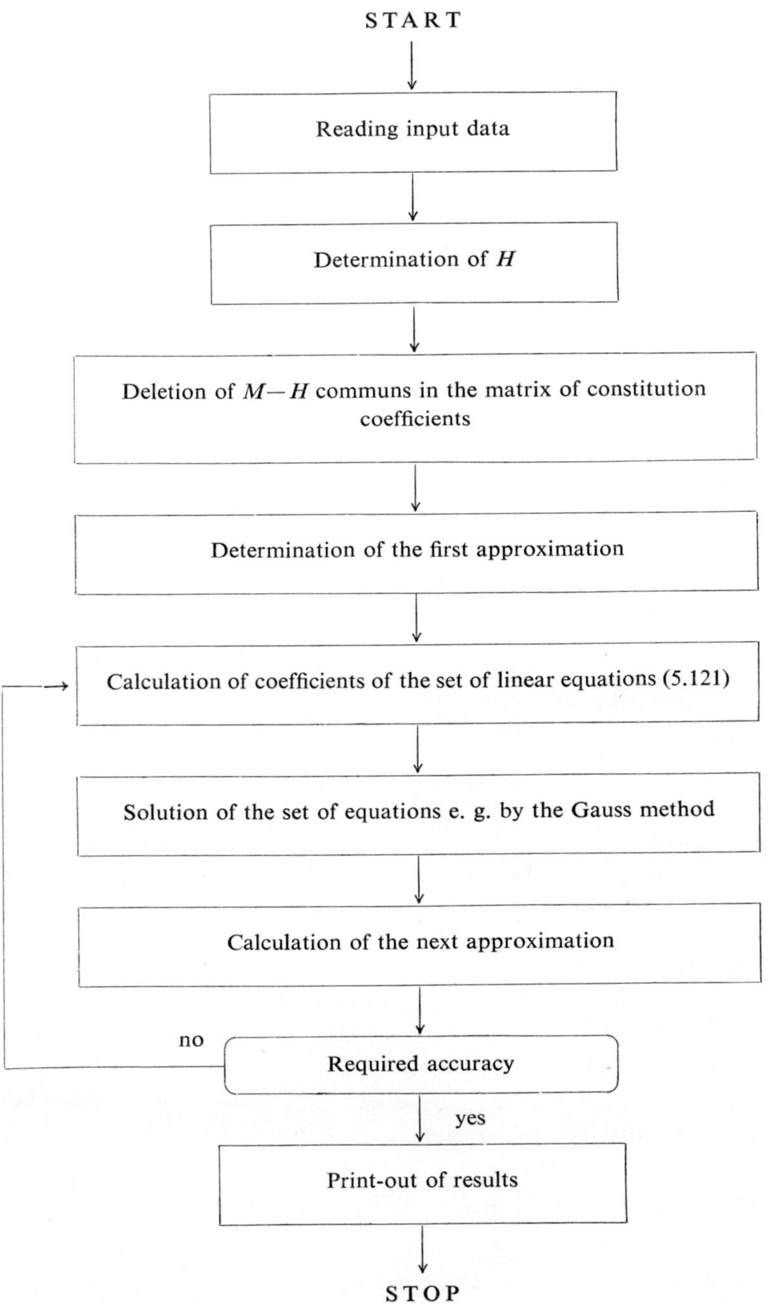

Fig. 21. Block diagram of the modified White-Johnson-Dantzig method.

fail unless the overall number of moles is invariable in the system in question. Thus the use of both conditions will reliably guarantee the result of the calculation with the required degree of accuracy in a minimum number of iteration steps.

Accuracy of calculation

The White-Johnson-Dantzig method may be regarded as one of non-linear programming. Dorn[48,49] has shown, that every problem that is based on minimalisation of a convex function has a dual form. Mutual linkage of the two formulations of the dual theorem can only be used to determine the overall free enthalpy in the individual steps, and thus also to establish the accuracy of the solution achieved. When minimalisation is expressed as

$$\min \left[\sum_{i=1}^{N} c_i n_i + \sum_{i=1}^{N} n_i \ln \frac{n_i}{n} \right], \tag{5.137}$$

with

$$\sum_{i=1}^{N} a_{ij} n_i = b_j \quad j = 1, 2, \ldots, M$$

and

$$n_i \geqq 0 ,$$

then the dual problem can be formulated as

$$\max \sum_{j=1}^{M} b_j e_j , \tag{5.138}$$

where

$$\sum_{j=1}^{M} a_{ij} e_j \leqq \ln \frac{n_i}{n} + c_i = f_i \quad i = 1, 2, \ldots, N .$$

The relation (5.138) can be interpreted thus: In analogy to equation (5.58) where of have considered the dimensionless function of overall free enthalpy as the sum of contributions by individual constituents, we may now ascribe to each element or basic particle species in the system a certain value of the free enthalpy function e_j, by which this element contributes to the overall free enthalpy of the system. Evidently, there should apply in the state of equilibrium

$$\sum_{i=1}^{M} b_j e_j - \sum_{i=1}^{N} f_i n_i = 0 . \tag{5.139}$$

Conversely, when the system is moving away from equilibrium, the difference of the two values rises continuously, the first term of equation (5.139) representing the upper, the second term the lower limit. The difference of the two terms defines for every iteration step an interval which corresponds to the distance of the system from equilibrium.

148

Abbreviation of the calculation

Use of a computer provides the possibility of performing calculations for large sets of data including temperature, pressure, changes in the initial mass balance and different numbers of constituents in the state of equilibrium. It is usually advantageous to arrange the data in such a manner that the initial set of data for numbers of moles of the individual constituents should only be determined for changes in the initial mass balance. When calculating the equilibrium of this system for different temperatures or pressures it is better to use the equilibrium composition data from the preceding calculation as initial data for the new situation. Very often the number of required iterations can be diminished considerably in this way.

5.6.4   The method of Lagrangian multipliers

The principle and procedure of the Lagrangian multiplier method was described in section 5.3.3. It has been shown that the numerical algorithm is based on solving a set of non-linear equations

$$\sum_{i=1}^{N} q_{ij} \exp \left(\sum_{k=1}^{M} a_{ik}\lambda_k - c_i \right) = d_j \quad j = 1, 2, ..., M , \qquad (5.140)$$

where the following holds:

$$q_{i1} = 1 , \quad d_1 = 1 \qquad\qquad i = 1, 2, ..., N$$

$$q_{ij} = a_{ij} - a_{i1}b_j/b_1 \quad d_j = 0 \quad j = 2, ..., M . \qquad (5.141)$$

Using the known multipliers λ_k $k = 1, 2, ..., M$ we now easily determine the equilibrium values of molar fractions, which are given by

$$y_i = \exp \left(\sum_{k=1}^{M} a_{ik}\lambda_k - c_i \right) \quad i = 1, 2, ..., N . \qquad (5.142)$$

It is evident that, like in the White-Johnson-Dantzig method, we must first determine the rank of the matrix of constitution coefficients, and in cases of $H < M$ delete $(M - H)$ linearly dependent mass balance equations (5.69). This formally decreases the number of elements in the system.

Another important factor in determining chemical equilibria by the Lagrangian multiplier method is determination of the first approximation of the multipliers λ_k $k = 1, 2, ..., M$. This first approximation is required in order to enable solution of the set of equations (5.140) by one of the differentiation methods (e.g. Newton's method with reduction parameter) or, for the application of one of the methods in which differentiation is not performed. Let us assume that the first approximation

$\lambda_k^{(1)}$ $k = 1, 2, \ldots, M$ is known, and let us describe the process of solving the set (5.140) by Newton's method with reduction parameter. We develop the left-hand side of the equations (5.140) into a Taylor series in the point $\lambda_k = \lambda_k^{(1)}$ $k = 1, 2, \ldots, M$ neglecting second- and higher-order terms. This leads to a set of linear equations

$$\sum_{s=1}^{M} r_{js} \Delta\lambda_s^{(1)} = d_j - \sum_{k=1}^{M} q_{ij} y_i^{(1)} \quad j = 1, 2, \ldots, M , \tag{5.143}$$

where

$$y_i^{(1)} = \exp\left(\sum_{k=1}^{M} a_{ik}\lambda_k^{(1)} - c_i\right) \quad i = 1, 2, \ldots, N$$

$$r_{js} = \sum_{i=1}^{N} q_{ij} a_{is} y_i^{(1)} \quad \begin{matrix} j = 1, 2, \ldots, M \\ s = 1, 2, \ldots, M \end{matrix} \tag{5.144}$$

This set of equations can be treated by e.g. the Gauss method. The second approximation of $\lambda_k^{(1)}$ values is obtained from the relation

$$\lambda_k^{(2)} = \lambda_k^{(1)} + \eta\,\Delta\lambda_k^{(1)} \quad k = 1, 2, \ldots, M , \tag{5.145}$$

where η is a reduction parameter. The choice of the reduction parameter may be based on the follwoing thoughts. The relation (5.142) for the equilibrium value of the molar fraction leads to a set of inequalities

$$\sum_{k=1}^{M} a_{ik}\lambda_k - c_i < 0 \quad i = 1, 2, \ldots, N . \tag{5.146}$$

Assume that the first approximation $\lambda_k^{(1)}$ $k = 1, 2, \ldots, M$ satisfies the set of inequalities (5.146). The reduction parameter $\eta \in (0, 1\rangle$ is then chosen such as to make the second approximation also satisfy the inequalities (5.146). In other words, we guarantee validity of the condition $y_i \in (0, 1)$ $i = 1, 2, \ldots, N$ through the numerical process. According to our experience this choice of the reduction parameter always guarantees convergence of the numerical process.

The overall number of moles may be determined in every iteration step as follows. Dividing the left- and right-hand sides of the mass balance equations

$$\sum_{i=1}^{N} a_{ij} n_i = b_j \quad j = 1, 2, \ldots, M \tag{5.147}$$

by the overall number of moles n and adding the left- and right-hand sides of the set of equations thus obtained, we arrive at a relation for the value of n

$$n = \frac{\displaystyle\sum_{j=1}^{M} b_j}{\displaystyle\sum_{j=1}^{M} \sum_{i=1}^{N} a_{ij} y_i} . \tag{5.148}$$

Using this relation we may determine the overall number of moles, and thus also numbers of moles of individual compounds from the known approximative molar fraction values. Therefore we can calculate the expression D in every iteration step:

$$D = \sum_{j=1}^{M} \left| 1 - \sum_{i=1}^{N} a_{ij} n_i / b_j \right| , \qquad (5.149)$$

D having been obtained by addition of the absolute values of differences of the left- and right-hand sides of the mass balance equations (5.147), the j-th equation having been divided by b_j. The iteration process is stopped on validity of the condition $D < \varepsilon$, where ε is the accuracy required. A value of $\varepsilon = 0.0001$ is generally selected. A low value of the expression D signals that values of n_i $i = 1, 2, ..., N$ determined from

$$n_i = n y_i = n \exp \left(\sum_{k=1}^{M} a_{ik} \lambda_k - c_i \right) \quad i = 1, 2, ..., N , \qquad (5.150)$$

practically satisfy the mass balance equations and, therefore, chemical equilibrium is established for these values of n_i $i = 1, 2, ..., N$. Division of the j-th mass balance equation by b_j is essential because values of b_j $j = 1, 2, ..., M$ may mutually differ by as much as several orders of magnitude.

To describe the method completely, there remains the determination of the first approximation $\lambda_k^{(1)}$ $k = 1, 2, ..., M$ such as to satisfy the inequalities (5.146). The following technique may be employed. Let $y_i^{(1)}$ $i = 1, 2, ..., N$ be the first approximation of the equilibrium molar fractions. Unless we have apriori information on the equilibrium composition we select $y_i^{(1)} = 1/N$. From relation (5.142) follows

$$\ln y_i = \sum_{k=1}^{N} a_{ik} \lambda_k - c_i \quad i = 1, 2, ..., N . \qquad (5.151)$$

Developing the left-hand side of the relation (5.151) into a Taylor series in the point $y_i = y_i^{(1)}$ leads to

$$\sum_{k=1}^{N} a_{ik} \lambda_k - c_i = \ln y_i \approx \ln y_i^{(1)} + \frac{1}{y_i^{(1)}} \left(y_i - y_i^{(1)} \right) . \qquad (5.152)$$

From relation (5.152) furthermore follows

$$y_i \approx y_i^{(1)} (1 - \ln y_i^{(1)}) + y_i^{(1)} \left(\sum_{k=1}^{N} a_{ik} \lambda_k - c_i \right) . \qquad (5.153)$$

The mass balance equations expressed in terms of molar fractions take the form of

$$\sum_{i=1}^{N} q_{ij} y_i = d_j \quad j = 1, 2, ..., M . \qquad (5.154)$$

Substitution of relation (5.153) into relation (5.154) gives a set of M linear equations for M unknown variables λ_k $k = 1, 2, \ldots, M$. The calculated values are substituted into the inequalities (5.146). If these are satisfied, we have obtained the first approximation $\lambda_k^{(1)}$ $k = 1, 2, \ldots, M$. In the reverse case we multiply negative values of λ_k by a positive number α and positive values of λ_k by a negative number $-\alpha$. Since constitution coefficients are non-negative numbers, there exists a positive number α, which can be applied in the above manner to secure validity of the expressions (5.146). If chemical equilibrium is being calculated for different conditions, i.e. different initial compositions, pressures and temperatures, it is better to employ values of multipliers resulting from the preceding calculation as the first approximation in the following calculation.

Knowing the value of the multipliers, we are able to determine whether inclusion of one or several more compounds into the system will markedly alter the equilibrium composition (assuming that the set of elements remains unaltered). From the relations (5.140) and (5.142) follows, that if the value of

$$\beta = \exp\left(\sum_{k=1}^{M} a_{N+1,k}\lambda_k - c_{N+1}\right), \tag{5.155}$$

is low enough, addition of another compound with the ordinary number $N + 1$ will not alter the equilibrium composition of the system.

Notes on construction of the programme

The numerical algorithm of the method of Lagrangian multipliers is shown in the block diagram in Fig. 22. The algorithm is composed of the following blocks:

a) Reading input data. Reading input data is identical with that of the White-Johnson-Dantzig method.

b) Determination of the rank of the matrix of constitution coefficients. In this part of the programme, $(M - H)$ linearly depend rows of the mass balance equations must be deleted. However, in technical practice the case of $H < M$ occurs very rarely.

c) Determination of the first approximation $\lambda_k^{(1)}$ $k = 1, 2, \ldots, M$. The procedure leading to the first approximation was described in this section. With the selection of the reduction parameter suggested by us, care must be taken that the first approximation satisfies the inequalities (5.146).

d) Solution proper of the set of non-linear equations (5.140) by means of Newton's method with reduction parameter.

e) Print-out of results.

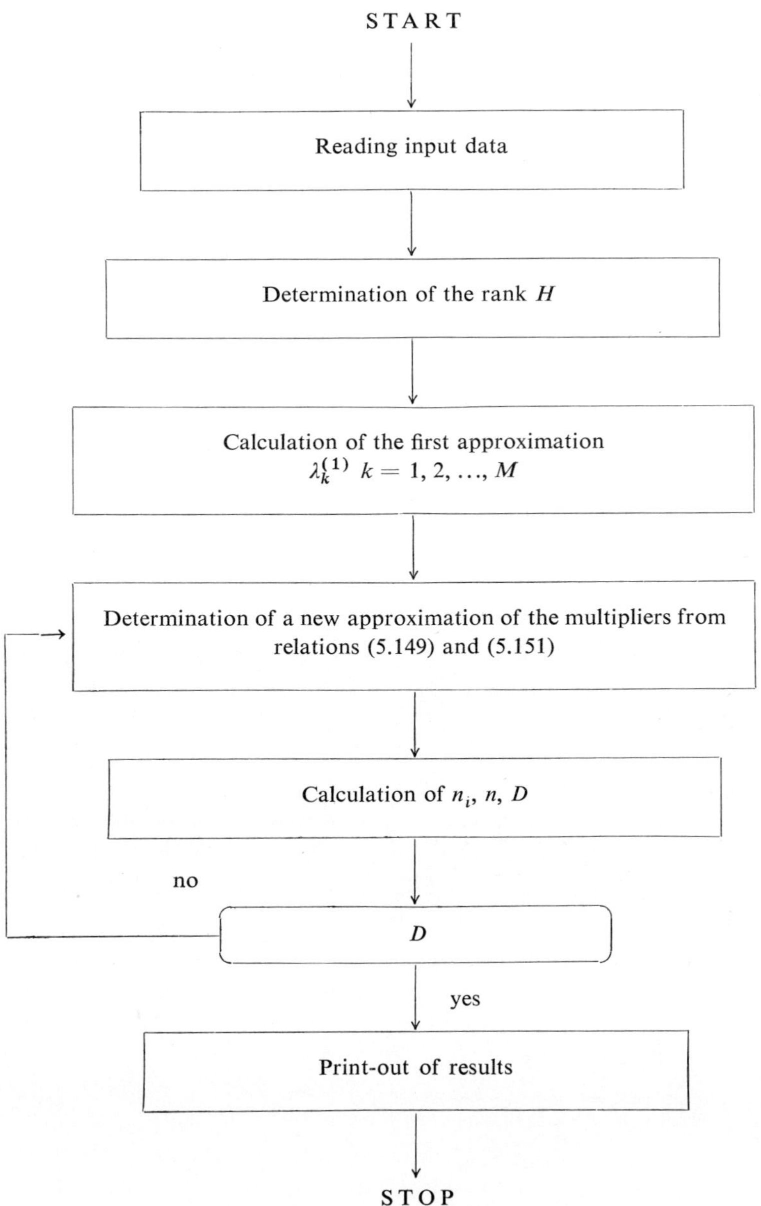

Fig. 22. Block diagram of the method of Lagrangian multipliers.

5.6.5 Mutual comparison of the methods

To compare the methods in terms of accuracy, number of steps required and time of calculation, oxidation of propane with a stoichiometric amount of air at 2200 K and 40 atm according to the scheme

$$C_3H_8(g) + 5\,O_2(g) + 20\,N_2(g) \rightarrow$$
$$\rightarrow \quad H, H_2, O, O_2, OH, H_2O, CO, CO_2, N_2, NO(g)\,,$$

was treated by three different methods. The results as well as thermochemical data employed are given in Table 3. Results obtained by Damköhler and Edse and by Kandiner and Brinkley are included for the sake of comparison.

It will be seen from the Table, that the accuracy of methods described in this chapter is sufficient equal in terms of practical application. Rather large deviations of these data from those reported by Damköhler and Edse and by Kandiner and Brinkley can be explained by errors in abbreviating data and by the lesser number of iteration steps performed.

The overall time needed for calculation obviously depends mainly on the number of elements and constituents in the system, on the form of the overall free enthalpy vs. composition relationship, which also directly influences the number of required iteration steps. In general, the time needed did not considerably differ between the systems calculated up to now.

On the other hand, differences in the time of calculation required for the system by different methods are mainly due to the method employed to determine the initial set of numbers of moles. Nett times of calculations after substitution of the same set of numbers of moles were practically equal for all three methods.

The above comparison supports the view that for medium sized systems the application of the above-described three methods will be equal in value. For large systems $(N \gg M)$, time can be saved by using methods based on minimalisation of the Gibbs function.

5.7 CALCULATION OF THE ENTHALPY BALANCE IN COMPLICATED SYSTEMS

Chemical equilibrium calculations are frequently accompanied by determinations of enthalpy changes involved in the overall chemical conversion process. Since such calculations are important for successful industrial application of newly developed processes, we are presenting at least the basic relationships required for judging enthalpy changes. The procedure to be described in the following is in fact a development of the NASA method of determining the coordinates (H, P).

We shall again consider a gaseous system of N compounds and M elements. Let us use ΔH to denote the amount of heat accepted or liberated by the system at con-

Table 3. Comparison of the results of chemical equilibrium calculations for the case of propane oxidation with air at 2200 K and 40 atm.

Thermochemical data			Numbers of moles of individual constituents in equilibrium mixture, by method of:				
Constituent	$-\left(\dfrac{G^{\circ} - H_0^{\circ}}{T}\right)$	ΔH_0°	Damköhler-Edse	Kandiner-Brinkley	Brinkley	White-Johnson-Dantzig	Lagrangian multipliers
H	32.353	51 620	0.0007	0.000 72	0.000 667	0.000 667	0.000 668
H_2	38.364	0	0.0222	0.022 15	0.020 807	0.020 807	0.020 808
O	43.488	58 586	0.0004	0.000 42	0.000 446	0.000 445	0.000 446
O_2	56.586	0	0.0310	0.031 14	0.033 718	0.033 712	0.033 727
OH	51.129	10 000	0.0332	0.033 20	0.015 709	0.015 710	0.015 709
H_2O	54.263	−57 104	3.9587	3.961 00	3.971 005	3.971 006	3.971 000
CO	54.826	−27 202	0.0844	0.084 20	0.082 314	0.081 294	0.081 328
CO_2	63.012	−93 969	2.9140	2.915 60	2.918 698	2.918 706	2.918 688
N_2	53.220	0	19.9741	19.986 30	19.986 649	19.986 644	19.986 655
CO	58.005	21 477	0.0274	0.027 41	0.026 712	0.026 712	0.026 712
n			27.0261	27.062 14	27.056 725	27.056 371	27.055 741
G/RT					−77.4960	−77.4938	−77.4969
Calculation time (Elliott 503)			Including 1. aproxim.		28 s	14 s	34 s
			1. approximation taken preceding calculation		12 s	10 s	8 s

stant pressure in the course of the reaction. Let us assume that the system includes p initial compounds. Let us denote the temperature of the input compounds T_S, and the temperature of the compounds at the end of the reaction T_E. There evidently applies

$$\Delta H = H_{END} - H_{START}, \tag{5.156}$$

where H_{START} and H_{END} is the enthalpy of the system at the beginning and end of the reaction. The following relationships apply to the values of H_{START} and H_{END}

$$H_{START} = \sum_{i=1}^{p} n_i^o (\Delta H_f^o)_{i,T_S}$$

$$H_{END} = \sum_{i=1}^{N} n_i (\Delta H_f^o)_{i,T_E}, \tag{5.157}$$

where $n_i^o\ i = 1, 2, \ldots, p$ is the number of moles of input compounds at the beginning of the reaction, $n_i\ i = 1, 2, \ldots, N$ is the number of moles of all compounds in equilibrium. As long as standard enthalpy of formation values ΔH_f^o of the individual compounds are known for the temperatures T_E and T_S, the relation (5.157) may be used directly for the calculation. Generally, however, these values are tabulated for 298.15 K. Assume therefore that the values of ΔH_f^o are known for all compounds at the temperature T_1. Using the thermodynamic relationship

$$\left(\frac{\partial \Delta H_f^o}{\partial T} \right)_P = C_P^o \tag{5.158}$$

equation (5.165) may be rewritten

$$H_{START} = \sum_{i=1}^{p} n_i^o \left[(\Delta H_f^o)_{i,T_1} + \int_{T_1}^{T_S} C_{P,i}^o(T)\, dT \right]$$

$$H_{END} = \sum_{i=1}^{N} n_i \left[(\Delta H_f^o)_{i,T_1} + \int_{T_1}^{T_E} C_{P,i}^o(T)\, dT \right]. \tag{5.159}$$

To apply the equations (5.159), however, we must know the C_P^o vs. temperature relationship. This can usually be expressed in a satisfactory manner by means of

$$C_P^o = a_0 + a_1 T + a_2 T^2$$

$$C_P^o = a_0 + \frac{a_1}{a_2 + T}$$

$$C_P^o = a_0 + \frac{a_1}{a_2 + T} + a_3 T + a_4 T^2 + a_5 \ln T, \tag{5.160}$$

where the first two relations are used for temperature intervals of less than 500 K.

5.7.1 Calculation of adiabatic temperature

Up to now, all chemical conversions were studied under isothermic conditions. In some cases, particularly for exothermic reactions, calculations are also required for adiabatic conditions. The reason is, that it is often difficult to keep strongly exothermic reactions isothermic under industrial-scale technologic conditions. Equilibrium conditions determined for isothermic and adiabatic conditions are limiting cases of the general course of processes involved.

Under adiabatic conditions the heat of reaction liberated is consumed for heating up the gas mixture, whereby the temperature of the system rises, influencing by feed-back the degree of conversion. The resulting temperature of reaction products at the end of an adiabatic reaction may be calculated by simultaneous solution of equilibrium relationships and the enthalpy balance, balance relationships being conserved during the process.

Let us first describe the procedure for a system with one chemical reaction

$$\sum_{i=1}^{N} v_i A_i = 0 . \tag{5.161}$$

Assume that the dependence of the equilibrium constant on temperature is known. This relationship can be obtained by means of one of the methods described in Chapter 3. The equilibrium degree of conversion ξ and temperature T must satisfy a set of two non-linear equations

$$K_a(T) - \left(\frac{P}{n^\circ + v\xi}\right)^v \prod_{i=1}^{N} (n_i^\circ + v_i\xi)^{v_i} = 0$$

$$H_{\text{END}} - H_{\text{START}} = 0 . \tag{5.162}$$

The first equation expresses the equilibrium condition for a general temperature T, the second equation represents the condition for adiabatic processes. Denoting the required temperature of the products T_E, we may write equation (5.162) in a more detailed form

$$K_a(T_E) - \left(\frac{P}{n^\circ + v\xi}\right)^v \prod_{i=1}^{N} (n_1^\circ + v_i\xi)^{v_i} = 0$$

$$\sum_{i=1}^{N} (n_i^\circ + v_i\xi) \left[(\Delta H_f^\circ)_{i,T_1} + \int_{T_1}^{T_E} C_{P_i}^\circ(T)\, dT \right] -$$

$$- \sum_{i=1}^{p} n_i^\circ \left[(\Delta H_f^\circ)_{i,T_1} + \int_{T_1}^{T_S} C_{P_i}^\circ(T)\, dT \right] = 0 , \tag{5.163}$$

where T_S is the initial temperature of the initial compounds, T_1 is the temperature at which standard enthalpy of formation values are known and T_E is the temperature required. The set of two non-linear equations (5.163) for the unknown variables ξ and T_E may be solved by eliminating ξ from the second equation and substituting into the first equation. Calculation of the adiabatic temperature involved one main difficulty. Before the calculation, the dependence of the equilibrium constant K_a and heat capacity values $C^o_{P,i}$ on temperature must be known for the temperature interval T_1, T_S, T_E, the values of T_E being obviously not known beforehand. Therefore considerable attention must be given to the selection of temperature relationship of the above quantities. When the calculated value of the temperature T_E lies outside the temperature range in which the relationships $K_a = K_a(T)$, $C^o_{P,i} = C^o_{P,i}(T)$ are valid, the result will not be very reliable.

Example 23

Calculate the temperature for adiabatic CO combustion by a theoretical oxygen amount at atmospheric pressure according to

$$CO(g) + \tfrac{1}{2}O_2(g) = CO_2(g).$$

Log $(K_a)_{2500} = 1.436$, values of $C^o_{P_i}$ are expressed for the sake of simplicity by a quadratic relationship

$$C^o_{P_{CO}} = 6.25 + 2.091 \times 10^{-3}T - 0.66 \times 10^{-6}T^2,$$

$$C^o_{P_{O2}} = 6.13 + 2.99 \times 10^{-3}T - 0.706 \times 10^{-6}T^2,$$

$$C^o_{P_{CO2}} = 6.85 + 8.533 \times 10^{-3}T - 2.475 \times 10^{-6}T^2$$

and $(\Delta H^o_f)_{298K}$ CO and CO_2 values are $-27\,201.9$ cal mole^{-1} and $-94\,259.8$ cal mole^{-1} resp., the dependence of the equilibrium constant on temperature defined by a procedure shown in Example 8, will be

$$\log K_a = \frac{14\,615}{T} - 1.244 \log T + 0.541 \times 10^{-3}T - 0.514 \times 10^{-7}T^2 - 1.242.$$

Starting out from 1 mole CO, the equilibrium relationship will be

$$K_a = \frac{\xi}{(1-\xi)}\left(\frac{3-\xi}{(1-\xi)}\right)^{0.5}.$$

Joining together the two equations we obtain

$$\log \frac{\xi}{(1-\xi)}\cdot\left(\frac{3-\xi}{1-\xi}\right)^{0.5} =$$

$$= \frac{14\,615}{T} - 1.244 \log T + 0.541 \times 10^{-3}T - 0.514 \times 10^{-7}T^2 - 1.242,$$

which, when plotted, gives curve A in Fig. 23.

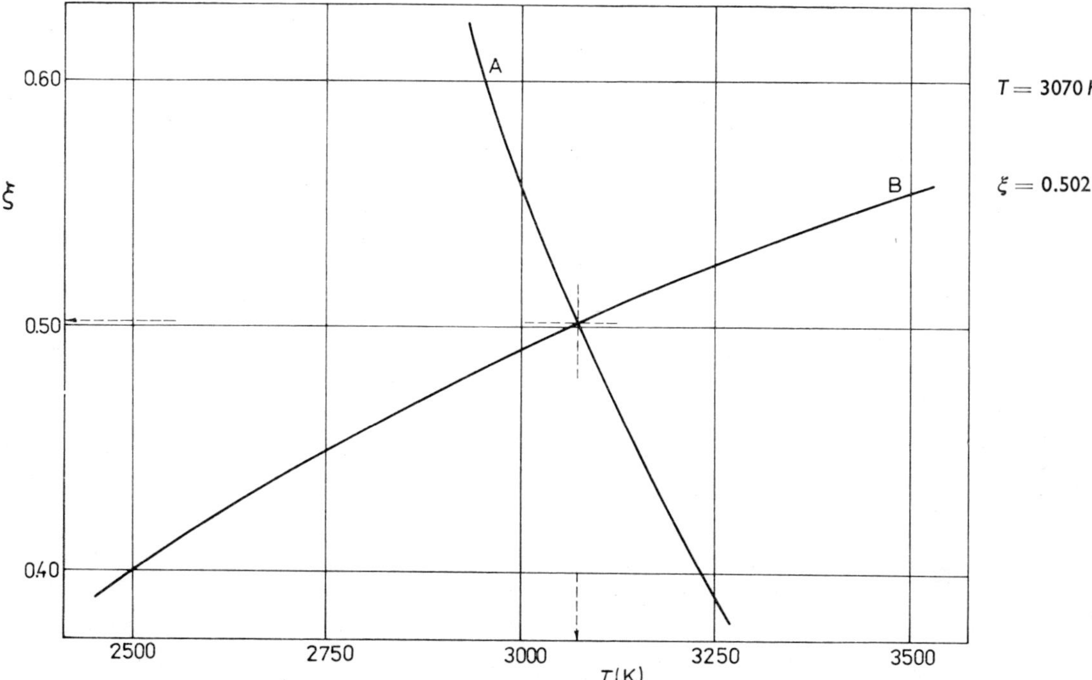

Fig. 23. Graphical determination of chemical equlibrium in the oxidation of carbon monoxide by the theoretical amount of oxygen under adiabatic conditions.

(A — equilibrium curve, B — enthalpic balance curve)

Calculation of the second relationship (5.163) leads to

$$\xi = \frac{2\,856 - 9.32T - 1.794 \times 10^{-3}T^2 + 0.354 \times 10^{-6}T^3}{-66\,888 - 2.47T + 2.473 \times 10^{-3}T^2 - 0.471 \times 10^{-6}T^3}.$$

This relation characterises the curve *B* in Fig. 23. Since chemical equilibrium as well as enthalpic balance must be satisfied in equilibrium in the case of an adiabatic process, the solution is obtained from the point of intersection of the two curves.

This procedure can be extended to the case of a system in which R linearly independent reactions are taking place. In this case, a set of $R + 1$ non-linear equations are solved for $R + 1$ unknown variables $\xi_1, \xi_2, \ldots, \xi_R, T_E$

$$(K_a)_r(T_E) - \left(\frac{P}{n}\right)^{\nu_r} \prod_{i=1}^{N} n_i^{\nu_{ri}} = 0 \quad r = 1, 2, \ldots, R$$

$$H_{END} - H_{START} = 0, \tag{5.164}$$

where

$$n_i = n_i^o + \sum_{r=1}^{R} v_{ri}\xi_r \quad i = 1, 2, \ldots, N$$

$$v_r = \sum_{i=1}^{N} v_{ri} \qquad r = 1, 2, \ldots, R . \tag{5.165}$$

To solve the set (5.164), the temperature relationship of the equilibrium constants $(K_a)_r$, $r = 1, 2, \ldots, R$ and of heat capacity values $C_{P,i}^o$, $i = 1, 2, \ldots, N$ must be known.

It is more advantageous to apply methods based on minimalisation of the Gibbs function. In this case we may proceed as follows: We determine the equilibrium composition, e.g. by means of the method of Lagrangian multipliers for several temperature values in the vicinity of the expected T_E value. To do this, we must know the dependence of $c_i = G_i^o/RT + \ln P$ values on temperature. At every temperature, for which we have calculated the equilibrium composition, we determine the values of $\Delta H = H_{END} - H_{START}$. Let $\Delta H < 0$ apply for the temperature $T_E^{(1)}$ and $\Delta H > 0$ for the temperature $T_E^{(2)}$. The required temperature T_E will then lie in the interval $(T_E^{(1)}, T_E^{(2)})$. Furthermore we can apply e.g. the interval halving method or the regula falsi method (see Appendix 3). The $c_i = c_i(T)$ $i = 1, 2, \ldots, N$ relationship is determined as follows: values of $-(G_T^o - H_T^o)/T$, $H_{T_1}^o$ are tabulated in the literature for various values of T. The standard temperature T_1 is usually 0 K or 298.15 K. The quantity $H_{T_1}^o$ is independent of temperature, and polynomial development to at most the third or fourth degree will usually suffice to elucidate the value of $-(G_T^o - H_T^o)/T$.

5.7.2   Calculation of the initial mass balance

A similar problem is involved in determining the initial set of numbers of moles in adiabatic chemical conversions, when it is required that a predetermined temperature should be maintained. This can be achieved by either adding an inert constituent, which by increasing its own temperature lowers the reaction temperature to the required value or, by adding another reactive constituent. The constituent added must cause at least one endothermic reaction to take place, thus consuming the excess heat of reaction. The result is a process, in which the overall enthalpy change is zero for the temperature given, i.e. reaction is enthalpically autonomous.

Both cases can be solved by any of the methods described for calculating isothermic chemical equilibria, selecting the ratio of input constituents by trial and error until the condition of zero overall enthalpy according to relation (5.147) is satisfied. This procedure, however, is laborious, although the use of the above-described procedures is a certain advantage.

A more expedient procedure requires construction of a new algorithm, again using the theory of Lagrangian multipliers. The procedure is similar to e.g. the Lagrangian multiplier method. The mass balance equation is rewritten as

$$\sum_{i=1}^{p-1} a_{ij} n_i^o + a_{pj} n_p^o = \sum_{i=1}^{N} a_{ij} n_i \quad j = 1, 2, ..., M , \tag{5.166}$$

where the right-hand side of relation (5.166) is equal to b_j $j = 1, 2, ..., M$ and the value of p defines the number of initial compounds. The number of moles $(p - 1)$ of initial compounds can be chosen beforehand, and the number of moles n_p^o of the remaining compound (the one which is to influence the enthalpy balance in the desired manner) is calculated from the relations (5.156), (5.159) and the condition $\Delta H = 0$

$$n_p^o = \alpha/\beta , \tag{5.167}$$

where

$$\alpha = \sum_{i=1}^{N} n_i \left[(\Delta H_f^o)_{i,T_1} + \int_{T_1}^{T_E} C_{P,i}^o(T)\, \mathrm{d}T \right] - \sum_{i=1}^{p-1} n_i^o \left[(\Delta H_f^o)_{i,T_1} + \int_{T_1}^{T_S} C_{P,i}^o(T)\, \mathrm{d}T \right] \tag{5.168}$$

$$\beta = (\Delta H_f^o)_{p,T_1} + \int_{T_1}^{T_S} C_{P,p}^o(T)\, \mathrm{d}T .$$

The remaining part of the calculation is identical with that of the Lagrangian multiplier method.

6 Chemical equilibria in real gas systems

The preceding chapter described methods of calculating complicated chemical equilibria in ideal gas systems. Since most organic reactions in the gas phase proceed at temperatures high enough for virtually all constituents to be in the above-critical region, the assumption of ideal behaviour is well satisfied at normal or slightly elevated pressure. Recently, however, high-pressure syntheses are finding increasing application in the chemical industry, some of which (methanol synthesis, hydro-formylation of olephines, synthesis of amines from olephins and ammonia, oxidative splitting of hydrocarbons, hydrogenation processes etc.) achieve technological interest only at pressures high enough for individual constituents and their mixtures to exhibit deviations from ideal behaviour. In such cases, corrections for the influence of non-ideal behaviour on equilibrium composition must be taken[72].

The description of state behaviour of real gases and their mixtures is an extensive branch of thermodynamics. It is not practically possible (neither would it be purpose-ful) to attempt a detailed discussion of the topic, the more as it has been treated in several outstanding publications[68,69,113,114,117,127,159] in great detail. Since highly complex, multi-component systems must be expected to originate in calculations of chemical equilibria, where the state behaviour in dependence on composition of the mixture usually is not known, it seems to be useful to limit the discussion to the determination of the real behaviour of mixtures from the known properties of their pure constituents.

The following section describes an attempt to introduce into the calculation corrections for the non-ideal character of the gas phase so as to allow the methods described in the preceding chapter to be applied to all techniques of calculating chemical equilibria. For practical reasons, moreover, only those procedures of des-cribing real behaviour were employed, the solution of which is quite general in character.

6.1 FUNDAMENTAL THERMODYNAMIC NOTIONS

6.1.1 Fugacity of a constituent in a mixture

From relations (3.20) and (3.33) follows a relationship between chemical potential and fugacity of the i-th constituent of a mixture in the form of

$$\mu_i = \mu_i^o + RT \ln \frac{f_i}{f_i^o} , \tag{6.1}$$

where μ_i^o and f_i^o is the chemical potential and fugacity, resp., of the i-th constituent in the standard state. For the standard state of so-called unit fugacity (temperature of the system, pure constituent, pressure 1 atm., ideal gas) the expression (6.1) may be rewritten as

$$\mu_i = \mu_i^o + RT \ln f_i , \tag{6.2}$$

where μ_i^o is the chemical potential of the pure constituent (Gibbs molar function) in the above standard state. Considering the linking condition between fugacity of the i-th constituent and its partial pressure in the ideal system $P_i = y_i P$, i.e.

$$\lim_{P \to 0} \frac{f_i}{y_i P} = 1 , \tag{6.3}$$

we may rewrite equation (6.2) for the ideal system as

$$\mu_i = \mu_i^o + RT \ln y_i P . \tag{6.4}$$

Since, furthermore,

$$\left(\frac{\partial \mu_i}{\partial P} \right)_{T,y} = \bar{v}_i , \tag{6.5}$$

where $\bar{v}_i$ is the partial molar volume of the i-th constituent, differentiation of the left- and right-hand sides of equation (6.2) with respect to pressure will lead to

$$\left(\frac{\partial \ln f_i}{\partial P} \right)_{T,y} = \frac{\bar{v}_i}{RT} . \tag{6.6}$$

The dependence of fugacity on temperature, pressure and composition y is obtained by integration of relation (6.6)

$$\ln f_i(T, P, y) = \ln f_i(T, P^*, y) + \int_{P^*}^{P} \frac{\bar{v}_i}{RT} \, dP , \tag{6.7}$$

where P^* is an arbitrary, non-zero pressure (due to the existence of the integral on

163

the righ-hand side of relation (6.7)). Let us now add and substract the expression $\ln y_i P^*$ on the right-hand side of the relation (6.7). We now obtain

$$\ln f_i(T, P, y) = \ln \frac{f_i(T, P^*, y)}{y_i P^*} + \ln y_i P + \int_{P^*}^{P} \left(\frac{\bar{v}_i}{RT} - \frac{1}{P} \right) dP . \tag{6.8}$$

This equality between the left- and right-hand sides of equation (6.8) must hold for all values of P^* and therefore also for $P^* \to 0$. Denoting the ratio $f_i/(y_i P)$ the fugacity coefficient φ_i, equations (6.8) and (6.3) result in

$$\ln \varphi_i(T, P, y) = \frac{1}{RT} \int_{0}^{P} \left(\bar{v}_i - \frac{RT}{P} \right) dP . \tag{6.9}$$

Thus, the original relationship for the chemical potential in a mixture of ideal gases

$$\mu_i = \mu_i^{\circ} + RT \ln P y_i , \tag{6.10}$$

will convert, for real systems, to

$$\mu_i = \mu_i^{\circ} + RT \ln P y_i + \int_{0}^{P} (\bar{v}_i - RT/P) \, dP \tag{6.11}$$

or, also, to

$$\mu_i = \mu_i^{*} + \int_{0}^{P} (\bar{v}_i - RT/P) \, dP , \tag{6.12}$$

where μ_i^{*} is the chemical potential of the constituent in the ideal gas state at the temperature and pressure of the system.

It is clear, therefore, that the partial molar volume of every constituent present in the system under study must be determined in order to define the real behaviour of the system.

6.1.2   Partial molar volume

Out of the many procedures of determining partial molar volume values, the selection of which depends on the quality of available data and required degree of accuracy, only those shall be discussed here which provide a general solution and which, at present, give best agreement.

6.1.2.1   *Virial expansion*

Virial expansion represents a manner of expressing behaviour of state which, on the basis of kinetic gas theory and statistical mechanics characterises deviations from ideal behaviour by means of a power series, the coefficients of which represent the force influence of pairs, groups of three and larger groups of molecules. Out of the possible forms[68] virial pressure-explicit expansion is used most often in the form of

$$P = \frac{nRT}{V}\left(1 + \frac{nB_m}{V} + \frac{n^2 C_m}{V^2} + \frac{n^3 D_m}{V^3} + \ldots\right), \tag{6.13}$$

where

$$B_m = \sum_{j,k=1}^{N} y_j y_k B_{jk}, \tag{6.14}$$

$$C_m = \sum_{j,k,l=1}^{N} y_j y_k y_l C_{jkl}, \tag{6.15}$$

$$D_m = \sum_{j,k,l,p=1}^{N} y_j y_k y_l y_p D_{jklp}, \tag{6.16}$$

$$n = \sum_{i=1}^{N} n_i \tag{6.17}$$

and

$$V = n v_m. \tag{6.18}$$

The equation (6.13) may also be expressed in the form of

$$z = \left[1 + \left(\frac{n}{V}\right)B_m + \left(\frac{n}{V}\right)^2 C_m + \left(\frac{n}{V}\right)^3 D_m + \ldots\right], \tag{6.19}$$

or, in an identical manner,

$$V = \frac{nRT}{P} + \frac{n^2 RT}{PV}B_m + \frac{n^3 RT}{PV^2}C_m + \frac{n^4 RT}{PV^3}D_m + \ldots \tag{6.20}$$

Putnam and Kilpatrick[130] have shown, that equation (6.20) may be converted to a volume-explicit form,

$$V = \frac{nRT}{P} + nB_m + \frac{n(C_m - B_m^2)\,P}{RT} + \frac{n(D_m - 3B_m C_m + 2B_m^3)\,P^2}{(RT)^2} + \ldots \tag{6.21}$$

or, formally,

$$V = \frac{nRT}{P} + nB_m + \frac{nC_m'}{RT}P + \frac{nD_m'}{(RT)^2}P^2 + \ldots \tag{6.22}$$

Since the following applies,

$$\bar{v}_i = \left(\frac{\partial V}{\partial n_i}\right)_{n_j \neq i, TP}, \tag{6.23}$$

we obtain

$$\bar{v}_i = \frac{RT}{P} + \left[B'_m + n\left(\frac{\partial B'_m}{\partial n_i}\right)_{n_{j\neq i},T,P} \right] + \left[C'_m + n\left(\frac{\partial C'_m}{\partial n_i}\right)_{n_{j\neq i},T,P} \right]\left(\frac{P}{RT}\right) +$$

$$+ \left[D'_m + n\left(\frac{\partial D'_m}{\partial n_i}\right)_{n_{j\neq i},T,P} \right]\left(\frac{P}{RT}\right)^2 + \ldots \tag{6.24}$$

Evidently, therefore, relations for the partial molar volume cannot be devised without expressing partial derivatives of virial coefficients of mixtures by means of constants for the individual constituents. For the sake of simplicity, let us calculate the derivative of the second virial coefficient.

Since $B'_m = B_m$, relation (6.14) may be modified to give

$$B_m = \frac{\sum\limits_{j,k=1}^{N} n_j n_k B_{jk}}{n^2} . \tag{6.25}$$

Assuming, that $B_{jk} = B_{kj}$, differentiation leads to

$$\left[B'_m + n\left(\frac{\partial B'_m}{\partial n_i}\right)_{n_{j\neq i},T,P} \right] = -\frac{\sum\limits_{j,k=1}^{N} n_j n_k B_{jk}}{n^2} + \frac{2\sum\limits_{j=1}^{N} n_j B_{ij}}{n} . \tag{6.26}$$

Equation (6.26) can be modified to

$$\left[B'_m + n\left(\frac{\partial B'_m}{\partial n_i}\right)_{n_{j\neq i},T,P} \right] = 2\sum\limits_{j=1}^{N} y_j B_{ij} - \sum\limits_{j,k=1}^{N} y_j y_k B_{jk} \tag{6.27}$$

or

$$\left[B'_m + n\left(\frac{\partial B'_m}{\partial n_i}\right)_{n_{j\neq i},T,P} \right] = \sum\limits_{j,k=1}^{N} y_j y_k (2B_{ij} - B_{jk}) , \tag{6.28}$$

which can be then substituted for the expression in square brackets in the relation (6.24).

When a similar procedure is followed with the other virial coefficients, the respective partial derivatives will take the form of

$$\left[C'_m + n\left(\frac{\partial C'_m}{\partial n_i}\right)_{n_{j\neq i},T,P} \right] = \sum\limits_{j,k,l=1}^{N} y_j y_k y_l (6C_{ijk} - 3C_{ijj} - 2C_{jkl}) -$$

$$- B_m \sum\limits_{j,k=1}^{N} y_j y_k (4B_{ij} - 3B_{jk}) , \tag{6.29}$$

166

$$\left[D_{\mathrm{m}}' + n \left(\frac{\partial D_{\mathrm{m}}}{\partial n_i} \right)_{nj \neq i,T,P} \right] = \sum_{j,k,l,p=1}^{N} y_j y_k y_l y_p (24 D_{ijkl} - 12 D_{ijjk} - 8 D_{jijj} - 3 D_{jklp}) -$$

$$- 3 B_{\mathrm{m}} \sum_{j,k,l=1}^{N} y_j y_k y_l (6 C_{ijk} - 3 C_{ijj} - 4 C_{jkkl}) +$$

$$+ 2 B_{\mathrm{m}}^2 \sum_{j,k=1}^{N} y_j y_k (6 B_i - 5 B_{jk}) - 6 C_{\mathrm{m}} \sum_{j=1}^{N} y_j B_{ij} , \tag{6.30}$$

etc.

Substitution of all these expressions for the respective terms of relation (6.24) leads to a relationship which expresses the partial molar volume of the i-th constituent on the basis of virial coefficients of the individual constituents.

6.1.2.2   The Redlich-Kwong equation

The Redlich-Kwong[132] equation is considered to be the most accurate two-constant equation of state. For the pure constituent, it is

$$P = \frac{RT}{v - b} - \frac{a}{\sqrt{(T)}\, v(v + b)} , \tag{6.31}$$

a, b being constants for a given substance. For mixtures, the authors of the equation (6.31) recommend constants to be determined from the relations

$$a_{\mathrm{m}} = \sum_{i=1}^{N} \sum_{j=1}^{N} y_i y_j a_{ij} , \tag{6.32}$$

where

$$a_{ij} = \sqrt{(a_{ii} a_{jj})} \quad i \neq j$$

and

$$b_{\mathrm{m}} = \sum_{i=1}^{N} y_i b_i . \tag{6.33}$$

The following then applies for the partial molar volume $\bar{v}_i$

$$\bar{v}_i = \alpha/\beta , \tag{6.34}$$

where

$$\alpha = \frac{RT}{v - b_{\mathrm{m}}} + \frac{RT b_i}{(v - b_{\mathrm{m}})^2} - \frac{(v + b_{\mathrm{m}})(\bar{a}_i + a_{\mathrm{m}}) - a_{\mathrm{m}} b_i}{Tv(v + b_{\mathrm{m}})^2}$$

$$\beta = \frac{RT}{(v - b_{\mathrm{m}})^2} - \frac{a_{\mathrm{m}}(2v + b_{\mathrm{m}})}{Tv^2(v + b_{\mathrm{m}})^2} . \tag{6.35}$$

a_m, b_m being constants of the mixture, defined by relations (6.32) and (6.33), constants a_i, b_i are parameters characterising the pure constituent and values of $\bar{a}_i$ can be obtained from

$$\bar{a}_i = -a_m + 2\sqrt{(a_m a_i)} \quad i = 1, 2, \ldots N . \tag{6.36}$$

6.1.2.3   The Beattie-Bridgman equation[8]

This equation of state, which very accurately expresses the behaviour of state of most gases up to high density values is usually written, for pure constituents,

$$P = \frac{RT(1-\varepsilon)}{v^2}(v + B) - \frac{A}{v^2}, \tag{6.37}$$

where

$$A = A_0\left(1 - \frac{a}{v}\right),$$

$$B = B_0\left(1 - \frac{b}{v}\right),$$

and

$$\varepsilon = \frac{c}{vT^3} .$$

In these cases A_0, B_0, c, a, b are individual constants.

When some prerequisites[9] are satisfied, this equation can be converted to a useful though less accurate form, identical with the three-constant virial development with pressure being the independent variable. For a mixture of gases the equation will take the form of

$$V = \frac{nRT}{P} + \frac{n\beta_m}{RT} + \frac{n\gamma_m}{(RT)^2}P + \frac{n\delta_m}{(RT)^3}P^2 , \tag{6.38}$$

where

$$\beta_m = RTB_{0m} - A_{0m} - \frac{Rc_m}{T^2}, \tag{6.39}$$

$$\gamma_m = -RTB_{0m}b_m + A_{0m}a_m - \frac{RB_{0m}c_m}{T^2} \tag{6.40}$$

and

$$\delta_m = \frac{RB_{0m}b_m c_m}{T^2}, \tag{6.41}$$

A_{0m}, B_{0m}, c_m, a_m, b_m are constants of the mixture: they are functions of the molar fractions and individual constants A_{0i}, B_{0i}, c_i, a_i, b_i of the individual constituents.

168

The partial molar volume of the i-th constituent according to relation (6.37)

$$\bar{v}_i = \frac{RT}{P} + \left[\frac{\partial(n\beta_m)}{\partial n_i}\right]_{n_{j\neq i},T,P} \frac{1}{RT} + \left[\frac{\partial(n\gamma_m)}{\partial n_i}\right]_{n_{j\neq i},T,P} \frac{P}{(RT)^2} +$$

$$+ \left[\frac{\partial(n\delta_m)}{\partial n_i}\right]_{n_{j\neq i},T,P} \frac{P^2}{(RT)^3} , \tag{6.42}$$

converts, after determination of partial derivatives of the mixture constants, to

$$\bar{v}_i = \frac{RT}{P} + \frac{[\beta_i + D_i(\beta)]}{RT} + \frac{[\gamma_i + D_i(\gamma) - \gamma_m]}{(RT)^2} P + \frac{[\delta_i + D_i(\delta) - 2\delta_m]}{(RT)^3} P^2 , \tag{6.43}$$

where

$$D_i(\beta) = RTD_i(B_0) - D_i(A_0) - R/T^2 D_i(c) , \tag{6.44}$$

$$D_i(\gamma) = -RT\{b_m[B_{0i} + D_i(B_0)] + b_i[B_{0m} - B_{0i}]\} +$$

$$+ \{a_m[A_{0i} + D_i(A_0)] + a_i[A_{0m} - A_{0i}]\} -$$

$$- \frac{R}{T^2} \{c_m[B_{0i} + D_i(B_0)] + c_i[B_{0m} - B_{0i}] + B_{0m}D_i(c)\} \tag{6.45}$$

and

$$D_i(\delta) = \frac{R}{T^2} \{B_{0m}b_m[c_i + D_i(c)] + b_m c_m[B_{0i} + D_i(B_0)] + b_i[B_{0m}c_m - B_{0i}c_i]\} ; \tag{6.46}$$

$$D_i(A_0) = f(y_i, A_{0i}) ,$$

$$D_i(B_0) = f(y_i, B_{0i}) ,$$

$$D_i(c) = f(y_i, c) . \tag{6.47}$$

6.1.2.4 The Benedict-Webb-Rubin equation

The Benedict-Webb-Rubin equation[12] was derived particularly to describe the behaviour of state of gas systems of relatively low molecular weight. It is, in fact, a perfected Beattie-Bridgman equation with 8 individual constants and is usually used in the pressure-explicit form as a function of reciprocal molar volume σ, i.e.

$$P = RT\sigma + \left(B_{0m}RT - A_{0m} - \frac{C_{0m}}{T^2}\right) \sigma^2 + (b_mRT - a_m) \sigma^3 +$$

$$+ a_m\alpha_m\sigma^6 + \frac{c_m\sigma^3}{T^2} (1 + \gamma_m\sigma^2) e^{-\gamma_m\sigma^2} . \tag{6.48}$$

It is advantageous to modify the relationship which defines fugacity of the constituent (equation (6.8)) to the form

$$RT \ln f_i = RT \ln RT \sigma_i + \int_0^\sigma \left[\left(\frac{\partial(PV)}{\partial n_i} \right)_{n_{j \neq i}, T, V} - RT \right] d \ln \sigma . \tag{6.49}$$

where

$$\sigma_i = n_i / V \quad \sigma = n / V . \tag{6.50}$$

Differentiation of the product of pressure and volume in equation (6.50) with respect to the number of moles of the constituent i with constant T, V and $n_{j \neq i}$ leads to:

$$\left(\frac{\partial(PV)}{\partial n_i} \right)_{n_{j \neq i}, T, V} = RT + 2 \left(B_{0m} RT - A_{0m} - \frac{C_{0m}}{T^2} \right) \sigma + 3 (b_m RT - a_m) \sigma^2 +$$

$$+ 6 a_m \alpha_m \sigma^5 + \frac{c_m \sigma^2}{T^2} \left\{ [3(1 + \gamma_m \sigma^2) - 2(\gamma_m \sigma^2)] e^{-\gamma_m \sigma^2} \right\} +$$

$$+ \left[RT \left(\frac{\partial B_{0m}}{\partial n_i} \right)_{n_{j \neq i}, T, V} - \left(\frac{\partial A_{0m}}{\partial n_i} \right)_{n_{j \neq i}, T, V} - \frac{1}{T^2} \left(\frac{\partial C_{0m}}{\partial n_i} \right)_{n_{j \neq i}, T, V} \right] \sigma +$$

$$+ \left[RT \left(\frac{\partial b_m}{\partial n_i} \right)_{n_{j \neq i}, T, V} - \left(\frac{\partial a_m}{\partial n_i} \right)_{n_{j \neq i}, T, V} \right] \sigma^2 +$$

$$+ \left(\frac{\partial (a_m \alpha_m)}{\partial n_i} \right)_{n_{j \neq i}, T, V} \sigma^5 + \frac{\sigma^2}{T^2} (1 + \gamma_m \sigma^2) e^{-\gamma_m \sigma^2} \left(\frac{\partial c_m}{\partial n_i} \right)_{n_{j \neq i}, T, V} -$$

$$- \frac{c_m \sigma^4}{T^2} (\gamma_m \sigma^2) e^{-\gamma_m \sigma^2} \left(\frac{\partial \gamma_m}{\partial n_i} \right)_{n_{j \neq 1}, T, V} . \tag{6.51}$$

The partial derivate can again be expressed only when the functional relationships of the dependence of mixture constants on individual constants of the constituents have been accurately defined.

6.2 DETERMINATION OF EQUILIBRIUM COMPOSITION

6.2.1 The generalized solution

The preceding section shows, that discussions of real behaviour of systems lead to relatively complicated expressions. These relationships can be employed practically only when simplifying assumptions have been introduced, limiting the need of a detailed description of the system in terms of equations of state.

Similar difficulties arise from attempts at deriving generalized equilibrium relationships. Before embarking upon a discussion of the simplifications which will enable a maximum application of the relationships which were developed for ideal systems, let us briefly indicate a more generalized solution.

6.2.1.1 Procedure based on expressing the equilibrium constants of individual reactions

Let us assume for the sake of simplicity, that the state behaviour of the mixture can be expressed by an equation of state with the second virial coefficient, e.g. in the form of

$$z = 1 + B_m(Y) P, \tag{6.52}$$

where the dependence of the virial coefficient of the mixture on individual coefficients of the pure constituents is defined by the relation

$$B_m(Y) = \sum_{i,k=1}^{N} y_i y_k B_{ik} \quad [T]. \tag{6.53}$$

Relations which describe the stoichiometry of the chemical processes continue to apply, the equilibrium condition is given as

$$\sum_{i=1}^{N} v_{ri}\mu_i = 0 \quad (r = H + 1, \ldots, N). \tag{6.54}$$

Substituting for the chemical potential in the form of

$$\mu_i = \mu_i^o + RT\ln P + RT\ln y_i + P\left[2\sum_{k=1}^{N} y_k B_{ik} - B_m(Y)\right], \tag{6.55}$$

into the equation (6.54) we obtain the equilibrium relation

$$\sum_{i=1}^{N} v_{ri} \left\{ \left(\frac{\mu_i^o}{RT}\right) + \ln P + \ln y_i + P/RT\left[2\sum_{k=1}^{N} y_k B_{ik} - B_m(Y)\right]\right\} = 0$$

$$(r = H + 1, \ldots, N). \tag{6.56}$$

The second set of equations, obtained from the mass balance, will have the same shape as for ideal systems

$$n_i = q_i - \sum_{r=H+1}^{N} v_{ri} n_r \quad (i = 1, 2, \ldots, H). \tag{6.57}$$

The relations (6.56) and (6.57) describe a set of N equations. While in ideal systems this set can be reduced to H equations, because numbers of moles of derived constituents are but functions of values for basic constituents, in real systems all N equations must be solved simultaneously. To this effect it is advantageous to modify the equilibrium as well as the balance relationships into a form in which numbers of moles are replaced by molar fractions. The relationships are then converted to the functional relations (see 5.116)

$$F_r = \sum_{i=1}^{N} v_{ri} \left\{ \left(\frac{\mu_i^o}{RT} \right) + \ln P + \ln y_i + \frac{P}{RT} \left[2 \sum_{i=1}^{N} y_k B_{ik} - B_m(Y) \right] \right\}$$

$$(r = H + 1, ..., N), \tag{6.58}$$

$$F_i = y_i + \sum_{r=H+1}^{N} a_{ir} y_r - \bar{q}_i \quad (i = 1, 2, ..., H), \tag{6.59}$$

where

$$a_{ir} = v_{ri} - \bar{q}_i \left(\sum_{k=1}^{H} v_{rk} - 1 \right). \tag{6.60}$$

The partial derivatives of the two functions with respect to molar fractions of the individual constituents will be

$$\left(\frac{\partial F_r}{\partial y_j} \right) = \sum_{k=1}^{N} v_{rk} \left\{ \frac{\delta_{jk}}{y_k} + \frac{2P}{RT} \left[B_{jk} - \sum_{l=1}^{N} y_l B_{lj} \right] \right\} \quad \begin{array}{l} (r = (H+1), ..., N) \\ (j = 1, 2, ..., N) \end{array} \tag{6.61}$$

$$\left(\frac{\partial F_i}{\partial y_j} \right) = \delta_{ij} \quad \begin{array}{l} (i = 1, 2, ..., H) \\ (j = 1, 2, ..., H) \end{array} \tag{6.62}$$

$$\left(\frac{\partial F_i}{\partial y_j} \right) = a_{ij} \quad \begin{array}{l} (i = 1, 2, ..., H) \\ (j = (H+1), ..., N). \end{array} \tag{6.63}$$

The equations $(6.58) - (6.63)$ suffice to solve the problem by Newton's method.

6.2.1.2 *Procedure based on the White-Johnson-Dantzig method*

Let us again assume that a description of real behaviour is achieved by means of the virial equation of state, limited to the second virial coefficient.

$$z = 1 + B_m(Y) P. \tag{6.64}$$

Continuing to express the overall free enthalpy of the system in dimensionless form, we find that for real systems the equation (5.58) converts to

$$Q(n^{(1)}, T) = \frac{G^o(T)}{RT} + \ln P + \sum_{i=1}^{N} n_i^{(1)} \ln y_i^{(1)} + \frac{Pn^{(1)}}{RT} B_m(T, Y). \tag{6.65}$$

The method is based on replacing overall free enthalpy of the system by a function which results from developing the original relationship (6.65) into a Taylor series, limited to second-power approximation. With constant temperature, therefore, the following relation applies

$$\Gamma(n) = Q(n^{(1)}) + \sum_{i=1}^{N} \frac{\partial Q(n^{(1)})}{\partial n_i} \Delta_i + \frac{1}{2} \sum_{i=1}^{N} \sum_{k=1}^{N} \frac{\partial^2 Q(n^{(1)})}{\partial n_i \, \partial n_k} \Delta_i \Delta_k \, . \tag{6.66}$$

The first derivative of overall free enthalpy of the system with respect to the number of moles of the constituent i is identical with chemical potential,

$$\left(\frac{\partial Q(n)}{\partial n_i} \right)_{n_j \neq i} = \left(\frac{G^\circ}{RT} \right)_i + \ln P + \ln y_i + \frac{2P}{RT} \sum_{l=1}^{N} y_l B_{il} - \frac{P}{RT} B_m(Y) \, . \tag{6.67}$$

The second derivative can be expressed as

$$\left(\frac{\partial^2 Q(n)}{\partial n_i \, \partial n_k} \right)_{n_j \neq i,k} = RT \left[\frac{\delta_{ik}}{n_i} - \frac{1}{n} + \frac{2B_{ik}P}{nRT} - \frac{2P}{RT} \sum_{l=1}^{N} \frac{(B_{il} + B_{kl}) n_l}{n^2} + \right.$$
$$\left. + \frac{2P}{RT} \frac{B_m(Y)}{n} \right] , \tag{6.68}$$

where δ_{ik} is Kronecker's delta (a diagonal matrix, the non-zero elements of which are equal to one) and $B_m(Y)$ is likewise determined by relations (6.53), (6.68). The second-power approximation of the original function $Q(n)$ will achieve the following form for a trial-and-error set of values $n = (n_1, n_2, \ldots, n_N)$

$$\Gamma(n) = \sum_{i=1}^{N} n_i \left\{ \frac{\mu_i^\circ}{RT} + \ln y_i^{(1)} - \frac{P}{RT} \left[2 \sum_{k=1}^{N} y_k^{(1)} B_{ik} - n_k^{(1)} B_m(Y^{(1)}) \right] - \right.$$
$$- \sum_{i=1}^{N} \sum_{k=1}^{N} [n_i - n_i^{(1)}] [n_k - n_k^{(1)}] \left[\frac{\delta_{ik}}{n_k^{(1)}} - \frac{1}{n^{(1)}} + \frac{2PB_{ik}}{n^{(1)}RT} - \right.$$
$$\left. \left. - \frac{2P}{RT} \sum_{l=1}^{N} \frac{(B_{ik} + B_{kl}) n_k^{(1)}}{n^2} + \frac{2P}{RT} \frac{B_m(Y)}{n} \right] \right\} . \tag{6.69}$$

Assuming that the balance relations are satisfied

$$\sum_{i=1}^{N} a_{ij} n_i = b_j \quad (j = 1, 2, \ldots, M) \tag{6.70}$$

a new function can be defined,

$$F(n) = \Gamma(n) + \sum_{j=1}^{M} \lambda_j \left(b_j - \sum_{i=1}^{N} a_{ij} n_i \right) , \tag{6.71}$$

where λ_j again are indefinite Lagrangian multipliers.

From this point onward, the procedure is identical with the original one, as the calculation of equilibrium composition depends on the minimum of the function $F(n)$. This minimum can be found by solving a set of equations obtained when partial derivatives of the function $F(n)$ with respect to the individual numbers of moles are made equal to zero.

To calculate n_i, the following expressions can be defined on the basis of Lagrangian multipliers

$$\sum_{i=1}^{N} n_i f_{ik} = H_k + \sum_{j=1}^{M} a_{jk} n_k^{(1)} \quad (k = 1, 2, ..., N),$$ (6.72)

where

$$f_{ik} = \frac{n_i^{(1)}}{RT} \left(\frac{\partial^2 G}{\partial n_i^{(1)} \partial n_i^{(1)}} \right)_{j \neq i,k},$$ (6.73)

therefore,

$$f_{ik} = \delta_{ik} - \frac{n_i^{(1)}}{n^{(1)}} \left\{ 1 - \frac{2B_{ik}P}{RT} + \frac{2P}{RT} \sum_{l=1}^{N} \frac{n_i^{(1)}}{n^{(1)}} [B_{il} + B_{kl}] - \frac{P}{RT} B_m(Y^{(1)}) \right\}$$ (6.74)

and

$$H_k = n_k^{(1)} \left\{ 1 - \frac{\mu_k}{RT} - \sum_{l=1}^{N} [f_{kl} - \delta_{kl}] \right\}.$$ (6.75)

It will be seen that in the case of an ideal system the variables n_i, defined by relations (6.73), are separable. The number of moles of each constituent can be expressed by means of only $n_i^{(1)}$, λ_j and the ratio $n/n^{(1)}$. In other words, the procedure is reduced to a set of $(M + 1)$ equations for M Lagrangian multipliers and the ratio $n/n^{(1)}$. In the case of non-ideal systems the situation is far more complicated, since the number of equations to be solved simultaneously is $N + M$.

It would appear that designing relations for constant temperature and volume will be simpler, as the second derivative of free energy is simpler than the analogue relation for free enthalpy

$$\left(\frac{\partial^2 F}{\partial n_i \partial n_k} \right)_{n_j \neq i,k} = RT \left[\frac{\delta_{ik}}{n_i} + \frac{2RT}{V_m} B_{ik} \right].$$ (6.76)

The simplification, however, is unsubstantial, as the number of equations to be solved remains unchanged.

6.2.1.3   Method of Lagrangian multipliers

Chemical equilibria in real gas systems can also be treated by the gradient method, as the function $F(\lambda_1, \lambda_2, ..., \lambda_M, n_1, n_2, ..., n_N)$ discussed in section 5.2.3.1 continues to be valid. This also applies to the linking condition (5.60). To determine the equilibrium composition we must again find the minimum of the function F by means of

$$\left(\frac{\partial F}{\partial n_i}\right)_{n_j \neq i, T, P} = 0 \quad (i = 1, 2, \ldots, N) . \tag{6.77}$$

When real behaviour is again described by means of the virial equation limited to the second virial coefficient in the form of (6.58), the function $Q(n)$ is changed from the form of (5.58) to

$$Q(n) = \sum_{i=1}^{N} n_i\left[c_i + \ln y_i + \frac{Pn}{RT} B_m(Y)\right] \quad [T] , \tag{6.78}$$

where

$$c_i = \left(\frac{G^o}{RT}\right)_i + \ln P ,$$

which is formally identical with the relation (6.59).

Differentiating the equation (5.113) in the sense of (6.77) (after substitution of the equation (6.78) for Q(n)), we obtain a set of equations

$$\left(\frac{\partial F}{\partial n_i}\right)_{n_j \neq i, T, P} = c_i + \ln y_i + \frac{2P}{RT} \sum_{l=1}^{N} y_l B_{il} - \frac{P}{RT} B_m(Y) + \sum_{j=1}^{M} \lambda_j a_{ij} , \tag{6.79}$$

from which, differing from relations obtained by differentiation of the equation (5.113) for the ideal system, molar fractions cannot be separated.

The result will again be a set of $N + M$ equations

$$c_i + \ln y_i + \frac{2P}{RT} \sum_{l=1}^{N} y_l B_{il} - \frac{P}{RT} B_m(Y) + \sum_{j=1}^{M} \lambda_j a_{ij} = 0 ,$$

$$\sum_{i=1}^{N} a_{ij} y_i - b_j t = 0 , \tag{6.80}$$

where

$$\sum_{i=1}^{N} y_i = 1$$

and

$$t = \frac{1}{n} ,$$

which must be solved simultaneously. Evidently this is the simplest of the three procedures described. Compared to the procedure 6.2.1.1 its convergence is independent of the selection of basic constituents, while compared to the procedure

6.2.1.2 it has the obvious advantage that it does not contain second partial derivatives of the overall free enthalpy vs. composition relationship. While these terms are simple in the case of an ideal system, deviations from ideal behaviour lead to such complicated relationships that the White-Johnson-Dantzig method becomes unsuitable for these purposes.

We may summarize, that the application of all three procedures is laborious and time-consuming even though sophisticated computers are used, if deviations from ideal behaviour are defined in the above-described manner by means of the second virial coefficient. The fact must also be taken into account, that the equation of state of ideal gases, limited to the second virial coefficient, gives rather uncertain results at high pressure. On the other hand, use of virial development with more coefficients or, of equations of state with more constants, results in relationships of such complexity, that the length of time needed for the calculation will usually not be proportional to the accuracy of the results.

6.2.2. The iterative procedure

Difficulties involved in calculating chemical equilibria in real systems with exact formulation of the problem can be overcome by making use of the advantages of the iterative procedure. Starting out from the assumption, that the influence of real behaviour will be considerably lesser, in relative terms, than the influence of the change in concentration due to conversion of the system from the initial conditions to equilibrium in the ideal gas state, the entire calculation may be viewed as a process of gradually improving approximation, from the determination of the equilibrium composition of the ideal mixture, through an ideal mixture of real gases to a real mixture of real gases. The fact that the significance of the three influences named decreases in the order given leads to the idea of using fixed fugacity coefficients for the equilibrium composition in the ideal gas system, making these coefficients independent of composition in the next step. Since this process can be repeated with every following approximation, a state will be achieved in which the concentration will not change in two successive steps (i.e. state of chemical equilibrium in the real system), the dimension of the set of algebraic equations remaining unchanged. The reason is, that the calculation was in fact divided into two separate blocks, equilibrium composition being calculated in the one and fugacity coefficients of the individual constituents for the corresponding composition being calculated separately in the other block. A great advantage consists in the fact, that the procedure can be employed for all algorithms derived for calculations of ideal systems as well as for all means of expressing fugacity coefficients of constituents in the mixture.

The procedure will be seen from the following application to methods involving minimalization of overall free enthalpy of the system (the application to methods based on stoichiometric analysis is quite similar).

176

In a real gas system at constant temperature and pressure, chemical equilibrium is achieved in the minimum point of the function

$$G = \sum_{i=1}^{N} n_i \left(c_i + \ln \varphi_i + \ln \frac{n_i}{n} \right) \qquad (6.81)$$

on a set of points $(n_1, n_2, \ldots, n_N)$ which satisfy the conditions of mass balance

$$\sum_{i=1}^{N} a_{ij} n_i = b_j \quad j = 1, 2, \ldots, M , \qquad (6.82)$$

where φ_i is the fugacity coefficient of the i-th constituent of the mixture, which generally depends on temperature, pressure and composition of the mixture. The other symbols have the same meaning as in Chapter 5. Since $\left| \ln \varphi_i \right| \ll \left| c_i \right|$ $i = = 1, 2, \ldots, N$, the following iterative procedure can be employed:

a) Select $\varphi_i \equiv 1$ $i = 1, 2, \ldots, N$, corresponding to the assumption of ideal behaviour of the gas mixture.

b) Determine the equilibrium composition by means of the White-Johnson-Dantzig method, method of Lagrangian multipliers or another method based on minimalization of the Gibbs function, denoting the equilibrium composition $(n_1^{(1)}, n_2^{(1)}, \ldots, n_N^{(1)})$.

c) Determine values of $\varphi_i^{(1)} = \varphi_i (T, P, n_1^{(1)}, n_2^{(1)}, \ldots, n_N^{(1)})$ $i = 1, 2, \ldots, N$.

d) Repeat the stage sub b), with the difference that the form of the function Q will now be

$$G = \sum_{i=1}^{N} n_i \left(c_i + \ln \varphi_i^{(1)} + \ln \frac{n_i}{n} \right) .$$

Since the expression $c_i + \ln \varphi_i^{(1)}$ $i = 1, 2, \ldots, N$ is a constant, any of the above-mentioned methods may be used. Denote the equilibrium composition obtained $(n_1^{(2)}, n_2^{(2)}, \ldots, n_N^{(2)})$.

e) Determine values of $\varphi_i^{(2)} = \varphi_i(T, P, n_1^{(2)}, n_2^{(2)}, \ldots, n_N^{(2)})$ etc.

This iterative procedure is repeated, until the required accuracy of the equilibrium composition has been achieved. According to the author's experience, two to three steps usually suffice.

In the stage sub a), $\varphi_i \equiv \varphi_i^{\circ}$ may be chosen instead of $\varphi_i \equiv 1$, φ_i° being the fugacity coefficient of the pure i-th constituent at the temperature and pressure of the system. This choice corresponds to an approximation of an ideal mixture of real gases.

The procedure used to calculate chemical equilibrium by means of equilibrium constants is the analogue of the one described. In this case, the set of non-linear equations to be solved is

$$K_r = (K\varphi)_r \, f_r(\xi_1, \xi_2, \ldots, \xi_R) \quad r = 1, 2, \ldots, R ,$$

where $(K\varphi)_r$ is obtained from a combination of fugacity coefficients (see relation (3.37)).

To illustrate the influence of non-ideal nature of the gas phase, calculations of two rather simple systems are shown, namely addition of ammonia to ethylene and formation of ethanethiol and diethylsulphide from ethanol, hydrogen and gaseous sulphur.

The equilibrium composition was calculated with the aid of the method of Lagrangian multipliers, fugacity coefficients φ_i were determined by the Redlich-Kwong

Table 4. Pressure dependence of diethylamine formation

T (K)		P (atm)									
		1	2	5	10	20	50	100	200	500	1000
600	i-i	0.1257	0.1861	0.2263	0.2391	0.2453	0.2489	0.2501	0.2057	0.2510	0.2511
	i-r	0.1257	0.1863	0.2263	0.2389	0.2444	0.2445	0.2479	0.2560	0.2580	0.2528
	r-r	0.1257	0.1863	0.2266	0.2397	0.2464	0.2514	0.2550	0.2584	0.2582	0.2540
700	i-i	0.0026	0.0097	0.0471	0.1115	0.1800	0.2323	0.2500	0.2586	0.2637	0.2653
	i-r	0.0026	0.0098	0,0476	0.1128	0.1819	0.2335	0.2504	0.2615	0.2670	0.2644
	r-r	0.0026	0.0098	0.0473	0.1121	0.1815	0.2352	0.2540	0.2637	0.2674	0.2649

Table 5. Equilibrium composition in ethanethiol synthesis at 80 atm pressure

T (K)		C_2H_5OH	H_2	$S_2(g)$	C_2H_5SH	Con-
400	i-i	0.1782×10^{-4}	0.3340×10^{-8}	0.5876×10^{-12}	0.2068	
	i-r	0.3901×10^{-5}	0.1563×10^{-8}	0.5933×10^{-5}	0.2634	
	r-r	0.8520×10^{-5}	0.1189×10^{-9}	0.3430×10^{-12}	0.3070	
500	i-i	0.6374×10^{-4}	0.4306×10^{-6}	0.1419×10^{-11}	0.1854	
	i-r	0.7233×10^{-4}	0.4390×10^{-6}	0.2296×10^{-7}	0.2075	
	r-r	0.6572×10^{-4}	0.2891×10^{-6}	0.7029×10^{-11}	0.2126	
600	i-i	0.1130×10^{-3}	0.1150×10^{-4}	0.2265×10^{-11}	0.1256	
	i-r	0.1190×10^{-3}	0.1137×10^{-4}	0.1831×10^{-8}	0.1396	
	r-r	0.1162×10^{-3}	0.9480×10^{-5}	0.5122×10^{-11}	0.1388	
700	i-i	0.1246×10^{-3}	0.1586×10^{-3}	0.2950×10^{-11}	0.6665×10^{-1}	
	i-r	0.1628×10^{-3}	0.1236×10^{-3}	0.2943×10^{-9}	0.8901×10^{-1}	
	r-r	0.1600×10^{-3}	0.1109×10^{-3}	0.4724×10^{-11}	0.8782×10^{-1}	

equation in the generalized form (see section 6.4.2.2). This choice is justified by several reasons.

1. The accuracy of the determined equilibrium composition is limited by the accuracy of the thermodynamic data. It is to be seen from the eq. (6.81), that there is no sense in demanding that the fugacity coefficient logarithm be determined to an accuracy better than that of the expression G_i^o/RT obtained from experimental data. Since values of G_i^o/RT are usually of the order of tens while $\ln \varphi_i$ values are usually of the order of tenths, excessive accuracy of determination of the fugacity coefficient is unnecessary.

2. A simple equation of state should be chosen. The deterioration of accuracy of fugacity coefficients obtained will be negligible. Moreover, in the effort at elucidating the chemical equilibrium in a broad temperature and pressure range, extrapolation outside the limits of validity may be subject to greater risk in the case of multi-constant equations as opposed to simple relationships.

3. Among equations of state with few constants, the Redlich-Kwong equation appears to be one of the best and most frequently employed ones.

4. The generalized form of this equation is well suited to the purpose of expressing non-ideal behaviour on the basis of critical data, which are easily available for a wide range of substances.

Table 5 (Continued)

stituent (n_i)

$(C_2H_5)_2S$	H_2S	$(C_2H_5)_2O$	CH_3CHO C_2H_6	C_2H_4	H_2O
0.2476	0.5465	0.2976×10^{-8}	0.1494	0.3471×10^{-5}	0.8505
0.3310	0.4055	0.2160×10^{-9}	0.3724×10^{-1}	0.1572×10^{-5}	0.9627
0.3160	0.3770	0.2657×10^{-9}	0.3048×10^{-1}	0.4651×10^{-6}	0.9695
0.7925×10^{-1}	0.7354	0.1613×10^{-7}	0.3279	0.2786×10^{-3}	0.6720
0.1238	0.6687	0.1346×10^{-7}	0.2724	0.2073×10^{-3}	0.7275
0.9807×10^{-1}	0.6893	0.1440×10^{-7}	0.2955	0.2551×10^{-3}	0.7045
0.2046×10^{-1}	0.8539	0.3025×10^{-7}	0.4147	0.3927×10^{-2}	0.5851
0.2948×10^{-1}	0.8307	0.3046×10^{-7}	0.3990	0.3623×10^{-2}	0.6009
0.2395×10^{-1}	0.8372	0.2955×10^{-7}	0.4047	0.3694×10^{-2}	0.5952
0.3468×10^{-2}	0.9294	0.2835×10^{-7}	0.4480	0.2939×10^{-1}	0.5518
0.7620×10^{-2}	0.9033	0.4559×10^{-7}	0.4369	0.2202×10^{-1}	0.5630
0.6797×10^{-2}	0.9051	0.4424×10^{-7}	0.4382	0.2214×10^{-1}	0.5616

In the first case, formation of ethyl-, diethyl- and triethylamine by addition of ammonia to ethylene was assumed to proceed in a sequence of reactions according to the scheme:

$$C_2H_4 \; + \; NH_3 \qquad\qquad = \quad C_2H_5NH_2$$

$$C_2H_4 \; + \; C_2H_5NH_2 \quad = \quad (C_2H_5)NH$$

$$C_2H_4 \; + \; (C_2H_5)_2NH \quad = \quad (C_2H_5)_3N \, ,$$

with an initial ratio of $C_2H_4 : NH_3 = 1 : 1$. The pressure dependence of equilibrium numbers of moles of diethylamine formed at 600 and 700 K is plotted in Table 4. Three cases are considered, ideal mixture of ideal gases (denoted i-i), ideal mixture of real gases (i-r) and real mixture of real gases (r-r).

For the formation of ethanethiol and diethylsulphide from ethanol, hydrogen and gaseous sulphur the reaction mixture is assumed to contain the initial constituents as well as ethanethiol, diethylsulphide, hydrogen sulphide, diethylether, acetaldehyde, ethylene, ethane and water. The results of calculating the equimolar composition of the initial mixture $\left(C_2H_5OH:H_2:S_2(g) = 1:1:\tfrac{1}{2}\right)$ at 80 atm pressure are also included in Table 5 for the above three types of mixtures (i-i, i-r, r-r) in the form of numbers of moles of individual constituents at the respective temperatures.

6.2.3   Approximate solution

Another considerable simplification is achieved by the assumption, that numbers of moles (molar fractions), being variables of the functions (6.58), (6.59) are mutually separable. Since in most methods non-linear relations are converted to sets of linear equations by means of Taylor series development, this assumption means that the first or higher partial derivatives do not contain terms which would mutually functionally bind individual components. The advantage then consists in the fact, that procedures derived for ideal gas systems can be employed with only slight modifications, without limiting the degree of complexity of the equation of state employed.

6.2.3.1   Determination of the equilibrium composition from known equilibrium constants of the individual reactions

The equilibrium condition for a set system of N constituents and M elements in which R independent reactions are taking place is expressed by equation (6.54), which is easily converted to the form of

$$(K_a)_r = \prod_{i=1}^{N} a_i^{\nu_{ri}} \quad (r = 1, 2, ..., R) \, . \tag{6.83}$$

For a real system, we may further modify to the form

$$(K_a)_r = \prod_{i=1}^{N} n_i^{\nu_{ri}} \left(\frac{P}{n}\right)^{\nu_r} \prod_{i=1}^{N} \varphi_i^{\nu_{ri}} \quad (r = 1, 2, \ldots, R) \tag{6.84}$$

or,

$$(K_a)_r = \prod_{i=1}^{N} n_i^{\nu_{ri}} \left(\frac{P}{n}\right)^{\nu_r} (K\varphi)_r \quad (r = 1, 2, \ldots, R). \tag{6.85}$$

In general, the term $(K\varphi)$ is a function of temperature, pressure and composition, so that the fugacity coefficient of each constituent must be calculated separately from the fugacity of the constituent in the mixture. When an ideal mixture of real gases is considered, however, the term $(K\varphi)$ will be independent of the composition of the system and may be included in the equilibrium constant. This procedure is very advantageous, as it allows the form of the equilibrium relationships to be retained. In this case, procedures described for ideal systems can be applied unchanged. Fugacity coefficient values of the individual constituents can be calculated by means of one of the methods to be discussed in the following section.

6.2.3.2 *Determination of equilibrium composition from the minimum of overall free enthalpy of the system*

In this case, relations derived for ideal systems require only slight modification by inclusion of a correction factor $RT \ln \varphi_i$ into the expression for the chemical potential of the i-th constituent, as follows from a combination of equations (6.9) and (6.11).

Therefore, in the White-Johnson-Dantzig method, all equations remain unchanged with the exception of equation

$$c_i' = \left(\frac{G^\circ}{RT}\right)_i + \ln P + \ln \varphi_i. \tag{6.86}$$

Similarly, the gradient method may also be employed unchanged, if c_i is replaced by the expression c_i' defined by relation (6.86). Other procedures discussed in the survey of methods of calculating in ideal gas systems are modified in a similar manner.

6.3 DETERMINATION OF THE FUGACITY COEFFICIENTS OF CONSTITUENTS IN THE SYSTEM

All relationships needed to calculate chemical equilibria in real systems were presented in the preceding action. Let us now turn back to the calculation of the partial molar volume. It will be seen from the relations given in section 6.1.2, that practical application is difficult in this form, as it is bound to a detailed description of the behaviour of state of the mixture involved, in dependence on its composition. For example, with virial developments we must know all interaction constants B_{jk}, C_{jkk}, C_{jjk}, C_{jkl} etc. Aside from the constants B_{jk}, which were measured and calculated for some gases on the basis of statistical mechanics and kinetic theory of gases, only the interaction constants C_{jkl} are known accurately for the square-well[69] type of potential. The other equations of state likewise cannot be used, unless the relationship between the constants of the mixture and the constants of the individual constituents is known. Considering the degree of complexity possible in chemically reacting systems which are interesting in terms of modern technology, accurate measurement of the behaviour of state would become very expensive. The most hopeful approach is therefore designing such relationships under certain simplifying assumptions, in which the behaviour of state of the mixture could be expressed on the basis of the behaviour of the pure constituents with sufficient accuracy.

Since the simplifications will depend on the pressure range to be considered for a given case, we shall consider separately equilibria calculations for high and for low pressures.

6.3.1 Determination of fugacity coefficients for high pressures

6.3.1.1 Virial expansion

Useful relationships for calculating real behaviour can be obtained by means of the following simplifications:

a) The power series will be limited to the 2., 3., and 4. virial coefficient,

b) interaction constants will be estimated from the virial coefficients of the pure constants by one of the three recommended methods.

Arithmetic mean

$$B_{jk} = \frac{B_{jj} + B_{kk}}{2} \; ; \quad C_{jjk} = \frac{2C_{jjj} + C_{kkk}}{3} \; ; \quad D_{jkkl} = \frac{D_{jjjj} + 2D_{kkkk} + D_{llll}}{4} \; . \tag{6.87}$$

Geometric mean:

$$B_{jk} = B_{jj}^{1/2} B_{kk}^{1/2} \; ; \quad C_{jjk} = C_{jjj}^{2/3} C_{kkk}^{1/3} \; ; \quad D_{jkkl} = D_{jjjj}^{1/4} D_{kkkk}^{1/2} D_{llll}^{1/4} \; . \tag{6.88}$$

Lorentz mean:

$$B_{jk} = \left(\frac{B_{jj}^{1/3} + B_{kk}^{1/3}}{2}\right)^3 \; ; \quad C_{jjk} = \left(\frac{2C_{jjj}^{1/3} + C_{kkk}^{1/3}}{3}\right)^3 \; ;$$

$$D_{jkkl} = \left(\frac{D_{jjjj}^{1/3} + 2D_{kkkk}^{1/3} + D_{llll}^{1/3}}{4}\right)^3 \; . \tag{6.89}$$

When the arithmetic mean is used, equations (6.14), (6.15) and (6.16) convert to

$$B_m = \sum_{i=1}^{N} y_i B_{ii}, \qquad (6.90)$$

$$C_m = \sum_{i=1}^{N} y_i C_{iii}, \qquad (6.91)$$

$$D_m = \sum_{i=1}^{N} y_i D_{iiii} \qquad (6.92)$$

and the relation (6.24) converts to

$$\bar{v}_i = \frac{RT}{P} + B_{ii} + \frac{P}{RT}\left(C_{iii} + B_m^2 - 2B_{ii}B_m\right) + \left(\frac{P}{RT}\right)^2 \times$$

$$\times \left(D_{iiii} - 3B_m C_{iii} - 3C_m B_{ii} + 3B_m C_m + 6B_m^2 B_{ii} - 4B_m^3\right); \qquad (6.93)$$

the final expression for φ_i, needed to express real behaviour, is therefore simplified to

$$RT\ln \varphi_i = B_{ii}P + \frac{1}{2}\frac{P^2}{RT}\left(C_{iii} + B_m^2 - 2B_{ii}B_m\right) +$$

$$+ \frac{1}{3}\frac{P^3}{(RT)^2}\left(D_{iiii} - 3B_m C_{iii} - 3C_m B_{ii} + 3B_m C_m + 6B_m^2 B_{ii} - 4B_m^2\right), \qquad (6.94)$$

where B_m, C_m, D_m are defined by relations (6.90), (6.91) and (6.92).

Similarly, expressions obtained by application of the two remaining approximations of interaction constants may also be derived. The relationships, however, are complicated to such a degree that their use is questionable with respect to the possible increase in precision. It seems better to use equation (6.94), substituting values calculated according to relations (6.88) and (6.89) for the constants B_m, C_m, D_m.

6.3.1.2   *The Redlich-Kwong equation*

When the behaviour of the gas mixture is described by the Redlich-Kwong equation of state (6.31) — (6.33) the dependence of the fugacity coefficient on temperature, pressure and composition takes the form of

$$\ln \varphi_i = \ln \frac{RT}{P(v - b_m)} + \frac{b_i}{v - b_m} - \frac{a_m b_i}{b_m RT^{1.5}(v + b_m)} +$$

$$+ \frac{1}{RT^{1.5}b_m^2}\left(\bar{a}_i b_m - a_m b_i + a_m b_m\right)\ln \frac{v}{v + b_m}, \qquad (6.95)$$

where the constants a_m, b_m, $\bar{a}_i$, b_i have the same significance as in section 6.1.2.2. The following is the best procedure of calculating the fugacity coefficient by means of the Redlich-Kwong equation:

a) Determine the values of a_i, b_i, which characterize the i-th constituent.

b) Determine the constants of the mixture a_m, b_m, from the known composition of the mixture and with the use of relations (6.32), (6.33).

c) Using the Redlich-Kwong equation of state, determine the molar volume in the mixture for the temperature T and pressure P.

d) Using relation (6.95) determine fugacity coefficients of all compounds in the mixture.

6.3.1.3 The Beattie-Bridgman equation

Partial molar volumes can be expressed by means of the Beattie-Bridgman equation similarly to the case in which virial expansion was used[11], functional relationships being defined with the use of relation (6.47). Either one of the three approximations defined by relations (6.87) – (6.89) is used for all constants or, as recommended by Beattie et al., the geometrical mean is taken for the constants A_{0m} and c_m, the Lorentz mean for B_{0m} and the arithmetic mean for a_m, b_m, so that

$$A_{0m} = \left(\sum_{i=1}^{N} y_i A_{0i}^{1/2}\right)^2 ,$$

$$B_{0m} = \tfrac{1}{4}\left(\sum_{i=1}^{N} y_i B_{0i}\right) + \tfrac{3}{4}\left(\sum_{i=1}^{N} y_i B_{0i}^{1/3}\right)\left(\sum_{i=1}^{N} y_i B_{0i}^{2/3}\right) ,$$

$$c_m = \left(\sum_{i=1}^{N} y_i c_i^{1/2}\right)^2 ,$$

$$a_m = \sum_{i=1}^{N} y_i a_i ,$$

$$b_m = \sum_{i=1}^{N} y_i b_i . \tag{6.96}$$

The required values of partial molar volumes will then be

$$\bar{v}_i = \frac{RT}{P} + \frac{[\beta_i + D_i(\beta)]}{(RT)} + \frac{[\gamma_i D_i(\gamma) - \gamma_m]}{(RT)^2} P + \frac{[\delta_i + D_i(\delta) - 2\delta_m]}{(RT)^3} P^2 , \tag{6.97}$$

where

$$D_i(\beta) = RTD_i(B_0) - D_i(A_0) - \frac{R}{T^2} D_i(c) , \tag{6.98}$$

184

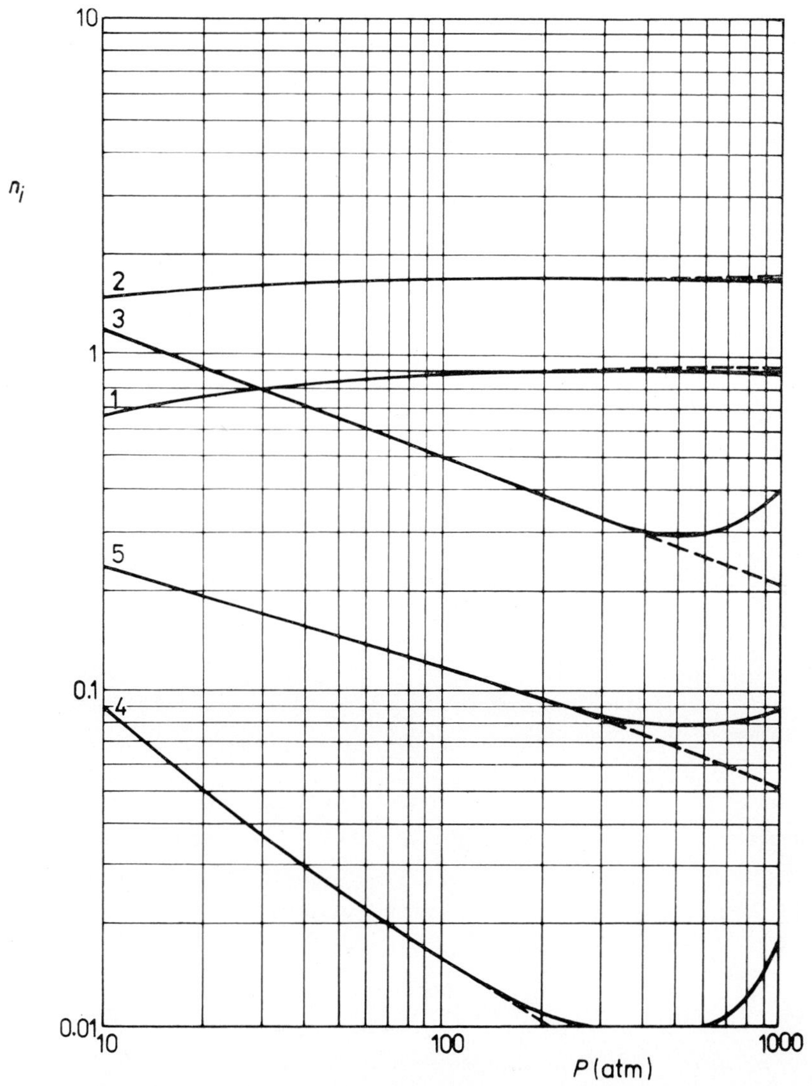

Fig. 24. Comparison of equilibrium of the decomposition of methanol by steam in dependence on pressure, for the ideal system (interrupted lines) and in the real gas state (full curves) at 900 K.

(Order of constituents: $1 - CH_4$, $2 - H_2O$, $3 - H_2$, $4 - CO$, $5 - CO_2$)

$$D_i(\gamma) = -RT\{B_m[B_{0i} + D_i(B_0)] + b_i[B_{0m} - B_{0i}]\} +$$

$$+ \{a_m[A_{0i} + D_i(A_0)] + a_i[A_{0m} - A_{0i}]\} -$$

$$- \frac{R}{T^2}\{c_m[B_{0i} + D_i(B_0)] + c_i[B_{0m} - B_{0i}] + B_{0m}D_i(c)\}, \qquad (6.99)$$

185

$$D_i(\delta) = \frac{R}{T^2} \left\{ B_{0m}b_m[c_i + D_i(c)] + b_m c_m[B_{0i} + D_i(B_0)] + b_i[B_{0m}c_m - B_{0i}c_i] \right\},$$

$$(6.100)$$

$$D_i(A_0) = -\left(A_{0i}^{1/2} - \sum_{i=1}^{N} y_i A_{0i}^{1/2}\right)^2 , \qquad (6.101)$$

$$D_i(B_0) = -\tfrac{3}{4}\left(B_{0i}^{1/3} - \sum_{i=1}^{N} y_i B_{0i}^{1/3}\right)\left(B_{0i}^{2/3} - \sum_{i=1}^{N} y_i B_{0i}^{2/3}\right) ,$$

$$D_i(c) = -\left(c_i^{1/2} - \sum_{i=1}^{N} y_i c_i^{1/2}\right)^2 ,$$

and β_m, γ_m and δ_m are defined by the relationships (6.39) — (6.41). The expression for the fugacity coefficient of the i-th constituent is now given by

$$RT\ln \varphi_i = [\beta_i + D_i(\beta)]\left(\frac{P}{RT}\right) + \frac{1}{2}[\gamma_i + D_i(\gamma) - \gamma_m]\left(\frac{P}{RT}\right)^2 +$$

$$+ \frac{1}{3}[\delta_i + D_i(\delta) - 2\delta_m]\left(\frac{P}{RT}\right)^3 . \qquad (6.102)$$

For the sake of illustration we have calculated the equilibrium composition of a mixture, formed by decomposition of methane with steam at an initial ratio of $1 : 2$ moles, 900 K and in the 10 to 1000 atm range. Thermochemical data for CH_4, H_2O, H_2, CO will be found in Example 10, for CO_2 in Example 11. The calculation was performed according to relations (6.96) — (6.102), constants of the Beattie-Bridgman equation are given in Appendix 11 for the individual pure constituents. The results are plotted in Fig. 24. It will be seen from the plot, that for all constituents 900 K is a high enough temperature for deviations from ideal behaviour to become apparent only at elevated pressure.

6.3.1.4 *The Benedict-Webb-Rubin equation*

Proceeding similarly to the preceding case and introducing, as recommended by the authors of[12], the arithmetic mean for the interaction terms in constant B_0, the geometric mean for A_0, C_0, γ and the Lorentz mean for a, b, c, α, we obtain the following expressions for the constants of the mixture

$$A_{0m} = \left(\sum_{i=1}^{N} y_i A_{0i}^{1/2}\right)^2 , \qquad a_m = \left(\sum_{i=1}^{N} y_i a_i^{1/3}\right)^3 ,$$

$$B_{0m} = \sum_{i=1}^{N} y_i B_{0i} , \qquad b_m = \left(\sum_{i=1}^{N} y_i b_i^{1/3}\right)^3 ,$$

$$C_{0m} = \left(\sum_{i=1}^{N} y_i C_{0i}^{1/2}\right)^2, \quad c_m = \left(\sum_{i=1}^{N} y_i c_i^{1/3}\right)^3,$$

$$\gamma_m = \left(\sum_{i=1}^{N} y_i \gamma_i^{1/2}\right)^2, \quad \alpha_m = \left(\sum_{i=1}^{N} y_i \alpha_i^{1/3}\right)^3. \tag{6.103}$$

The equation (6.51) converts to

$$\left(\frac{\partial(PV)}{\partial n_i}\right)_{n_j \neq i, T, V} = RT + \left[(B_{0m} + B_{0i}) RT - 2(A_{0m}A_{0i})^{1/2} - 2T^2(C_{0m}C_{0i})^{1/2}\right] \sigma +$$

$$+ 3\sigma^2 \left[(b_m^2 b_i)^{1/3} RT - (a_m^2 a_i)^{1/3}\right] + 3\sigma^5 \left[a_m(\alpha_m^2 \alpha_i)^{1/3} + \alpha(a_m^2 a_i)^{1/3}\right] +$$

$$+ \left[\frac{3\sigma^2}{T^2} (c_m^2 c_i)^{1/3} (1 + \gamma_m \sigma^2) - \frac{2c_m \sigma^4}{T^2} (\gamma_m \gamma_i)^{1/2} (\gamma_m \sigma^2)\right] e^{-\gamma_m \sigma^2} \tag{6.104}$$

and the expression for the fugacity coefficient will be given by

$$RT \ln \varphi_i = \left[(B_{0m} + B_{0i}) RT - 2(A_{0m}A_{0i})^{1/2} - \frac{2}{T^2} C_{0m}C_{0i})^{1/2}\right] \sigma +$$

$$+ \tfrac{3}{2}\left[(b_m^2 b_i)^{1/3} RT - (a_m^2 a_i)^{1/3}\right] \sigma^2 + \tfrac{3}{5}\left[a_m(\alpha_m^2 \alpha_i)^{1/3} + \alpha_m(a_m^2 a_i)^{1/3}\right] \sigma^5 +$$

$$+ \frac{3\sigma^2}{T^2} (c_m^2 c_i)^{1/3} \left[\frac{1}{\gamma_m \sigma^2} (1 - e^{-\gamma_m \sigma^2}) - (\tfrac{1}{2} e^{-\gamma_m \sigma^2})\right] -$$

$$- \frac{2\sigma^2 c_m}{T^2} \left(\frac{\gamma_i}{\gamma_m}\right)^{1/2} \left[\frac{1}{\gamma_m \sigma^2} (1 - e^{-\gamma_m \sigma^2}) - \tfrac{1}{2}(2 + \gamma_m \sigma^2) e^{-\gamma_m \sigma^2}\right]. \tag{6.105}$$

6.3.2 Determination of fugacity coefficients at low pressures

At low pressures, generally up to $P_r = 0.6 \sim 0.8$, gaseous systems may be considered to be slightly non-ideal, which fact allows a considerable simplification of the non-ideality correction relations. The assumption that the system forms an ideal mixture is mostly satisfied in these cases, although each of the constituents exhibits real behaviour. Thus we may apply the Lewis-Randall rule according to which fugacity of a constituent in the mixture is equal to the overall molar fraction of the constituent. The proportionality constant is fugacity of the pure constituent, i.e.

$$f_i = f_i^\circ y_i. \tag{6.106}$$

Hence,

$$\left(\frac{\partial V}{\partial n_i}\right)_{P,T,n_j \neq i} = \left(\frac{V}{n_i}\right) \quad \text{and} \quad \bar{v}_i = v_i, \tag{6.107}$$

and thus the definition relation (6.11) converts to

$$\mu_i = \mu_i^o + RT \ln Py_i + \int_0^P \left(v_i - \frac{RT}{P} \right) dP . \tag{6.108}$$

6.3.2.1 Virial expansion

The equation (6.94) is furthermore simplified to

$$RT \ln \varphi_i = B_i P + \frac{C_i}{2} \left(\frac{P^2}{RT} \right) + \frac{D_i}{3} \left(\frac{P^3}{(RT)^2} \right) , \tag{6.109}$$

sufficient results being frequently obtained with the aid of the form with second virial coefficient,

$$RT \ln \varphi_i = B_i P . \tag{6.110}$$

6.3.2.2 The Redlich-Kwong equation

The equation (6.95) is simplified to

$$\ln \varphi_i = \ln \frac{RT}{P(v_i - b_i)} + \frac{b_i}{v_i - b_i} - \frac{a_i}{RT^{1.5}(v_i + b_i)} - \frac{a_i}{b_i RT^{1.5}} \ln \frac{v_i + b_i}{v_i} . \tag{6.111}$$

The fugacity coefficient of the *i*-th constituent in an ideal mixture of real gases can be calculated as follows:

a) Determine the constants of the Redlich-Kwong equation of state for each pure constituent.

b) Determine the molar volume of the pure *i*-th constituent $i = 1, 2, ..., N$ for the given temperature and pressure, denoting it v_i.

c) Use relation (6.111), substituting $v = v_i$, to determine the fugacity coefficient φ_i of the pure constituent which, assuming an ideal mixture of real gases, is equal to the fugacity coefficient of the *i*-th constituent in the mixture.

This procedure can again be extended for application to any other equation of state.

6.3.2.3 The Beattie-Bridgman equation

The equation (6.102) is also simplified, converting to

$$RT \ln \varphi_i = \beta_i \frac{P}{RT} + \frac{\gamma_i}{2} \left(\frac{P}{RT} \right)^2 + \frac{\delta_i}{3} \left(\frac{P}{RT} \right)^3 , \tag{6.112}$$

$$\beta_i = RT\,B_{0i} - A_{0i} - \frac{Rc_i}{T^2}, \tag{6.113}$$

$$\gamma_1 = -RTB_{0i}b_i + A_{0i}a_i - \frac{RB_{0i}c_i}{T^2}, \tag{6.114}$$

$$\delta_i = \frac{RB_{0i}b_ic_i}{T^2}. \tag{6.115}$$

In many cases the equation (6.102) can be simplified down to

$$RT\ln\varphi_i = [\beta_i + D_i(\beta)]\frac{P}{RT}, \tag{6.116}$$

where β_i is defined by relation (6.113) and $D_i(\beta)$ by relations (6.98) and (6.101).

6.4 DETERMINATION OF THE CONSTANTS OF EQUATIONS OF STATE OF PURE CONSTITUENTS

6.4.1 Tabulated data

The accuracy of determining deviations from ideal behaviour mainly depends on the kind and quality of available data. The optimum result may be expected when individual constants are known for every constituent present. Up to the present, however, constants of equations of state have only been published for a few elements and simple compounds (e.g. H_2, O_2, CO, CO_2 etc.) and for the lower members of homologic hydrocarbon series.

Constants of the Beattie-Bridgman equation are surveyed[11] in Appendix 11. Similarly, Appendix 12 lists constants of the Benedict-Webb-Rubin equation[117]. Hirschfelder[69] discusses the posibility of calculating virial coefficients by means of statistical mechanics, including also a list of values published since 1954.

When the values of the constants are not known, but the behaviour of state of a constituent has been measured, the constants can be calculated with sufficient accuracy by the least squares method. Novák[115,116] describes a very advantageous calculation, involving statistical weighing of individual data.

6.4.2 Determination of the constants of equations of state from critical data

6.4.2.1 Virial expansion

Quite a number of procedures are known for determining virial coefficients from data[117]. Recently, useful methods were published by McGlashan and Potter[104] and David and Hamann[41], who use a two-constant virial equation with reduced coefficients in the form of

$$z = \frac{Pv}{RT} = 1 + \frac{B_r}{v_r} + \frac{C_r}{v_r^2}, \qquad (6.117)$$

where

$$B_r = \frac{B}{v_c}, \quad C_r = \frac{C}{v_c^2}, \quad v_r = \frac{v}{v_c}.$$

The authors named express the temperature dependence of the reduced second virial coefficient by the series

$$B_r = B_1 + \frac{B_2}{T_r} + \frac{B_3}{T_r^2} + \frac{B_4}{T_r^3}, \qquad (6.118)$$

Table 6. Temperature coefficients of the reduced second virial coefficient*

Constituent	T_c (K)	v_c (ml^3 . mol^{-1})	interval (T_r)	B_1	B_2	B_3	B_4
Inert gases	—	—	0.80—2.70	+0.3867	− 0.7104	− 0.9768	+0.1706
Methane	191.1	*99	1.06—2.67	+0.4042	− 0.7517	− 1.0553	+0.2714
Ethane	305.5	148	0.89—1.67	+1.612	− 5.416	+ 5.099	−2.461
Propane	370.0	200	0.74—1.65	−0.7073	+ 2.7609	+ 4.3019	+1.0213
Butane	425.2	255	0.64—1.35	−5.0620	+15.277	−15.608	+4.1136
Pentane	469.8	311	0.66—1.22	+1.5881	− 3.5783	+ 1.5122	−0.7831
2-Methylpropane	408.1	263	1.01—1.40	+3.6915	−13.032	+14.652	−6.5373
2,2-Dimethyl-propane	433.8	303	0.75—1.26	+0.1226	+ 0.2519	− 1.7169	+0.0874
Ethylene	282.4	129	0.96—1.50	+4.912	−18.575	+23.892	−8.975
Propene	365.0	181	0.76—1.43	−0.8060	+ 3.1860	− 4.8766	+1.2500
1-Butene	419.1	240	0.90—1.25	+6.3190	−20.220	+20.609	−7.9224
2-Methylpropene	417.9	239	1.01—1.31	+0.0489	+ 0.1094	− 1.1562	−0.2422
Benzene	562.0	260	0.53—1.12	−1.5770	+ 4,5248	− 4.9866	+0.7948

* Critical values taken from Kobe and Lynn[91].

Table 7. Relationship between critical quantities, constants of equations of state and virial coefficients

Quantity	Type of equation			
	van der Waals	Berthelot	Dieterici	Redlich-Kwong
P_c	$a/27b^2$	$(aR/216b^3)^{0.5}$	$a/4e^2b^2$	$(0.0867/b)^5(a/0.4278)^2\ R^{1/2}$
T_c	$8a/27bR$	$(8a/27bR)^{0.5}$	$a/4bR$	$(a/bR)\ (0.0867/0.4278)^{2/3}$
c	$3b$	$3b$	$2b$	$3.847b$
$B(T)$	$b-a/RT$	$b-(a/RT^2)$	$b-(a/RT)$	$b-(a/RT^{3/2})$
$C(T)$	b^2	b^2	$b^2-(ab/RT)+$ $+\ (a^2/2R^2T^2)$	$b^2+(ab/RT^{3/2})$

the coefficients of which are given in Table 6 for some hydrocarbons. For the reduced third coefficient C_r, David and Hamann[41] propose the relation

$$C_r = 0.5419 - \frac{1.1249}{T_r} + \frac{1.0973}{T_r^2} . \tag{6.119}$$

Values obtained in this manner agree well for $v_r > 1.3$ and $T_r > 1$. In spite of that, values thus obtained are somewhat higher than constants calculated by means of the $(12:6)$ Lennard Jones potential.

Very often, satisfactory values of virial coefficients can be obtained from two-constant generalized equations according to relationships listed in Table 7. The process based on the van der Waals equation is simplest, while results obtained from the Dieterici and Redlich-Kwong equations are usually more accurate. Particularly the last-named equation provides remarkably good results even at elevated pressure. On the whole, however, accuracy is lower than that of results obtained on the basis of the Lennard-Jones potential.

6.4.2.2 The Redlich-Kwong equation

The authors of the Redlich-Kwong equation recommend the following relations for calculating the constants a, b:

$$a = 0.4278(RT_c)^2\ \sqrt{(T_c)}/P_c$$

$$b = 0.0867RT_c/P_c . \tag{6.120}$$

These relations were obtained on the basis of conditions applying to the critical point.

With substances which are influenced by quantum effects, better results can be achieved (except in the vicinity of the critical point) by introducing effective critical quantities

$$T_c^{ef} = T_c + 8$$

$$P_c^{ef} = P_c + 8 .$$

(6.121)

6.4.2.3 The Beattie-Bridgman equation

Generalization of the Beattie-Bridgman equation was suggested by Su and Chang[152,153] in the form of

$$P_r v_r = T_r + \frac{\beta_r}{v_r'} + \frac{\gamma_r}{v_r'^2} + \frac{\delta_r}{v_r'^3} ,$$

(6.122)

where

$$\beta_r = \frac{P_c}{(RT_c)^2} \beta = T_r B_{0r} - A_{0r} - \frac{c_r}{T_r^2} ,$$

$$\gamma_r = \frac{P_c^3}{(RT_c)^3} \gamma = - T_r B_{0r} b_r + A_{0r} A_r - \frac{B_{0r} c_r}{T^2} ,$$

$$\delta_r = \frac{P_c^3}{(RT_c)^4} \delta = \frac{B_{0r} b_r c_r}{T_r^2} ,$$

$$v_r = \frac{v}{RT_k} P_k ,$$

$$A_{0r} = \frac{A_0}{\dfrac{R^2 T_c^2}{P_c}} ; \quad B_{0r} = \frac{B_0}{\dfrac{RT_c}{P_c}} ; \quad c_r = \frac{c}{\dfrac{RT_c}{P_c}} ;$$

$$a_r = \frac{a}{\dfrac{RT_c}{P_c}} ; \quad b_r = \frac{b}{\dfrac{RT_c}{P_c}} .$$

Keyes[90] determined the numerical value of dimensionless ratios by comparing a number of non-polar substances, thus enabling the determination of individual constants from critical data:

$$A_0 = 0.4758 \left(\frac{RT_c}{P_c}\right)^2 P_c , \quad a = 0.1127 \left(\frac{RT_c}{P_c}\right) ,$$

$$B_0 = 0.18764 \left(\frac{RT_c}{P_c}\right) T_c^3 , \quad b = 0.03833 \left(\frac{RT_c}{P_c}\right)$$

(6.123)

and

$$c = 0.05 \left(\frac{RT_c}{P_c} \right) T_c^3 \ .$$

In some cases, where it is possible to limit the calculation to the first virial term of equation (6.123), i.e. to the expression

$$v = \frac{RT}{P} + \frac{\beta}{RT} \ , \tag{6.124}$$

the following form is obtained by generalization

$$v = \frac{RT}{P} + \frac{RT_c}{P_c} 0.18764 - \frac{0.4758}{T_r} - \frac{0.05}{T_r^3} \ . \tag{6.125}$$

At rather low pressures this equation gives better results than the Berthelot equation.

Deviations of the dimensionless ratios defined by the relations (6.123) led to efforts at further improvement. For example, Keyes compared experimental data in an empirical manner, finding that for non-polar gases the constant c takes the form of

$$c = 0.023 \ \frac{A_0^3}{R^3 B_0} \ . \tag{6.126}$$

Hirschfelder and Roseveare[67], using the $(6:12)$ Lennard Jones potential, obtained a value of 0.024 for the constant on the right-hand side of equation (6.126), while Corner[27], using the $(6:9)$ potential, arrived at a value of 0.013. Corner also tried to express the remaining constants in the form of

$$a = 0.45 B_0 \ ; \quad b = 0.1 B_0 \ . \tag{6.127}$$

For non-polar gases these values agree very well.

6.4.2.4   The Benedict-Webb-Rubin equation

The Benedict-Webb-Rubin equation may be generalized in a manner similar to that used for the Beattie-Bridgman equation. Modification then leads to

$$\frac{z}{z_c} = 1 + \left(B_{0r} - \frac{A_{0r}}{T_r} - \frac{C_{0r}}{T_r^3} \right) \frac{1}{v_r} + \left(b_r - \frac{a_r}{T_r} \right) \frac{1}{v_r^2} + \frac{\alpha_r a_r}{T_r v_r^5} +$$

$$+ \frac{c_r}{T_c^3 v_r^2} \left(1 + \frac{\gamma_r}{v_r^2} \right) \exp \left(- \frac{\gamma_r}{v_r^2} \right) , \tag{6.128}$$

where

$$A_{0r} = \frac{A_0 P_c}{(RT_c)^2}, \qquad b_r = \frac{bP_c^5}{(RT_c)^2},$$

$$B_{0r} = \frac{B_0 P_c}{RT_c}, \qquad c_r = \frac{cP_c^2}{(RT_c)^3},$$

$$C_{0r} = \frac{C_0 P_c}{T_c^2 (PT_c)^2}, \qquad \alpha_r a_r = \frac{aP_c^5}{(RT_c)^6},$$

$$a_r = \frac{aP_c^2}{(RT_c)^3} \quad \text{and} \quad \gamma_r = \frac{\gamma P_c^2}{(RT_c)^2}.$$

Values of dimensionless ratios, determined for hydrocarbons by Joffe[85], Darin[40] and Su and Viswanath[151] are listed in Table 8.

Table 8. Values of reduced constants of the Benedict-Webb-Rubin equation

Constant	Author		
	Joffe	Darin	Su-Viswanath
A_{0r}	0.31315	0.32230	0.24181
B_{0r}	0.13464	0.1306	0.076431
C_{0r}	0.16920	0.15343	0.21217
a_r	0.059748	0.053419	0.044071
b_r	0.04307	0.04162	0.037152
c_r	0.059416	0.053583	0.06448
$\alpha_r a_r$	0.0000961	0.0001028	0.00011369
γ_r	0.042113	0.0001913	0.06

It will be seen from the table that there are considerable differences between the values. Opfell[118] tried to eliminate this problem by expressing the constants in the form of linear functions of an acentric factor. For non-branched aliphatic hydrocarbons and isohydrocarbons, Canjar[22] and Griskey[61] proposed empirical relationships between the values of the constants and critical temperature, as well as the number of carbon atoms in the molecule.

Although these relationships are rather complex, the results will not always be proportional in quality to the labour in volved. Much better agreement may be achieved when constants of equations of state are calculated not from generalized relationships, but from values calculated with the aid of known individual constants of suitable reference constituents, similar in structure to the constituent in question. Brown[18] tried to develop this procedure for the Beattie-Bridgman equation. No-

194

vák[115,116] recently proposed a suitable procedure based on the Benedict-Webb-Rubin equation. The results achieved for pure constituents are, as a mean, two to three times more accurate than those of a generalized calculation using relations (6.122) and (6.125).

6.4.3   Estimation of constants

In order to find out rapidly whether real behaviour of gas phase must be taken into account when calculating a chemical equilibrium, the fugacity coefficients of pure constituents may be estimated directly from the generalized Gamson-Watson fugacity diagram[57,64]. When the calculation shows that the assumption of real behaviour is justified, values obtained in this way can be included in the input data at the start of the calculation.

7 Practical calculation procedure

7.1 SOURCES OF THERMOCHEMICAL DATA

Reliable calculations of chemical equilibria and their extensive application in industrial practice are linked to the availability of suitable thermochemical data. Although data are now available for most lower members of homologic hydrocarbon series, the precision of which allows equilibrium compositions to be determined accurately, there still remain many compounds which cannot be included in equilibrium investigations. The purpose of the following section is to give a basic orientation in the selection of data which might be employed for highly exacting and accurate determinations of the equilibrium composition of mixtures, as well as for an estimate of the feasibility of chemical conversions.

7.1.1 Accuracy of data

The fundamental requirement of reliable calculations is accuracy of the data employed. This is easily proved by differentiating the definition relations

$$\delta(\Delta G^\circ)/RT \approx \delta_r K_a \,, \tag{7.1}$$

$$\delta(\Delta G^\circ) \leqq \delta(\Delta H^\circ) + T\delta(\Delta S^\circ) \,, \tag{7.2}$$

which determine the relationship between the absolute error of the determination of the standard change of free enthalpy or change of entropy, and the relative error of the equilibrium constant. From these relations follows, that an error of 50% in determining the equilibrium constant corresponds at 300 K to an error of only 300 cal mole^{-1} in determining the heat of reaction or, of 1.0 cal K^{-1} mole^{-1} in determining the entropy value. It may be similarly shown that to determine, at 300 K, the equilibrium constant with a relative error of 10% requires that the heat of combustion be determined to 0.03 or, the entropy to 0.2 cal K^{-1} mole^{-1}. Although calorimetric techniques are highly refined nowadays, the high degree of precision required necessitates highly sophisticated experimentation. It is clear, however, that even this precision of determining the equilibrium constant causes unwanted scatter in the equilibrium composition values, as may be seen from several very simple reaction schemes (Figs. 9 to 12) and from the next example.

Example 24

Find the relationship between the error of determining equilibrium composition and the error of determining the equilibrium constant for an ideal system, in which a single chemical reaction is taking place. Apply the results to the second part of Example 10. Extend this procedure to the case of R linearly independent reactions.

From the equilibrium condition

$$K_a = \left(\frac{P}{n}\right)^v \prod_{i=1}^{N} n_i^{v_i},$$

follows, for the given temperature and pressure,

$$\frac{\mathrm{d}\ln K}{\mathrm{d}\xi} = \sum_{i=1}^{N} \frac{v_i^2}{n_i} - \frac{v^2}{n}$$

where $n_i = n_i^o + v_i\xi$ $i = 1, 2, \ldots, N$ are mass balance equations. Conversion from the derivative to the ratio of differences gives the resulting relationship

$$\delta_r\xi \approx \alpha\delta_r K_a$$

$$\alpha = \left[|\xi|\left(\sum_{i=1}^{N} \frac{v_i^2}{n_i} - \frac{v^2}{n}\right)\right]^{-1},$$

where δ_r is a symbol for the relative error $(\delta_r\xi = \delta\xi/|\xi|)$. The value of the parameter α is always positive, as follows from Appendix 3. This relationship may be practically applied as follows: first determine the equilibrium composition by means of a given equilibrium constant. Use the result to determine the stability coefficient α. From the known value of $\delta_r K_a$ now determine the absolute or relative error of the degree of conversion $\delta\xi$ or $\delta_r\xi$. Clearly, the following holds for the estimate of the absolute error of equilibrium composition

$$\delta n_i = |v_i|\,\delta\xi \quad i = 1, 2, \ldots, N\,.$$

The relationship for the coefficient α shows, that in reactions which take place with no change in the mole number $(v = 0)$, the calculated equilibrium composition is less sensitive to an error in the equilibrium constant value as compared to reactions which are accompanied by a distinct change of the mole number $(|v| \gg 0)$.

The second part of Example 9 involves calculation of the equilibrium composition of a system in which the following chemical reaction is taking place

$$CH_4 + H_2O = CO + 3H_2,$$

at 1 atm pressure and 900 K $(K_a = 2.053)$. The initial composition is 1 mole CH_4 and 2 moles H_2O. Solution of the respective equilibrium condition leads to these results:

$$\xi = 0.794$$

$$n_{CH_4} = 0.206$$

$$n_{H_2O} = 1.206$$

$$n_{CO} = 0.794$$

$$n_{H_2} = 2.382$$

$$n = 4.588$$

The numerical value of the stability coefficient α is

$$\alpha = \left[0.794 \times \left(\frac{1}{0.206} + \frac{1}{1.206} + \frac{1}{0.794} + \frac{9}{2.382} - \frac{4}{4.588}\right)\right]^{-1} \approx 0.12 \, .$$

Assuming an error of 1% in the determination of the equilibrium constant $(\delta_r K_a = 0.01)$, we find

$$\delta\xi = \delta_r\xi \times \xi = 0.794 \times 0.12 \times 0.01 \approx 0.001$$

and, therefore,

$$\delta n_{CH_4} = \delta\xi \approx 0.001$$

$$\delta n_{H_2O} = \delta\xi \approx 0.001$$

$$\delta n_{CO} = \delta\xi \approx 0.001$$

$$\delta n_{H_2} = 3\delta\xi \approx 0.003 \, .$$

The equilibrium constant may be subject to an error even larger than the one in this example. When the equilibrium constant is being determined from thermodynamic properties of the pure constituents (equation 7.13), the right-hand side of this equation will frequently include a difference of large numbers.

For a system in which R linearly independent reactions are taking place, the relation for the relative error of the degree of conversion may be rewritten as

$$\delta_r\xi_i \approx \sum_{s=1}^{R} |\alpha_{si}| \, \delta_r(K_a)_s \quad i = 1, 2, \ldots, N \, , \tag{7.3}$$

$$\alpha_{si} = \left[\xi_i \frac{d \ln (K_a)_s}{d\xi_i}\right]^{-1} = \left[\frac{d \ln (K_a)_s}{d \ln \xi_i}\right]^{-1} , \tag{7.4}$$

where $(K_a)_s$ is the equilibrium constant of the s-th reaction. The following procedure is quite analogue to the case of a system in which a single reaction is proceeding.

The dependence of the accuracy of determining equilibrium composition by means of methods which do not require stoichiometric analysis, on the accuracy of the parameters c_i $i = 1, 2, \ldots, N$ is best seen from the relationship (5.142). The exponent on the right-hand side of this relationship includes a difference of two large numbers. In the case of most compounds, the absolute value of the parameter c_i lies in the interval (20;150) at above-critical temperatures. Thus, if it is required that the value of the equilibrium molar fraction y_i be determined with an accuracy of $p\,\%$, the value of the thermodynamic parameter c_i must be determined to an accuracy of roughly $100p\,\%$. The reason consists in the fact, that the relation $\delta_r y_i \approx |c_i|\,\delta_r c_i$ $i = 1, 2, \ldots, N$ follows from the equation (5.142). For the sake of simplicity, the influence of errors of determining Lagrangian multipliers was neglected.

7.1.2 Methods of determining thermochemical data

Determination of the values of thermochemical quantities is a very extensive field of experimental techniques and calculation procedures. The following section should therefore be considered as nothing but a brief introduction.

It is well known that it is impossible to compute the absolute values of thermochemical functions. Entropy is an exception to this rule, but again entropy does not suffice to calculate the equilibrium composition. Consequently, experimental, i.e. calorimetric measurement is essential for every determination of an equilibrium constant. In spite of that, all procedures may be classified by the nature of the work involved as experimental and calculatory ones.

7.1.2.1 Experimental methods

The direct determination of an equilibrium by experimental means was previously the only possibility of determining the values of equilibrium constants. Nowadays, direct measurement is only seldom employed, as it is rather time-consuming and moreover the data obtained are of low precision. Moreover, chemical conversions which are nowadays investigated are too complicated for a reliable determination of the chemical equilibrium. Lastly, it is very difficult in some cases to arrange the reactions in a manner which would permit the necessary measurements to be carried out.

Determination of the equilibrium by analysis of the reaction mixture

The system is allowed to react under defined conditions until equilibrium is achieved. Then the concentration of individual constituents is determined and the equilibrium constant for the given temperature is calculated by substitution into relation (4.10). To determine the temperature course of the equilibrium constant, measurements must be done at several different temperatures. In this way, Rideal[134] determined the equilibrium of 2-propanol dehydrogenation to acetone as far back as 1922. Other examples of the experimental determination of the equilibrium constant of industrially significant reactions are described by Vvedenskii[166].

Determination of the change of standard free enthalpy from the electromotoric force of a galvanic element

In cases where the reaction to be studied can be carried out in a reversible galvanic element, the change of standard free enthalpy related to bringing the constituents into a state of equilibrium, is given by the relationship

$$\Delta G^\circ = -zFE^\circ, \tag{7.5}$$

where F is the number of Faraday charges which pass through the element, E° is the standard electromotoric force of the element and z is the number of equivalents which have reacted. When relation (7.5) is joined together with the reaction isotherm for a process described by the stoichiometric equation

$$\Delta G = \Delta G^\circ + RT \ln K_a, \tag{7.6}$$

we find, that

$$E = E^\circ - \frac{RT}{zF} \ln K_a. \tag{7.7}$$

It will now be seen, that

$$\ln K_a = \frac{E^\circ zF}{RT}. \tag{7.8}$$

Calorimetric measurement

Calorimetric measurement, the results of which can be made use of for determining the equilibrium constant, are twofold: low-temperature techniques and combustion techniques.

In the first case, the course of the molar heat of the pure substance is measured down to the very low temperature region, generally to 12 K. The value of S°_T may then be calculated from the third law of thermodynamics, Debye's relation[65] which extrapolates the interval from the lowest calorimetric data to 0 K, and enthalpic changes of phase transitions. The procedure is very well described by Scott[141], Vvedenskii[166] and others[66,150].

In the second case, either values of the heat of reaction at constant temperature are determined or, heats of combustion. These data are essential for all procedures in which statistical mechanics is employed. The technique is very exacting in experimental respects, and it is particularly difficult to achieve the required accuracy. For this reason, measurement of the heat of combustion is at present limited to a few best-equipped laboratories.

7.1.2.2   Calculation methods

These procedures may in principle be classified in two groups. In the first, values of thermochemical quantities are calculated by means of exact relations of statistical thermodynamics, the second group includes semiempirical and empirical estimates based on structural similarity or various additive rules, which are derived from analogies between atoms or groups of atoms in different types of molecules.

Statistical methods

Statistical methods are extensively employed at present. They make use of the postulates of statistical and quantum mechanics permitting, on the basis of the equipartition principle and of data concerning molecular structure, the calculation of the main energy contributions corresponding to individual types of motion. Structural data are obtained from spectra, recently predominantly from microwave spectra. Vibration frequency levels, which must be known if the contribution of vibrational motion is to be calculated, are likewise obtained from spectroscopic measurements.

The extent of this topic is considerably too wide for the present section. The reader will find more detailed information in the monographs[31,76,84,128,136].

Estimations

Lack of data relating to molecular structure, as well as a lack of spectroscopic data which are essential for the procedures of statistical mechanics, has led to the design of certain rules which then were verified on compounds, the thermochemical data of which are known. Use is made of the similarity of increments related to the same structural group in different compounds. Two kinds of methods are employed: either the respective value of a thermochemical quantity is obtained from several contributions which correspond to individual structural groups — we call these group contributions methods, or, we set out from the assumption that the composition of every compound may be made up of several similar substances (e.g., hexane may be made up by adding together two pentane molecules and substracting one butane molecule). This additive rule is then assumed to apply to individual thermochemical quantities too. These are called group equation methods.

Since there is a large number of procedures which may be employed to estimate values of thermochemical quantities, we have selected the ones which are used most often: these are listed in Table 9 with a brief characteristic and reference to the original literature. It should be noted, however, that these procedures seldom lead to data which could be employed for a serious calculation of the equilibrium composition. This is also shown in Tables 10 and 11, in which values of C_p^o and ΔG_f^o at 298.15 K are compared for several substances, the data having been calculated by means of statistical mechanics on the one hand and estimated by one of the empirical methods on the other.

Table 9. Survey of methods of estimating thermodynamic quantities

Method	Possibility of calculating	Mean relative deviation, %
1. Offtermat	ΔH_f^o	± 5
2. Anderson-Beyer-Watson	C_p^o, S^o, ΔH_f^o	± 1.7
3. Franklin	$((H^o - H_0^o)/T)$, $((G^o - H_0^o)/T)$, ΔH_f^o, ΔG_f^o	± 1.8
4. Souders-Matthews-Hurd	C_p^o, $((H^o - H_0^o)/T)$, ΔH_f^o, ΔS_f^o	± 1.4
5. van Krevelen-Chermin	G_f^o	± 3.5
6. Bennewitz-Rossner	C_p^o	± 2.7
7. Dobratz	C_p^o	± 2
8. Meghreblian-Crawford-Parr	C_p^o	± 2.4
9. Stull-Mayfield	C_p^o	± 2.6
10. Bruins-Czarnecki	ΔG_f^o	± 5
11. Bremner-Thomas	ΔG_f^o	± 4.2
12. Falkovskii	ΔG_f^o	± 7.3
13. Satoh	C_p^o	± 4.5
14. Group equations method	C_p^o, S^o, $((H^o - H_0^o)/T)$, $((G^o - H_0^o)/T)$, ΔH_f^o, ΔG_f^o, $\log K_a$	Bound to accuracy of data employed
15. Correlation method	C_p^o, S^o, $((G^o - H_0^o)/T)$, ΔG_f^o	Bound to accuracy of data employed
16. Similarity method	C_p^o, S^o, $((H^o - H_0^o)/T)$, $((G^o - H_0^o)/T)$	Only for vibrationally similar substances

Table 9. (Continued)

Application	Ref.	Note
All types of compounds	119	Not suitable for calculating K_a
All types of compounds	1	Better suited for hydrocarbons (4)
All types of compounds except those containing halogens	55, 163	For halogen-containing compounds (2)
For hydrocarbons only	143, 144	For non-hydrocarbons (2), (4), for halogen-containing compounds (2)
All types of compounds	92	For hydrocarbons, prefer (4)
All types of compounds	13	Not suitable for calculating K_a
All types of compounds	46	Not suitable for calculating K_a
All types of compounds	29, 105	Not suitable for calculating K_a
All types of compounds	149	Not suitable for calculating K_a
For hydrocarbons and compounds containing oxygen and nitrogen	21	Not suitable for calculating K_a
For hydrocarbons and oxygen-containing compounds only	14	Not suitable for calculating K_a
For hydrocarbons and oxygen-containing compounds only	51	Not suitable for calculating K_a
All types of compounds	139	Not suitable for calculating K_a
Higher members of homologic series of all types	158	Data for first members of homologic series must be known
Higher members of homologic series of all types	156, 157	Data for first members of homologic series must be known
All types of compounds	32—36	Not applicable to calculation of thermo-chemical quantities of formation

Table 10. Comparison of ΔG_f^o values at 298.15 K, calculated for several compounds by different methods (cal mole^{-1})

Compound	Values calculated from statistical mechanics		Anderson-Bayer-Watson		Franklin		Souders-Matthews-Hurd		van Krevelen-Chermin		Bruins-Czarnecki		Bremner-Thomas		Falkovskij	
	ΔG_f^o	Ref.	ΔG_f^o	Dev.*	ΔG_f^o	Dev.	ΔG_f^o	Dev.	ΔG_f^*	Dev.	ΔG_f^o	Dev.	ΔG_f^o	Dev.	ΔG_f^o	Dev.
Propane	− 5.61	137	− 5.44	0.17	− 5.82	−0.21	−5.96	−0.35	− 6.18	−0.57	−13.34	−7.73	− 6.0	−0.39	− 6.5	−0.89
Heptane	1.94	137	1.64	−0.30	2.37	0.43	2.10	0.16	1.87	−0.07	− 0.48	−2.42	2.0	0.06	0.0	−1.94
2,2,3-Trimethyl-butane	1.02	137	− 0.92	−1.94	1.09	0.07	1.44	0.42	1.56	0.54	− 0.48	−1.50	2.0	0.98	0.0	−1.02
Cyclopentane	9.23	137	6.09	−3.14	8.94	−0.29	9.31	0.08	8.86	−0.37	9.95	0.72	10.0	0.77	8.0	−1.23
Cyclohexane	7.59	137	6.29	−1.30	7.35	−0.24	7.94	0.34	7.16	−0.43	12.94	5.35	12.0	4.41	9.6	2.01
Ethylene	16.28	137	16.39	0.11	16.70	0.42	15.40	−0.88	16.97	0.69	12.85	−3.43	—	—	13.0	−3.28
2-Butene, cis	15.74	137	15.60	−0.14	16.05	0.31	15.31	−0.43	15.24	−0.50	16.80	1.06	16.0	0.26	16.2	0.46
1,3-Butadiene	36.1	137	36.54	0.53	59.47	23.46	35.33	−0.68	36.30	0.29	—	—	—	—	—	—
Acetylene	50.00	137	50.08	0.08	50.01	0.01	49.94	−0.06	52.23	2.23	—	—	—	—	—	—
Benzene	31.05	137	29.26	−1.79	30.71	−0.54	31.16	0.11	30.73	−0.32	—	—	—	—	—	—
Methanol	−38.7	83	−43.21	−4.50	−40.70	−2.00	—	—	—	—	−41.30	−2.60	−42.00	−3.30	−40.8	−2.10
1-Propanol	−39.5	45	−39.55	−0.05	−36.6	2.7	—	—	−38.00	1.50	−37.30	2.20	40.00	−0.05	−37.6	2.30
Dimethylether	−26.2	155	−28.90	−2.70	−32.10	−5.9	—	—	−25.9	0.3	—	—	−32.0	−5.8	−31.8	−5.6
Acetaldehyde	−31.96	126	−34.7	−2.74	−31.41	0.55	—	—	−31.31	0.55	—	—	−35.5	−3.54	−31.9	0.06
Acetone	−36.45	121	−37.16	−0.71	−36.5	−0.05	—	—	—	—	—	—	—	—	—	—

* calculated value − experimental (1. column).

Table 11. Comparison of C_P^o values at 298.15 K, calculated for several compounds by different methods (cal K^{-1} mole^{-1})

Compound	Values calculated from statistical mechanics		Dobratz		Stull-Mayfield		Meghreblian Crawford-Parr		Anderson-Beyer-Watson		Sounders-Matthews-Hurd		Satoh	
	C_P^o	Ref.	C_P^o	Dev.*	C_P^o	Dev.	C_P^o	Dev.	C_P^o	Dev.	C_P^o	Dev.	C_P^o	Dev.
Propane	17.57	137	16.59	−5.6	17.29	−2.6	18.73	−6.6	17.68	0.6	17.56	0.6	20.09	15
Heptane	39.57	137	33.55	−15	34.86	−12	36.99	−6.8	41.68	5.1	41.60	4.9	44.29	12
2,2,3-Trimethylbutane	39.33	137	33.53	−15	34.86	−12	36.99	−5.9	41.29	5.0	40.74	3.6	41.36	5.2
Cyclopentane	19.82	137	25.90	31	26.39	33	27.13	37	25.05	26	19.75	−0.3	22.98	16
Cyclohexane	25.40	137	30.22	19	30.82	21	31.72	25	29.26	15	25.35	−0.2	24.96	−1.7
Ethylene	10.41	137	9.41	−9.6	10.48	0.7	11.10	6.6	10.60	−1.8	10.39	−2.2	10.04	−3.5
2-Butene, cis	18.86	137	16.89	−10	19.16	1.6	20.17	7.0	19.34	2.5	18.73	0.7	22.92	22
1,3-Butadiene	19.01	137	14.58	−23	16.64	−12	16.58	−13	18.95	−0.3	19.09	0.4	18.92	−0.5
Acetylene	10.50	137	—	—	9.48	−9.7	9.96	−5.1	9.91	−5.6	10.53	−0.1	10.48	−0.2
Benzene	19.52	137	20.04	2.7	20.02	2.6	20.02	2.6	21.10	7.6	19.53	0.1	21.30	9.1
Methanol	10.76	83	10.93	1.6	—	—	10.63	−1.4	11.86	10	—	—	10.70	−0.6
1-Propanol	22.78	45	4.10	5.7	—	—	23.70	4.1	27.00	15	—	—	—	—
Dimethylether	15.2	155	16.5	8.5	—	—	14.9	−2.0	—	—	—	—	15.1	−0.7
Acetaldehyde	13.1	126	13.7	4.6	—	—	13.9	6.1	13.3	1.5	—	—	—	—
Acetone	17.9	121	17.5	−2.0	—	—	17.8	−0.5	17.5	−2.0	—	—	20.26	22

* % [(calculated value − experimental)/experimental] × 100.

7.1.3 Method of tabulating data

Two methods have come to be conventionally used when processing thermo-chemical data in the form of tables.

In the elder method, tables include values of ΔG_f° at 25 °C (298.15 K) and the constants a, b, c of the second-power relationship $C_P^\circ = f(T)$ for individual compounds. It is sometimes preferred to tabulate values of ΔH_f° and S° instead of ΔG_f°, again at 298.15 K, the two forms being identical in view of the validity of the equation

$$\Delta G_T^\circ = \Delta H_T^\circ - T \Delta S_T^\circ . \tag{7.9}$$

This method has the disadvantage that calculation of ΔG° values for temperatures different from those for which the data have been tabulated makes it necessary to employ empirical relations for the temperature dependence of molar heats, the precision of which varies — with the use of three constants — between 0.5 and 1%, which is usually insufficient for accurate work. Moreover such calculations are time-consuming.

For this reason, thermochemical data for individual substances are nowadays generally tabulated directly as functions of temperature. It was also found useful to directly tabulate values of $\log K_f$ which can be calculated according to the definition equation

$$\log K_f = - \frac{\Delta G_f^\circ}{2.303RT} \tag{7.10}$$

enabling very extensive and accurate additive calculation of the equilibrium constant in the form of

$$\log K_a = \sum_{i=1}^{N} v_i \log K_f . \tag{7.11}$$

Development of low-temperature calorimetry and of statistical mechanics has likewise made it possible to calculate standard enthalpy of formation at absolute zero temperature, $(\Delta H_f^\circ)_0$, also abbreviated H_0°. Since the third law of thermodynamics leads to

$$(\Delta H_f^\circ)_0 = (\Delta G_f^\circ)_0 , \tag{7.12}$$

it became possible to define the *H*-function and *G*-function,

$$\left(\frac{H_T^\circ - H_0^\circ}{T} \right) \quad \text{and} \quad \left(\frac{G_T^\circ - H_0^\circ}{T} \right).$$

Hence we can easily calculate the values of the equilibrium constant and heat of reaction of the processes in question, using the relations

$$-R \ln K_a \equiv \left(\frac{\Delta G^\circ}{T}\right) = \sum_{i=1}^{N} v_i \left[\left(\frac{G_T^\circ - H_0^\circ}{T}\right)_i + \left(\frac{\Delta H_0^\circ}{T}\right)_i\right] \ [T] \qquad (7.13)$$

and

$$\Delta H^\circ = \sum_{i=1}^{N} v_i \left[\left(\frac{H_T^\circ - H_0^\circ}{T}\right)_i + \left(\frac{\Delta H_0^\circ}{T}\right)_i\right] \ [T]. \qquad (7.14)$$

An example of data tabulated in this manner, which at present appears to be the most advantageous one, is shown in Table 12.

In order to allow equilibrium compositions to be calculated by means of methods involving minimalisation of the overall free enthalpy of a system, it should be advantageous to tabulate the function

$$\frac{G_T^\circ}{RT} = \left(\frac{(G_T^\circ - H_0^\circ)}{T}\right)\frac{1}{R} + \left(\frac{\Delta H_0^\circ}{RT}\right). \qquad (7.15)$$

Aston[4], who was the first to recognize the significance of this quantity, called it the simple thermodynamic function. Up to the present, however, tabulation of this function has not come into general use, and now it can be determined without difficulty from relation (7.13) or from values of $\log K_f$.

Table 12. Values of thermochemical functions of vinyl chloride
$$2C(g) + 3/2\,H_2(g) + 1/2\,Cl_2(g) = C_2H_3Cl(g) \quad \Delta H_0^\circ = 9300 \text{ cal/mole}$$

T (K)	C_P°	$\dfrac{(H^\circ - H_0^\circ)}{T}$	ΔH_f°	$\dfrac{-(G^\circ - H_0^\circ)}{T}$	S°	ΔG_f°	$\log K_f$
298.15	12.824	9.47	7500	53.59	63.06	11 400	-8.366
300	12.882	9.50	7500	53.64	63.14	11 400	-8.334
400	15.557	10.68	7000	56.54	67.22	12 800	-7.012
500	17.787	11.89	6500	59.05	70.94	14 400	-6.275
600	19.604	13.03	6200	61.32	74.35	16 000	-5.813
700	21.089	14.07	5800	63.41	77.48	17 600	-5.500
800	22.347	15.04	5600	65.35	80.39	19 300	-5.276
900	23.488	15.90	5400	67.18	83.08	21 000	-5.108
1000	24.344	16.70	5300	68.90	85.60	22 800	-4.978
1100	25.132	17.43	5200	70.49	87.92	24 600	-4.882
1200	25.843	18.11	5100	72.07	90.18	26 300	-4.789
1300	26.470	18.73	5000	73.54	92.27	28 100	-4.719
1400	27.004	19.30	5000	74.94	94.24	29 900	-4.668
1500	27.437	19.82	5000	76.30	96.12	31 600	-4.607

7.1.4 Data collections

The effort at extending the application of thermodynamic calculations in studies of chemical reactions has led some authors to collect all known data in the form of collections arranged in a unified manner. The best known, and most easily available are the following:

Bichovski F. R., Rossini F. D.: Thermochemistry of Chemical Substances, New York 1936.

Bricke E. V., Kapustinskii A. P.: Termicheskie konstanty neorganicheskikh veshchestv (Thermal constants of inorganic substances) Moscow — Leningrad 1949.

Bockris J. M., White J. L., Mackenzie J. D.: Physicochemical Measurement at High Temperatures, London 1959.

Fyzikálně-chemické tabulky (Physico-chemical Tables), Vol. II., SNTL, Prague 1954.

Kelley K. K.: The Entropies of Inorganic Substances, Bull. US. Bureau of Mines, Circ. 434 (1941), Circ. 476 (1949), Circ. 477 (1950).

Kharasch M. S.: Heats of Combustion of Organic Compounds, J. Research Natl. Bur. Standards 2, 359 (1929).

Kobe K. A.: Thermochemistry of Petrochemicals, Austin 1958.

Parks G. S., Hufmann H. M.: The Free Energies of Some Organic Compounds, Chemical Catalog Co., New York 1932.

Rossini F. D., Pitzer K. S., Arnett R. L., Braun R. M., Pimentel G. C.: Selected Values of Physical and Thermodynamic Properties of Hydrocarbons and Related Compounds, Carnegie Press, Pittsburgh 1953.

Rossini F. D., Wagman D. D., Evans W. H., Levine S., Jaffe I.: Selected Values of Chemical Thermodynamic Properties, Circ. 500, Natl. Bur. Standards, U. S. Government Printing Office, Washington, D. C., 1952.

Stull D. R., Carr I., Chao J., Dergazarian T. E., du Plessis L. A., Jostad R. E., Levine S., Oetting F. L., Petrella R. V., Prophet H., Sinke G. C.: JANAF Thermochemical Tables, Clearinghouse for Federal Scientific and Technical Information, Springfield, Va, 1966.

Stull D. R., Sinke G. C.: Thermodynamic Properties of Elements (Advances in Chemistry Series No 18), American Chemical Society, Washington, D. C., 1956.

Stull D. R., Westrum E. F. Jr., Sinke G. C.: The Chemical Thermodynamics of Organic Compounds, John Wiley & Sons, 1969.

Vvedenskii A. A.: Termodynamicheskie raschoty petrokhimicheskikh protsesov (Thermodynamic calculations of Petrochemical Processes), SNTL, Prague 1963.

Wagman D. D., Evans W. H., Halow I., Parker V. B., Bailey S. M., Schumm R. H.: Selected Values of Chemical Thermodynamic Properties, Circ. 270, Natl. Bur. Standards, Washington 1965.

Zeise H.: Thermodynamik (Thermodynamics). Vol. III/1, S. Hirzel Verlag, Leipzig 1954.

7.2 STRATEGY OF CALCULATION

There are two kinds of problems accessible by chemical equilibrium calculations, which may benefit research activities as well as industrial practice. The first kind includes cases in which the reaction or set of reactions studied are known and we want to determine the maximum yield. The information obtainable by calculating the chemical equilibrium is very valuable, as it represents the limit which can be approached in practical work, provided the reaction mechanism and kinetics are known. The significance of the calculation will become the more evident if we realize that in contemporary sophisticated large-scale production processes, a yield increment of as little as 0.1% may be of great economical importance.

The second kind of problems involves those cases in which the reaction is a new one, holding great technological promise but not worked out to any great experimental detail, and where we want to decide whether it is practical in thermodynamical respects, what maximum yield may be expected and how the individual variables will influence the degree of conversion.

In both cases the problem must be formulated very carefully and with a certain amount of professional sensitivity, particularly in terms of selection of the components. The results of the calculation must likewise be interpreted correctly and the right conclusions must be drawn from them. Only then may it be expected that the calculated results will be practically usable and that they will yield a maximum of information which, under suitable circumstances, may speed up the experimental investigation to a large degree.

Let us illustrate the problem by the following examples. Fig. 25 shows the synthesis of ammonia at 100 atm. The equilibrium curve has been plotted as the relationship between the molar fraction of ammonia formed, according to the equation

$$\tfrac{3}{2} H_2(g) + \tfrac{1}{2} N_2(g) = NH_3(g) \tag{7.16}$$

and the reaction temperature, based on data obtained in practical work with different catalysts at equal volume rates of the synthesis mixture. Similarly, Fig. 26 shows the dependence of the molar fraction of converted SO_2 on temperature according to the reaction

$$SO_2(g) + \tfrac{1}{2} O_2(g) = SO_3(g) \tag{7.17}$$

for several catalysts at different gas volume rates, compared with the equilibrium curve. It is clear in both cases, that technological research must be directed to development of a suitable catalyst, capable of operating at the lowest possible temperature. The optimum temperature is then selected with respect to the volume rate chosen, also considering such factors as heat transfer and the reactor geometry related to it, catalyst grain size etc.

When the process involved can be described by a single chemical reaction, as in the two examples mentioned, the situation is considerably facilitated. To judge totally new reactions, a criterion[47] based on the absolute value ΔG_T^o of the reaction has been in use for some time already. Three different cases may be distinguished:

$$\Delta G_T^o < -10\ 000 \ \text{cal}$$
$$-10\ 000 < \Delta G_T^o < 10\ 000 \ \text{cal} \tag{7.18}$$
$$\Delta G_T^o > 10\ 000 \ \text{cal} .$$

In the first case the reaction is hopeful, in the second case impractical and in the third case there is not much hope of practical application of the process, since it demands extreme conditions to force the reaction to run in the desired direction. When it has been found that the reaction is thermodynamically practical, it is not difficult to calculate the influence of individual reaction variables on the degree of conversion.

When judging the majority of organical reactions, however, different competing reactions must often be taken into account, the number and significance of which is

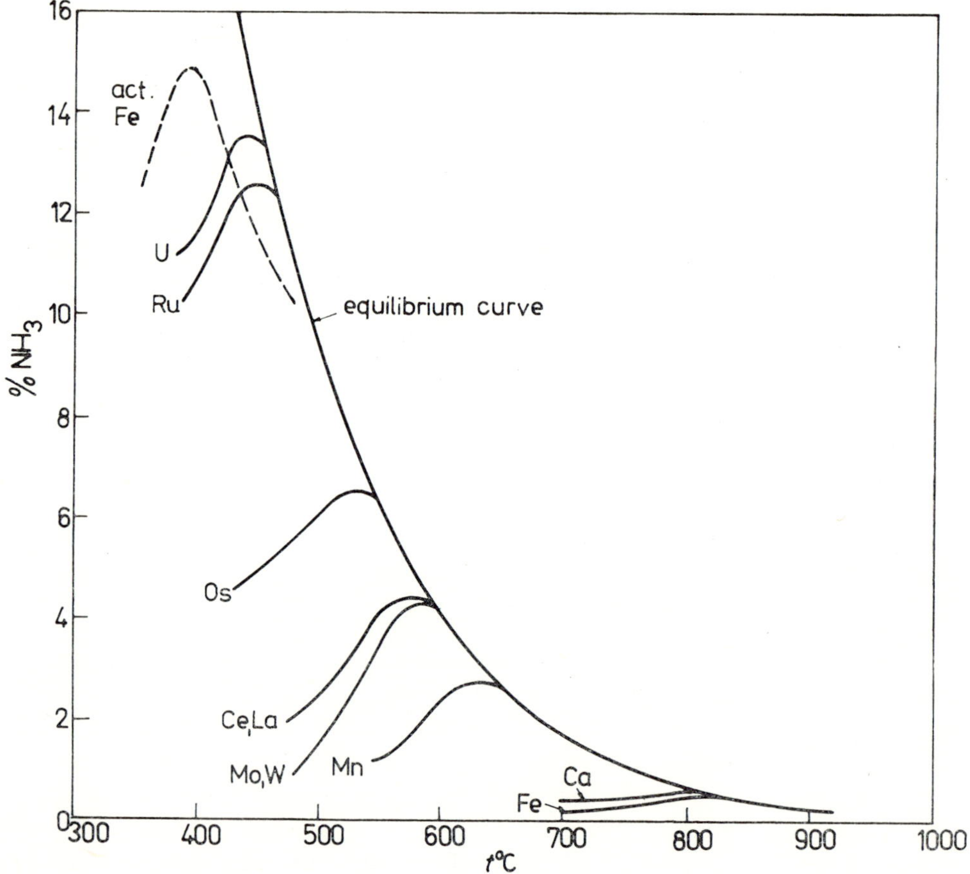

Fig. 25. Activity of catalysts in ammonia synthesis at 100 atm and equal volume velocities.

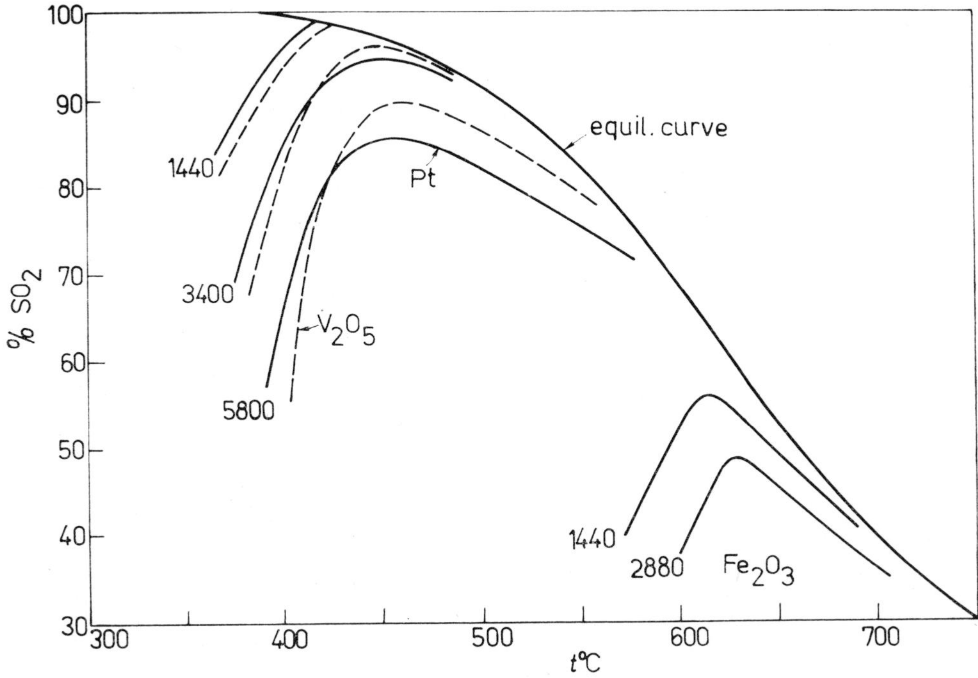

Fig. 26. Comparison of the efficiency of catalysts in oxidation of SO$_2$ at different gas volume rates.

(Initial gas mixture composition 7 molar % SO$_2$, 10 molar % O$_2$, 83 molar % N$_2$.)

usually affected by temperature. It is then far more difficult to arrange the calculation so as to be able to draw the correct conclusions for laboratory — or technological — scale research.

Considering that the contemporary level of computing methods as described in the preceding section does not in principle depend on the number of components considered, the following procedure is recommended for the calculations:

Start the calculation with maximum number of constituents to be considered and use this set to determine the equilibrium composition in a sufficiently broad temperature range. For every temperature the result will be an equilibrium mixture, in which the concentration of individual components however generally does not correspond to our expectation. The concentration of the required constituents, for which the calculation is in fact being performed, is suppressed by unwanted constituents which are thermodynamically more stable under these conditions. Therefore, before repeating the calculation, we divide all components into two groups depending on whether the constituent in question is a desired or an unwanted one. The calculation is then repeated in the same temperature interval without that unwanted constituent, the concentration of which was highest in the initial calculation. Comparison of these results with the preceding set of results for every temperature generally provides a good picture of the concentration change of the individual constituents. Usually there is a shift in favour of that constituent, the concentration of which was second highest in the initial mixture. If this is true, and the constituent involved again is unwanted, the calculation is repeated, disregarding this constituent. When the problem studied involves assessment of a new reaction system, the above procedure is followed until those constituents arrive at maximum concentrations which are required. When a known chemical process is investigated (e.g. in order to improve its yield) the calculation is repeated, gradually eliminating unwanted constituents until the best possible agreement is obtained between calculated and experimentally determined concentrations of the key constituents. Here the calculation is interrupted and all initially considered constituents are divided into three groups: required constituents, unwanted constituents (those which were eliminated from the calculation in successive repetitions) and thermodynamically unstable constituents (those present in the final calculation in very low concentrations). By constructing for every constituent the chemical reaction of its origin from the initial constituents, we can also divide the possible chemical conversion processes into three groups. The first group are required reactions, the preferred course, mechanism and kinetics of which must be given maximum attention in the following experimental study. The second group includes competing reactions, the course of which must be hindered by suitable selection of the reaction conditions, in the case of catalytical reactions by designing a suitable catalyst. In practical work the study of these processes is even more important than that of required processes, since the unwanted processes may frequently elucidate the cause of loss and depreciation of the initial constituents, thus contributing to preferential running of required reactions. Finally, thermodynamically unfeasible reactions may be eliminated from further consideration.

Final calculations are then carried out for the set of required constituents in a sufficiently dense network of individual variables, influencing the equilibrium composition.

7.3   PROCESSING THE CALCULATED RESULTS

With larger calculations done under varying initial conditions, the original table of results as obtained from the computer is difficult to understand. Therefore it is recommended to process the results either into concise tables or graphs. Graphical treatment is usually most suitable for further use, and may be done in two ways:

1. By plotting the numbers of moles of individual components vs. different values of one of the reaction variables (temperature, pressure or number of moles of the initial components), the other variables being constant. If the differences in concentration of individual constituents are extreme, it is better to plot numbers of moles on a logarithmic scale.

2. By plotting the number of moles of one component vs. different values of one reaction variable in parametric dependence on a second variable. Examples of this method are Figs. 28, 33 and 34.

Combination of the two methods will usually allow treatment even of voluminous sets of calculated data in a compact form, suited to subsequent use.

7.4 EXAMPLE

For an example of the solution of a practical technological problem we have selected the case of partial methane oxidation, which has recently become an important industrial process for making cheap acetylene. Besides a number of patents, the process has been described in several papers[23,81,87,99,124,167], the majority of which includes a thermodynamic description of the reaction system as well as conclusions which, however, are as yet incomplete, insufficiently comprehensive and sometimes even inaccurate.

7.4.1 Preliminary classification of reactions taking place

The possibility of preparing acetylene by dehydrogenation of methane was provided as early as the past century by Bethelot[160] by the reaction

$$2\,CH_4(g) \;=\; C_2H_2(g) + 3\,H_2(g), \quad \Delta H^{\circ}_{1000K} = 96\,164\ cal. \tag{7.19}$$

Since the dehydrogenation process takes place at elevated temperatures, the isothermal course of the process must be maintained either by introducing into the system a large amount of heat or by combustion of a part of the components of the reaction mixture with oxygen or air. Moreover, a number of unwanted reactions may take place, causing further loss of the raw material or of the acetylene already formed. The processes which may take place in the system may be classified beforehand as follows:

a) Main (required) reactions, expressed by relation (7.19)

b) Dehydrogenation of methane to other unsaturated hydrocarbons, e.g.

$$4\,CH_4(g) \;=\; C_4H_2(g) + 7\,H_2(g)$$

$$4\,CH_4(g) \;=\; C_4H_4(g) + 6\,H_2(g)$$

$$4\,CH_4(g) \;=\; C_4H_6(g) + 5\,H_2(g)$$

$$4\,CH_4(g) \;=\; C_4H_8(g) + 4\,H_2(g) \tag{7.20}$$

etc.

Table 13. Description of constituents in the equilibrium mixture

Serial number	Constituent	Reference to thermochemical data
1	Methane	50
2	Oxygen	137
3	Acetylene	59
4	Ethylene	137
5	Methylacetylene	137
6	Allene (propadiene)	137
7	Propene	137
8	Diacetylene	own calculation
9	Vinylacetylene	145, own calculation
10	Ethylacetylene	137
11	Dimethylacetylene	137
12	1,2-Butadiene	137
13	1,3-Butadiene	137
14	1-Butene	137
15	2-Butene, cis	137
16	2-Butene, trans	137
17	2-Methylpropene	137
18	Hydrogen	137
19	Water	137
20	Carbon monoxide	137
21	Carbon dioxide	137
22	Graphite	137

c) Methane combustion

$$CH_4(g) + \tfrac{1}{2}O_2(g) = CO(g) + 2\,H_2(g)$$
$$CH_4(g) + O_2(g) = CO(g) + H_2(g) + H_2O(g)$$
$$CH_4(g) + \tfrac{3}{2}O_2(g) = CO(g) + 2\,H_2O(g)$$
$$CH_4(g) + 2\,O_2(g) = CO_2(g) + 2\,H_2O(g) \tag{7.21}$$

d) Acetylene decomposition

$$C_2H_2(g) = 2\,C(s) + H_2(g) \tag{7.22}$$

e) Oxidation of the hydrogen formed

$$H_2(g) + \tfrac{1}{2}O_2(g) = H_2O(g). \tag{7.23}$$

For the initial calculation, a system was selected the constituents of which are listed in Table 13. Beside acetylene, higher acetylenic hydrocarbons, olephins and diolephins were considered. The majority of these was detected qualitatively in the reaction mixture[81,87,99,124].

7.4.2 Thermochemical data

The lowest temperature, at which satisfactory results can be obtained, is about 1150 °C. On the other hand, it is very difficult to perform a reaction in such a manner as to avoid formation of an excessive amount of combustion products. Therefore, the calculation must be limited to a temperature range of 1350 K to 1750 K. For this temperature range, the value of the G-function and the heat of formation at 0 K must be known for each constituent considered. For most constituents data are presented in the literature (see third column in Table 13) up to 1500 K only, for vinylacetylene up to 1000 K while data are missing for diacetylene. The missing data were obtained by means of statistical mechanics; the calculations were, however continued up to 1500 K only for fear of increasing deviations in the contribution of vibrational motion and unjustified neglection of electronic contributions. For the same reasons, extension of data from 1500 K to 1750 K was done for all constituents by extrapolating with the aid of a power series.

The following molecular structure data were used to calculate the values of the G-function of diacetylene[28]:

$$d_{C\equiv C} = 1.205 \pm 0.002 \text{ Å}$$

$$d_{C=C} = 1.379 \pm 0.001 \text{ Å}$$

$$d_{C\equiv H} = 1.064 \text{ Å} .$$

Table 14. Values of the *G*-function of diacetylene in the ideal gas state

T (K)	$-((G^o - H_0^o)/T)$ (cal K^{-1} mole^{-1})
298.15	48.245
300	48.317
400	51.922
500	55.076
600	57.917
700	60.498
800	62.960
900	65.053
1000	67.088
1100	68.992
1200	70.782
1300	72.470
1400	74.069
1500	75.588

Contributions of vibrational motion were calculated from the following values of basic vibrational frequencies[52,53]: 220(2), 482(2), 627(2), 630(2), 874(1), 2020(1) and 3329(2). Results of a calculation done with new values of universal constants[175] are shown in Table 14.

Since no thermochemical data was found which could be used to calculate the heat of formation at 0 K, the molar free enthalpy of formation was estimated for several different temperatures by means of Franklin's[55] and the van Krevelen-Chermin[92] methods. From these data, the probable value of $\Delta H_0^0 = 106.5$ kcal mole^{-1} obtained.

In the same way, using molecular structure data and basic vibration frequencies reported by Stamm, Halverson and Whalen[145] G-function values of vinylacetylene were calculated for the temperature range 1000 to 1500 K. The results are given in Table 15.

Table 15. Values of the *G*-function of vinylacetylene in the ideal gas state

T (K)	$-((G^o - H_0^o)/T)$ (cal K^{-1} mole^{-1})
1100	77.28
1200	79.40
1300	81.40
1400	83.38
1500	85.08

In analytical form the temperature relationship of G-functions of individual constituents was expressed with the use of a power series of the form

$$-\frac{G^o - H_0^o}{T} = A + BT + CT^2 + DT^3 + \ldots + \frac{Z}{T} + \frac{Y}{T^2} + \frac{V}{T^3} + \ldots \quad (7.24)$$

The number of constants to be employed is determined by the computer in the course of the calculation, depending on the accuracy required. In the case here described, the permitted deviation was 0.01 cal K^{-1} mole^{-1}. This demand can be satisfied by using only five constants, the series taking the form of

$$-\left(\frac{G^o - H_0^o}{T}\right) = A + BT + CT^2 + \frac{Z}{T} + \frac{Y}{T^2}. \quad (7.25)$$

For illustration, Table 16 shows the correlation of the G-function of ethylacetylene. The table shows good agreement between original and correlated values.

Other thermodynamic functions may likewise be expressed in this way with sufficient accuracy. The fact that we are justified in extrapolating through an interval

Table 16. Correlation of *G*-function values of ethylacetylene

$$A = 0.542\ 694 \times 10^2$$
$$B = 0.311\ 306 \times 10^{-1}$$
$$C = -0.375\ 686 \times 10^{-5}$$
$$Z = -0.271\ 801 \times 10^4$$
$$Y = 0.231\ 432 \times 10^6$$

T (K)	Tabulated values (cal K^{-1} mole^{-1})	Calculated values (cal K^{-1} mole^{-1})	Deviation (cal K^{-1} mole^{-1})
300	-56.7800	-56.7818	-0.0018
400	-60.7800	-60.7719	0.0081
500	-64.3800	-64.3851	-0.0051
600	-67.7000	-67.7081	-0.0081
700	-70.8400	-70.8093	0.0007
800	-73.7400	-73.7335	0.0065
900	-76.5100	-76.5095	0.0005
1000	-79.1600	-79.1565	0.0035
1100	-81.6900	-81.6875	0.0025
1200	-84.1100	-84.1119	-0.0019
1300	-86.4300	-86.4362	-0.0062
1400	-88.6600	-88.6654	-0.0054
1500	-90.8100	-90.8031	0.0069

Standard deviation (%) = 0,008

Table 17. Comparison of extrapolated and tabulated values of the *G*-function of methane

Temperature interval (K)	Calculated value for $T = 1500$ K	Tabulated value for T = 1500 K	Deviation
300—1000	52.883	52.893	0.010
300—1100	52.887		0.006
300—1200	52.890		0.003
300—1300	52.892		0.001
300—1400	52.893		0.000

of 250 K was verified using the G-function of methane as example: The value for a temperature of 1500 K, which is known from literature, was calculated from a five-constant series (7.25). The constants of the correlation equation were successively calculated from values in the temperature intervals 300—1000 K, 300—1100 K etc. The calculated values are to be seen in Table 17, which shows that the data agree very well.

Although small deviations cannot be excluded in values of the G-function thus obtained for temperatures of 1500 to 1750 K, we yet may expect that the deviations will be roughly equal for all constituents, and that therefore their influence on the finally determined equilibrium composition will be practically insignificant.

7.4.3   Calculation of the equilibrium composition

Firstly the heterogenous equilibrium of the system, which contained all constituents including solid graphite, was calculated by a modified gradient method in dependence on the number of moles of oxygen. The results are shown in Fig. 27. As expected, the unwanted constituents CO, CO_2 and graphite prevailed in the equilibrium mixture besides hydrogen and water. Therefore, these constituents may be left out of the further considerations.

The calculation continued with the determination of the equilibrium composition of the system of 19 constituents at 1350, 1450, 1550, 1650 and 1750 K in dependence on the number of moles of oxygen entering the reaction per 1 mole methane, with the purpose of characterising the influence of temperature on the course of side reactions and determining roughly the optimum ratio of initial constituents. In Figs. 28 to 32 the result of the calculation is presented in graphic form as the relationship of the logarithm of the number of moles of constituents which can be formed from 1 mole methanol.

Fig. 33 shows the temperature dependence for an input composition of 0.5 moles oxygen per 1 mole methane, which appears to be the optimum. The temperature dependence of the equilibrium at 1650 K and several different pressures is shown in Figs. 34 to 36. The pressure dependence for the ratio 0.5 moles O_2 per 1 mole CH_4 at 1350 and 1600 K is shown in Figs. 37 and 38.

From all relationships mentioned up to now follows, that the amount of diacetylene formed by the competing reaction is relatively very high. Due to this, the calculation may be distorted, i.e. less favoured reactions cannot participate fully. In order to verify this, one more informative calculation of the equilibrium composition without diacetylene was carried out for a temperature of 1650 K. The dependence on the number of moles of oxygen is expressed graphically in Fig. 39. The pressure dependence for 0.5 moles O_2 per 1 mole CH_4 is plotted in Fig. 40.

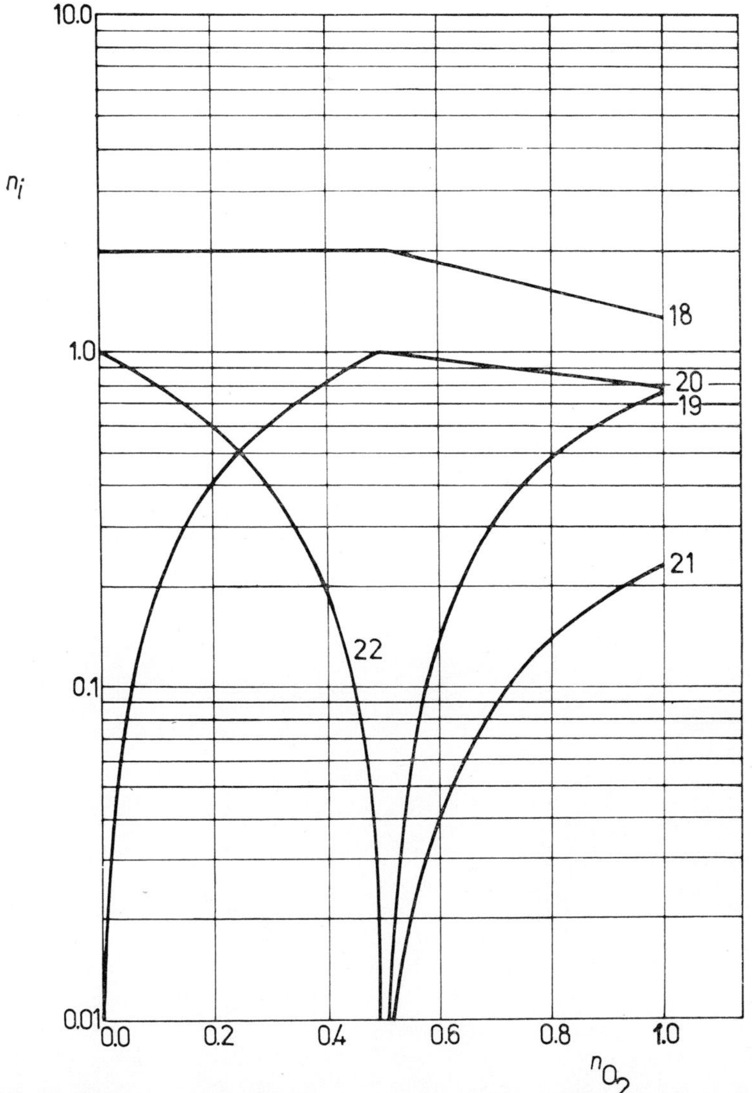

Fig. 27. Equilibrium in partial methane oxidation. The dependence of the number of moles of individual constituents formed from 1 mole methane on the number of moles of oxygen in the initial mixture, at 1650 K and 1 atm pressure. For notation of constituents, see Table 13.

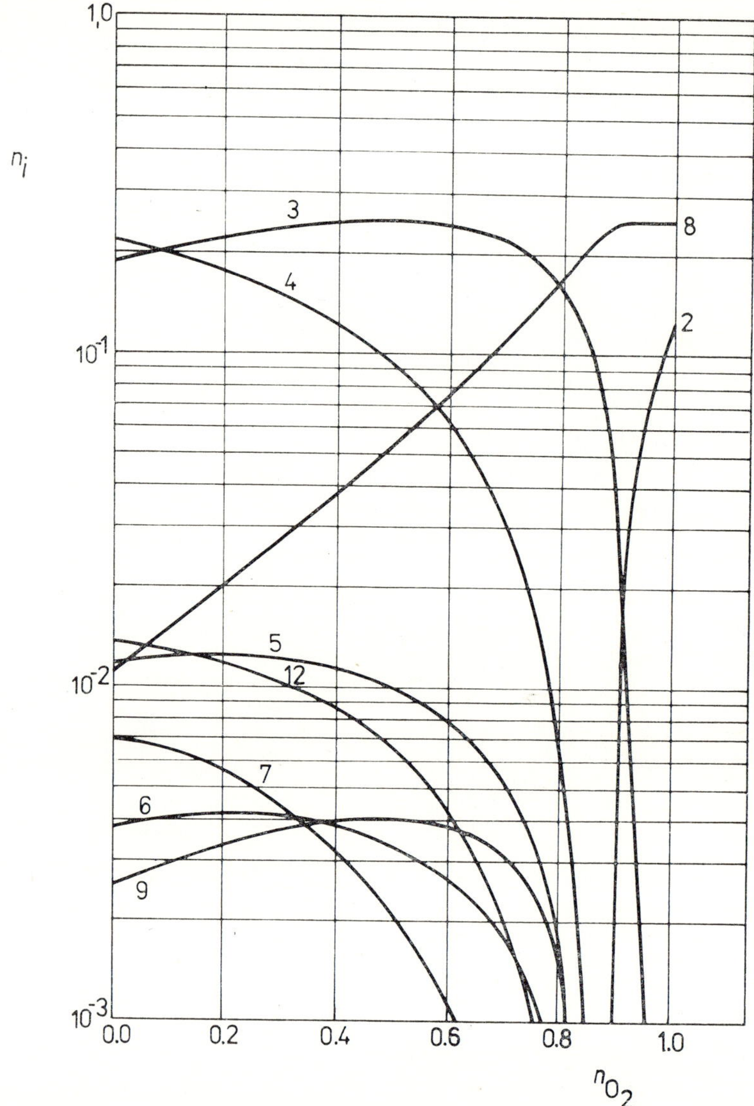

Fig. 28. Equilibrium in partial methane oxidation at 1350 K and 1 atm pressure. The dependence of numbers of moles of individual constituents formed from 1 mole methane, on the number of moles of oxygen in the initial mixture. For notation of constituents, see Table 13.

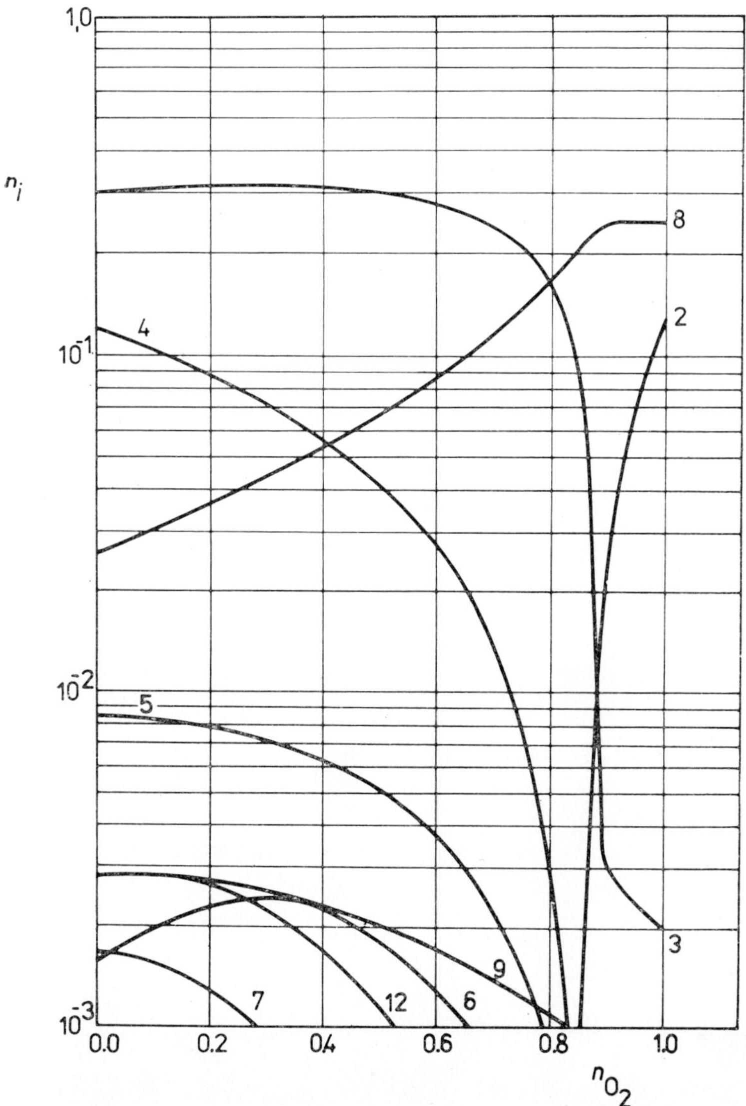

Fig. 29. Equilibrium in partial methane oxidation at 1450 K and 1 atm pressure. The dependence of numbers of moles of individual constituents formed from 1 mole methane, on the number of moles of oxygen in the initial mixture. For notation of constituents, see Table 13.

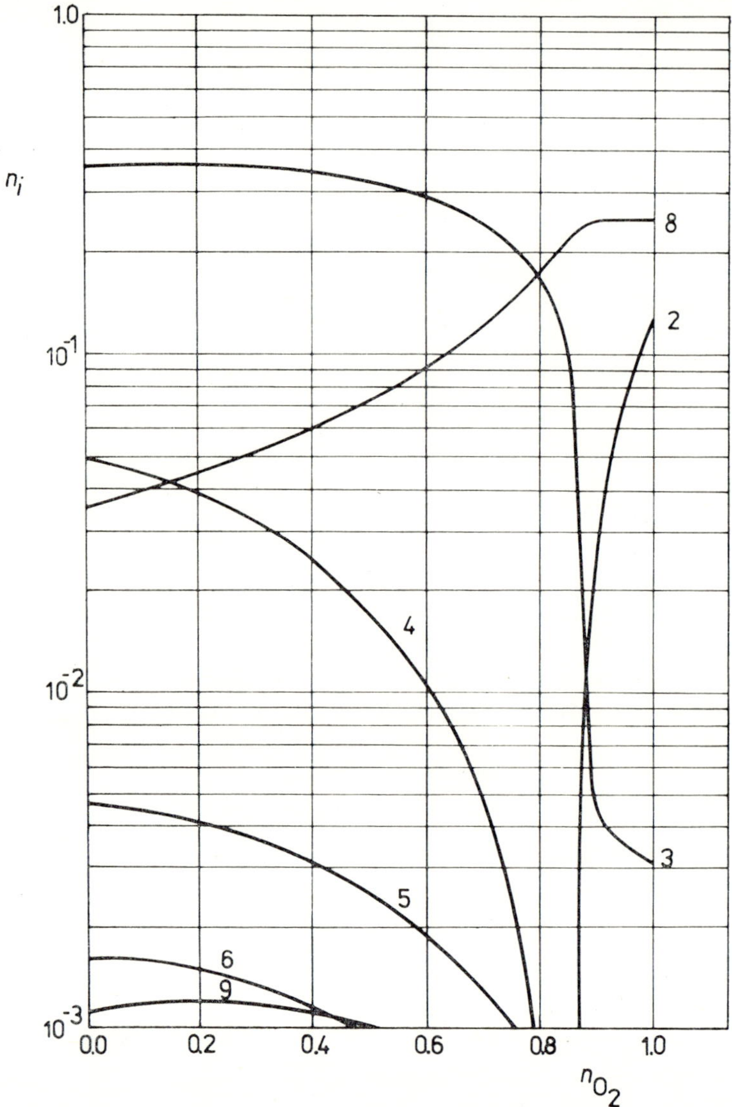

Fig. 30. Equilibrium in partial methane oxidation at 1550 K and 1 atm pressure. The dependence of numbers of moles of individual contituents formed from 1 mole methane, on the number of moles of oxygen in the initial mixture. For notation of constituents, see Table 13.

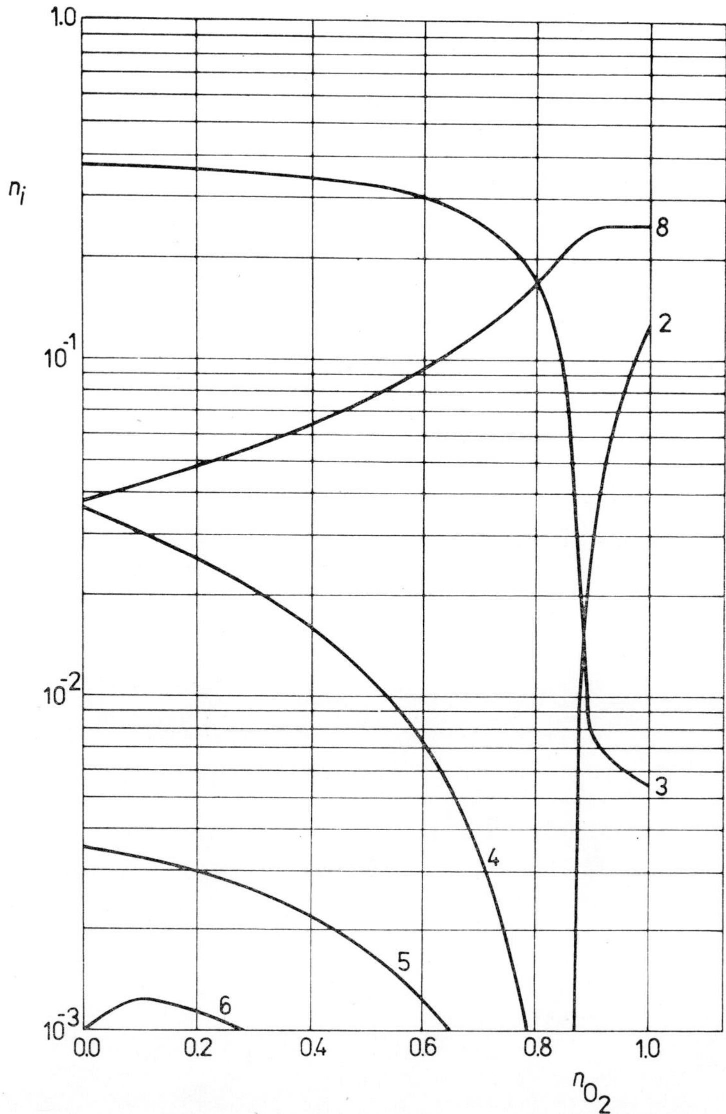

Fig. 31. Equilibrium in partial methane oxidation at 1650 K and 1 atm pressure. The dependence of numbers of moles of individual constituents formed from 1 mole methane, on the number of moles of oxygen in the initial mixture. For notation of constituents, see Table 13.

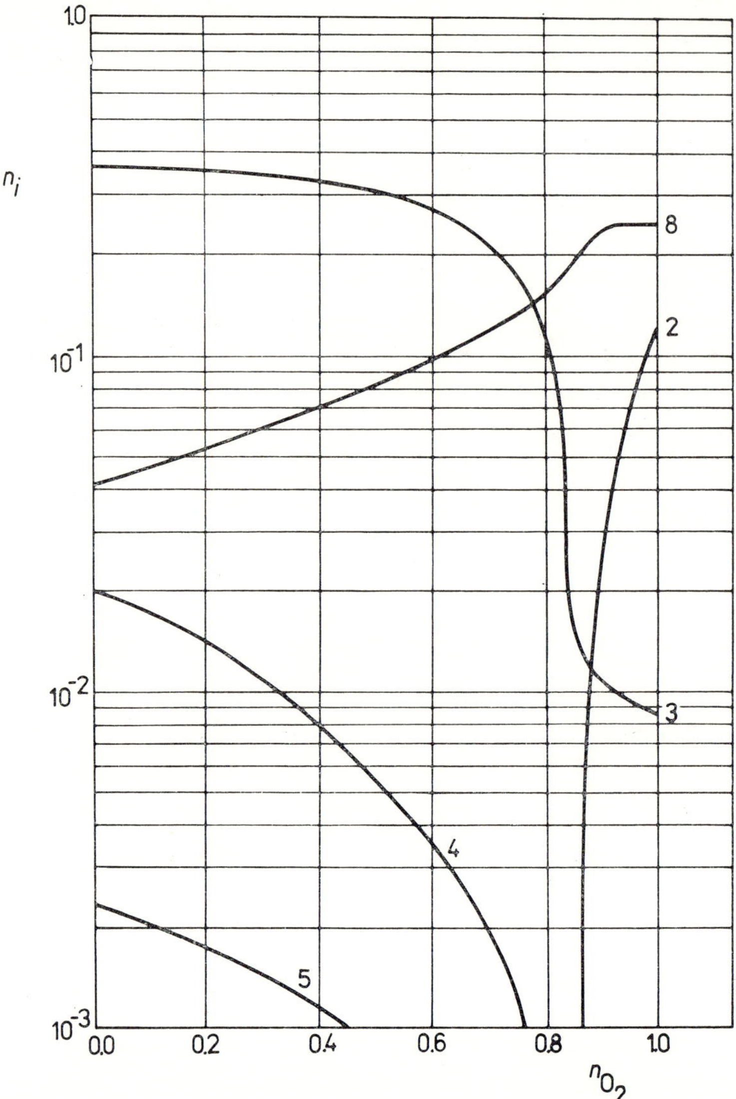

Fig. 32. Equilibrium in partial methane oxidation at 1750 K and 1 atm pressure. The dependence of numbers of moles of individual constituents formed from one mole methane, on the number of moles of oxygen in the initial mixture. For notation of constituents, see Table 13.

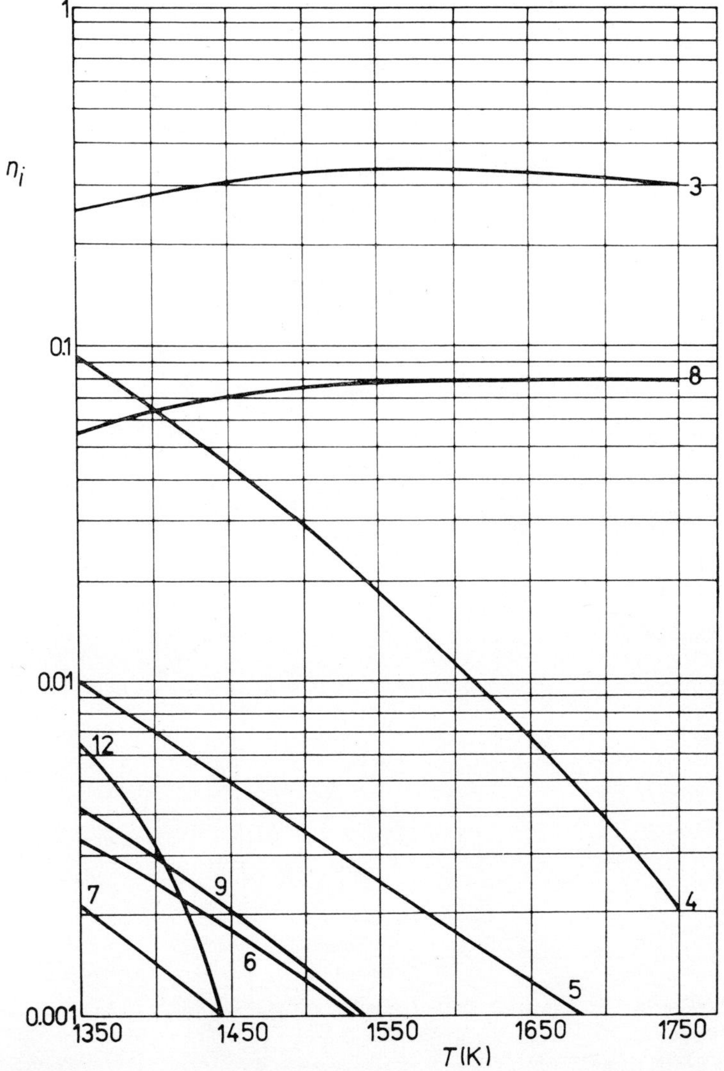

Fig. 33. Equilibrium in partial methane oxidation at 1 atm pressure. The dependence of numbers of moles of individual constituents formed from 1 mole methane and 0.5 mole oxygen, on temperature. For notation of constituents, see Table 13.

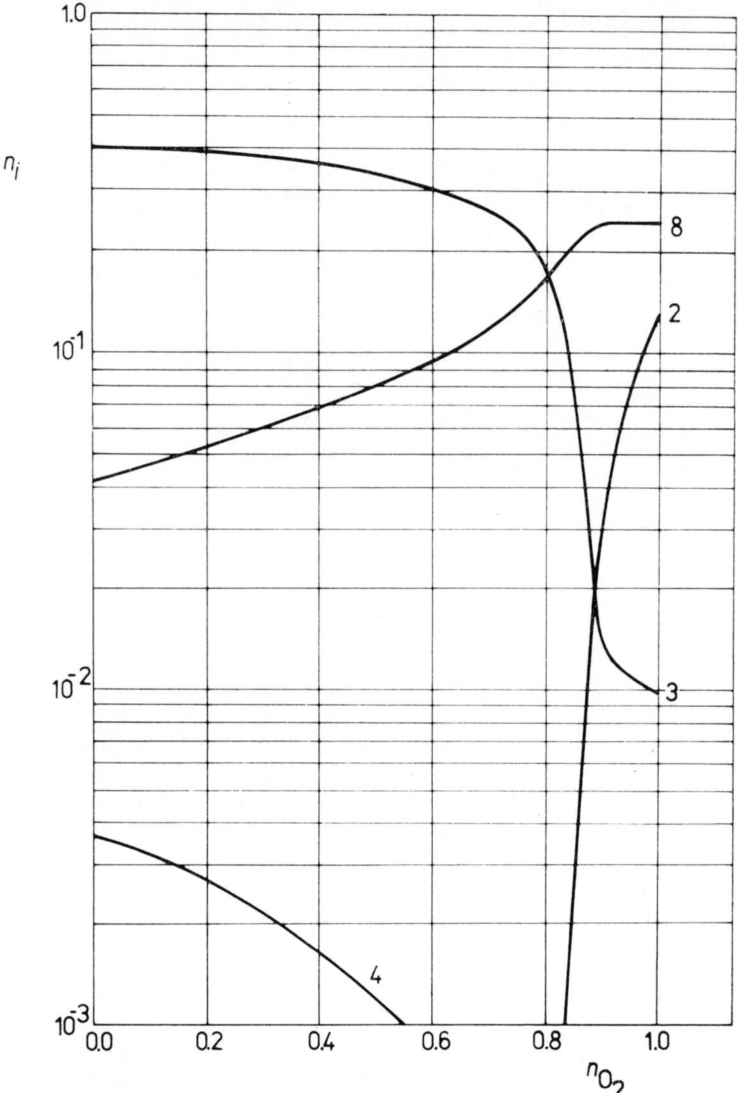

Fig. 34. Equilibrium in partial methane oxidation at 1650 K. The dependence of numbers of moles of individual constituents formed from 1 mole methane, on the number of moles of oxygen in the initial mixture at 0.1 atm pressure. For notation of constituents, see Table 13.

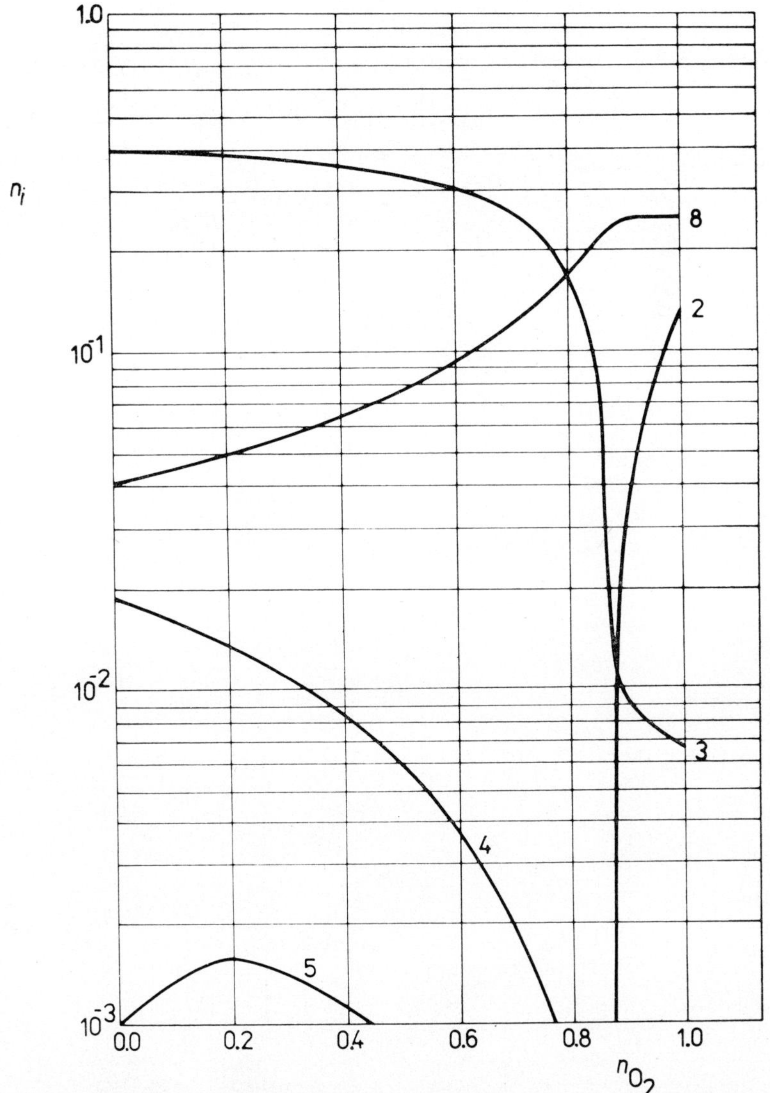

Fig. 35. Equilibrium in partial methane oxidation at 1650 K. The dependence of numbers of moles of individual constituents formed from one mole methane, on the number of moles of oxygen in the initial mixture at 0.5 atm pressure. For notation of constituents, see Table 13.

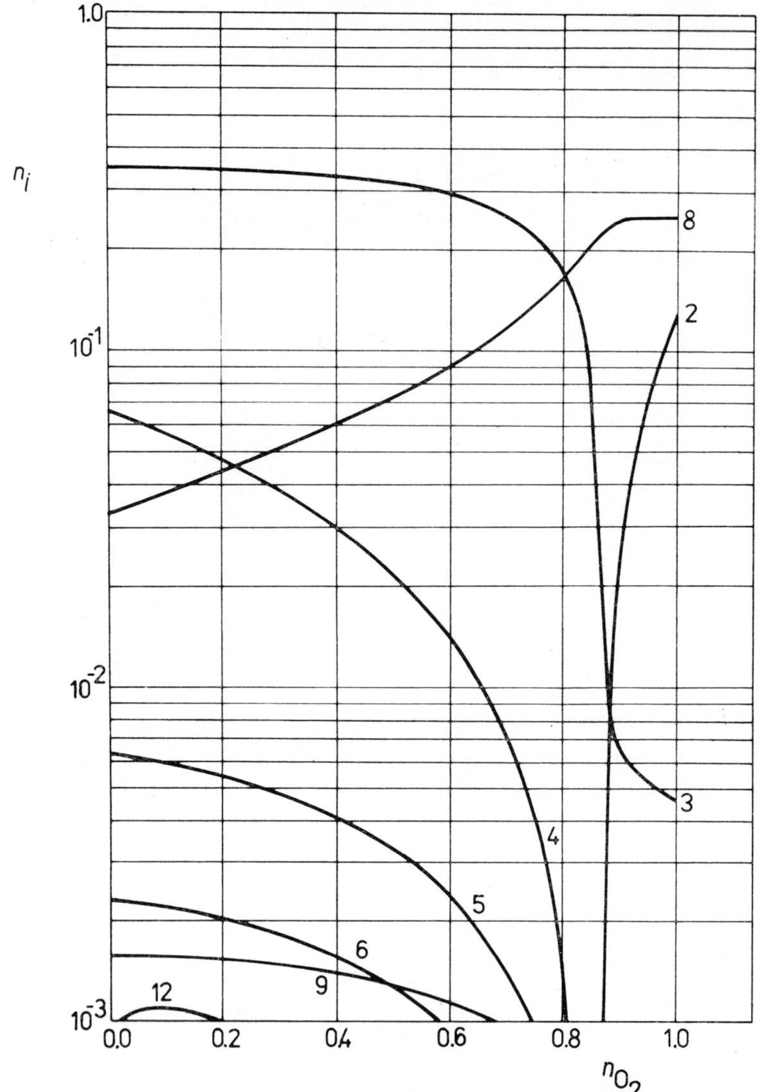

Fig. 36. Equilibrium in partial methane oxidation at 1650 K. The dependence of numbers of moles of individual constituents formed from one mole methane, on the numbers of moles of oxygen in the initial mixture at 2 atm pressure. For notation of constituents, see Table 13.

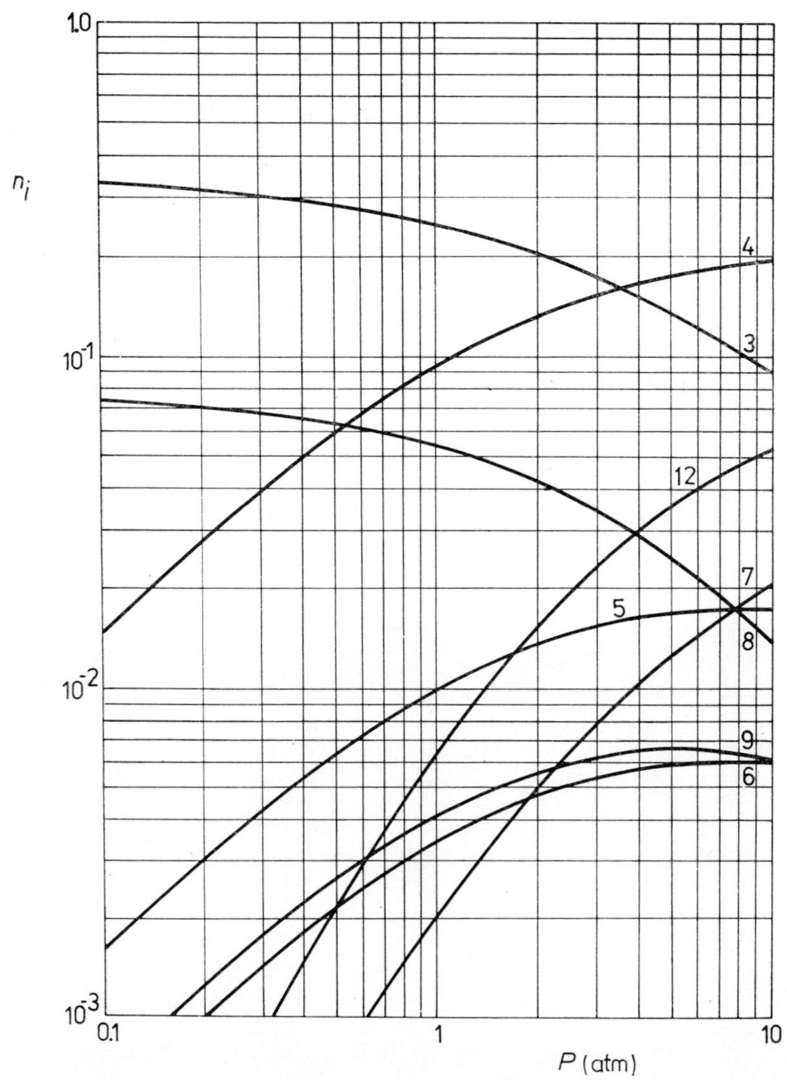

Fig. 37. Equilibrium in partial methane oxidation at 1350 K. The dependence of numbers of moles of individual constituents formed from 1 mole methane and 0.5 mole oxygen, on pressure. For notation of constituents, see Table 13.

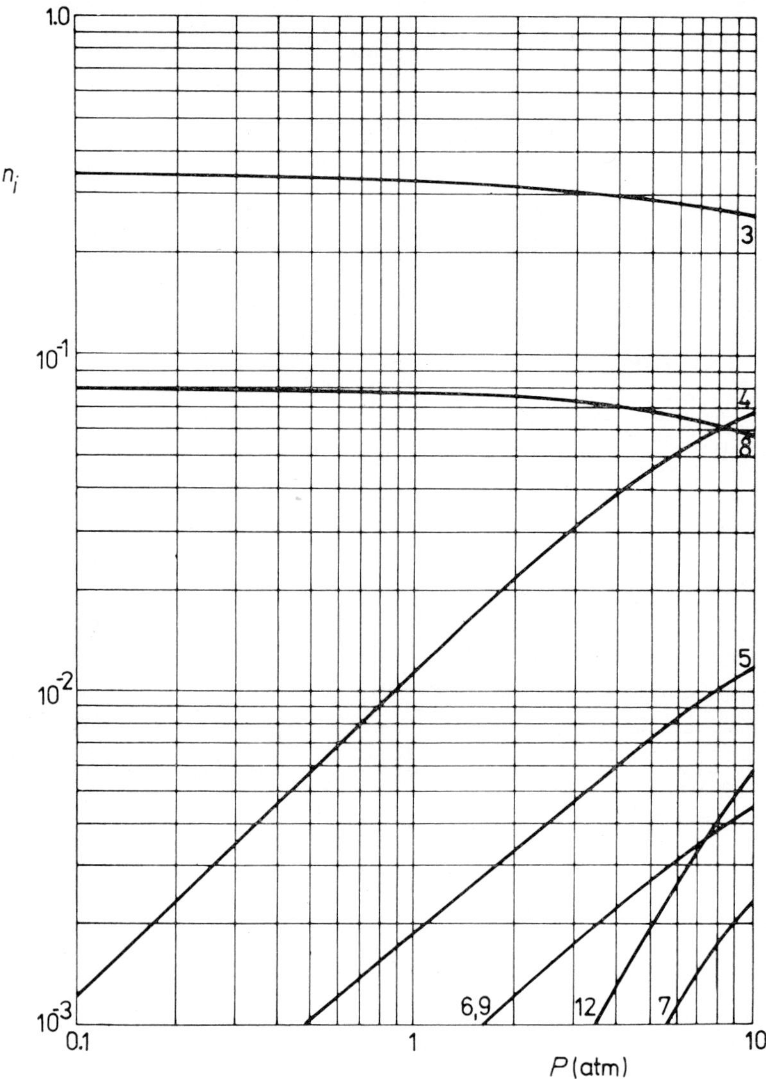

Fig. 38. Equilibrium in partial methane oxidation at 1600 K. The dependence of numbers of moles of individual constituents formed from 1 mole methane and 0.5 mole oxygen, on pressure. For notation of constituents, see Table 13.

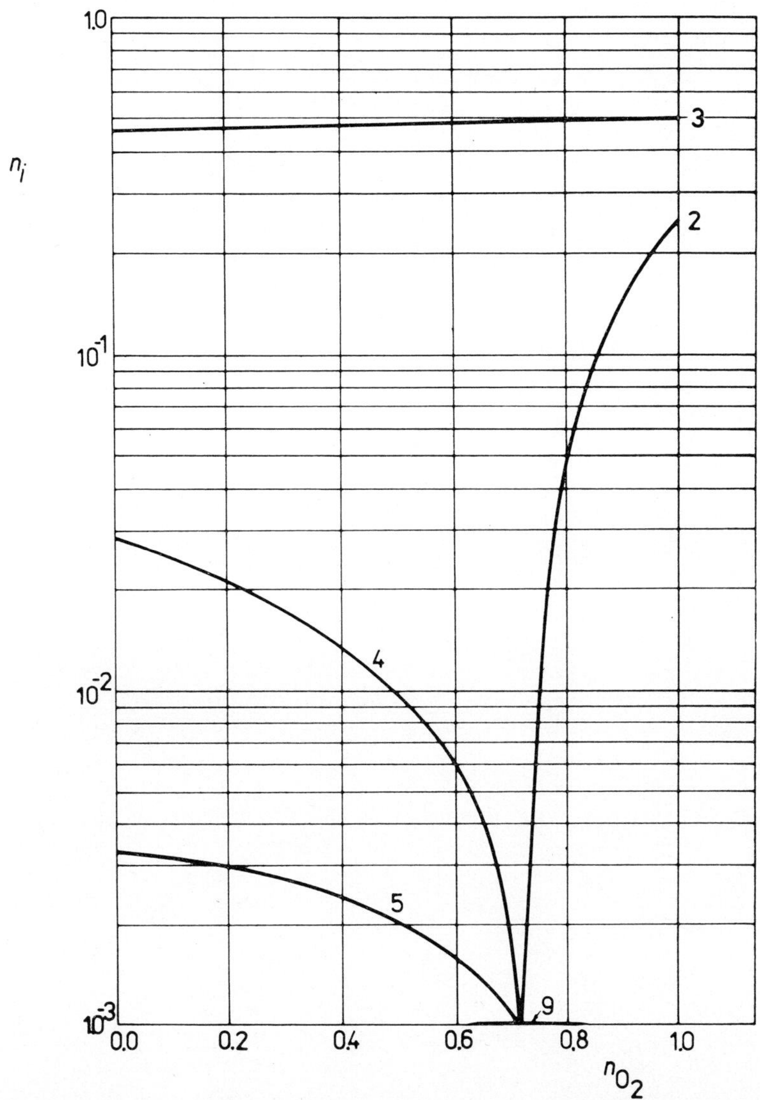

Fig. 39. Equilibrium in partial methane oxidation without formation of diacetylene. The dependence of numbers of moles of individual constituents formed from 1 mole methane, on the number of moles of oxygen in the initial mixture, at 1650 K and 1 atm pressure. For notation of constituents, see Table 13.

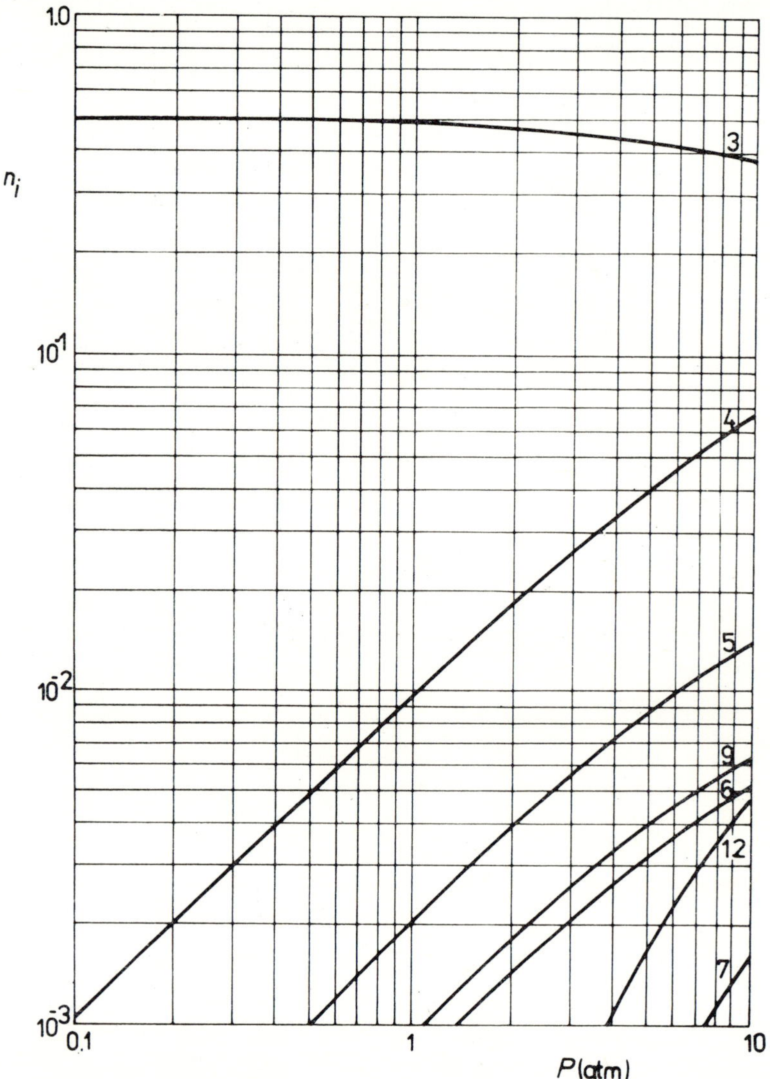

Fig. 40. Equilibrium in partial methane oxidation without formation of diacetylene. The dependence of numbers of moles of individual constituents formed from 1 mole methane and 0.5 mole oxygen, on pressure at 1650 K. For notation of constituents, see Table 13.

7.4.4   Technologic conclusions

1. Synthesis of acetylene from methane is thermodynamically feasible. In the range of reaction conditions considered, the equilibrium mixture contains predominantly acetylene.

2. The amount of acetylene in the equilibrium mixture rises with rising temperature and decreasing pressure.

3. The relationship between the amount of acetylene formed in the equilibrium mixture and the number of moles of oxygen in the initial mixture has a slow rising trend at low temperatures up to 0.8 moles per 1 mole methane, while it remains constant at higher temperature. Above this level the acetylene concentration decreases distinctly throughout the temperature range studied.

4. Unwanted constituents may be classified in two groups: the first includes higher unsaturated hydrocarbons formed by dehydrogenation, i.e. diacetylene, ethylene, methylacetylene, allene, propylene, vinylacetylene and 1,2-butadiene. With the exception of diacetylene there is no difficulty in suppressing the formation of these constituents with regard to the inverse dependence on the reaction conditions. This conclusion agrees with experimental results up to now obtained, according to which optimum yields are obtained at higher temperatures and low pressure. Suppression of diacetylene formation is more difficult.

5. At the reaction temperature, acetylene is unstable, having a tendency to decompose into carbon and hydrogen. There remains therefore the need of suitably selecting the technologic arrangement so as to counter only two reactions of the possible number of competing reactions, namely diacetylene formation and decomposition of acetylene to graphite and hydrogen. Since diacetylene formation is a sequential reaction,

$$2\,C_2H_2(g)\ =\ C_4H_2(g)\ +\ H_2(g) \tag{7.26}$$

its concentration should decrease with the decreasing reaction time. Rapid cooling of the reaction products should help to counter acetylene decomposition into carbon and hydrogen.

6. Oxygen present in the initial mixture is not consumed in the course of the reaction by hydrogen combustion, but by total oxidation of some of the methane. This conclusion is supported by the data plotted in Fig. 27, showing the termodynamic stability of hydrogen in the molecular form. If the initial constituent is methane only, the most stable constituents in equilibrium are graphite and hydrogen. When hydrogen is added to the initial mixture, it will react primarily with graphite (and, of course, gaseous methane) to form carbon monoxide until all the available carbon is exhausted. In this moment the entire amount of hydrogen continues to be present in the equilibrium mixture. Simultaneous oxidation of CO to CO_2 and

H$_2$ to H$_2$O takes place when another portion of oxygen is added. This conclusion agrees well with experimental data, which show a relatively high content of hydrogen in the reaction mixture. The fact that formation of CO and CO$_2$ cannot be totally suppressed is related to this finding. The process could only be arranged in such a manner that the mixture contained H$_2$, CO and CO$_2$ in a ratio suitable for use in methanol or ammonia synthesis.

7. Calculated results agree well with experimental data[81,87,99,124], the acetylene to ethylene ratio is roughly 10 : 1, diacetylene to vinylacetylene 4 : 1 and diacetylene to methylacetylene 3 : 1.

Appendix 1

SOLUTION OF ONE NON-LINEAR EQUATION

The non-linear equation

$$f(x) = 0 \, , \tag{A1.1}$$

where $f(x)$ is a continuous function in the given interval, capable of being differentiated to the required order, is solved by the iteration method which can be written in the general form of

$$x^{(n+1)} = \varphi\left(x^{(n)}, x^{(n-1)}, \ldots, x^{(n-k)}\right) \quad n = k, k+1, \ldots \tag{A1.2}$$

The value of the $(n + 1)$-th approximation to the solution of the non-linear equation (A1.2) is determined from the preceding $k + 1$ approximations $x^{(n)}, x^{(n-1)}, \ldots, x^{(n-k)}$. Most often employed are one-step $(k = 0)$ and two-step $(k = 1)$ methods. When there holds, starting from an index n_0 onward

$$\left|x^{(n+1)} - \bar{x}\right| \leqq \alpha \left|x^{(n)} - \bar{x}\right|^p \quad n > n_0 \, , \tag{A1.3}$$

where $\bar{x}$ is the solution of equation (A1.1) and $\alpha > 0$, $p \geqq 1$ are constants, then we say that the iteration method in question is a p-th order method. There must also hold for $p = 1$, that $\alpha < 1$, in order that the iteration process should converge. The higher the value of the parameter p, the more rapidly will the iteration process converge in the vicinity of the solution. When there are no derivatives of the function $f(x)$ on the right-hand side of the iteration relationship (A1.2), the iteration process is called a non-derivative method. In the reverse case, it is called a derivative method. Non-derivative methods are always first order methods. This, however, need not apply in the inverse sence. There also exist derivative methods which are first-order methods (e.g. the modified Newton method). In general, however, derivative methods which include derivatives of not higher than q-th order are $(q + 1)$-th order methods More details will be found e.g. in[25,42,94,131]. Let us now discuss the best-known iteration methods.

1.   Method of interval halving

The method of interval halving is a two-step, first-order method. Before it is applied, two approximations $x^{(0)}$, $x^{(1)}$ must be known, such that $\bar{x} \in (x^{(0)}, x^{(1)})$. It is furthermore assumed that there is only one solution of the equation (A2.1) in the interval $(x^{(0)}, x^{(1)})$. Evidently, $f(x^{(0)}) \times f(x^{(1)}) < 0$. The procedure now takes place according to the following scheme.

a) Select the approximation $x^{(2)}$ equal to the mean of the interval $(x^{(0)}, x^{(1)})$, i.e.

$$x^{(2)} = \frac{(x^{(0)} + x^{(1)})}{2}.$$

$$(A1.4)$$

b) Determine the sign of the product $f(x^{(2)}) \times f(x^{(0)})$. In the case of $f(x^{(2)}) \times f(x^{(0)}) < 0$ apply the following procedure to the interval $(x^{(0)}, x^{(2)})$. In the reverse case, the solution of equation (A1.1) will be found in the interval $(x^{(1)}, x^{(2)})$.

c) Obtain the approximation $x^{(3)}$ by halving the interval $(x^{(0)}, x^{(2)})$ in the case of $f(x^{(2)}) \times f(x^{(0)}) < 0$; in the reverse case, $x^{(3)}$ will be found in the interval $(x^{(1)}, x^{(2)})$. Repeat the procedure, until the condition is satisfied, that

$$\left| x^{(n+1)} - x^{(n)} \right| < \varepsilon,$$

$$(A1.5)$$

where ε is the required accuracy.

2.   The regula falsi method

The regula falsi method is likewise a two-step, first-order method, requiring the same conditions as the interval-halving method. Differing from the latter, where the interval is divided into two equal parts irrespective of the values of $f(x^{(0)})$ and $f(x^{(1)})$, in the regula falsi method this interval is divided in the ratio of $\left| f(x^{(0)})/f(x^{(1)}) \right|$. Thus the approximation $x^{(2)}$ can be obtained by constructing the point of intersection of the real axis with a straight line passing through the points $[x^{(0)}, f(x^{(0)})]$ and $[x^{(1)}, f(x^{(1)})]$. It can easily be proved, that

$$x^{(2)} = x^{(0)} - \frac{f(x^{(0)})}{f(x^{(1)}) - f(x^{(0)})} (x^{(1)} - x^{(0)}).$$

$$(A1.6)$$

As with the preceding method, we find out whether the solution of the equation (A1.1) lies in the interval $(x^{(0)}, x^{(1)})$ or in the interval $(x^{(1)}, x^{(2)})$, repeating this process. The numerical process is concluded as soon as the condition (A1.5) is satisfied. Although the two methods are of equal order, the regula falsi method usually converges more quickly.

The interval-halving and the regula falsi methods converge reliably, but rather slowly. In many fields of technology, the left-hand side of the equation (A1.1) often includes a number of parameters, the purpose being to find the solution $\bar{x}$ for different values of these parameters. When a chemical equilibrium is being studied, these parameters are temperature, pressure and initial composition, the required solution being the degree of conversion. Thus, in spite of the rapid computers available, first-order methods are too slow for chemical equilibrium studies with many variants of the initial conditions. Since the number of operations to be carried out in every step and the possibility of divergence of the method rise distinctly with the rising order of the method, it is usually best to employ second-order methods.

3.   Newton's method

Newton's method is a typical one-step, second-order method. Differing from the two methods discussed, only one approximation $x^{(0)}$ need be known before application of Newton's method. Developing the left-hand side of equation (A1.1) into a Taylor series in the point $x = x^{(0)}$ and neglecting the second- and higher-power terms, we obtain the equation

$$f(x^{(0)}) + f'(x^{(0)})(x - x^{(0)}) = 0 , \tag{A1.7}$$

where $f'(x^{(0)})$ is the first derivative of the function $f(x)$ in the point $x = x - x^{(0)}$. Denoting the solution of the equation (A1.7) the next approximation $x^{(1)}$ and generalising the resulting expression, we obtain the algorithm of Newton's iteration method

$$x^{(n+1)} = x^{(n)} - \frac{f(x^{(n)})}{f'(x^{(n)})} \quad n = 0, 1, \ldots . \tag{A1.8}$$

From relation (A1.7) follows, that the approximation $x^{(1)}$ is the point of intersection of a tangent to the function $f(x)$ in the point $x = x^{(0)}$ with the real axis. Newton's method is a second-order method, because

$$\left|x^{(n+1)} - \bar{x}\right| = \left|\varphi(x^{(n)}) - \bar{x}\right| = \left|\varphi(\bar{x}) + \varphi'(\bar{x})(x^{(n)} - \bar{x}) + \right.$$
$$+ \tfrac{1}{2}\varphi''(\bar{x})(x^{(n)} - \bar{x})^2 + \sigma(\left|x^{(n)} - \bar{x}\right|^3) - \bar{x}\Big| =$$
$$= \tfrac{1}{2}\left|\varphi''(\bar{x})(x^{(n)} - \bar{x})^2 (1 + \sigma(\left|x^{(n)} - \bar{x}\right|))\right| \leqq \alpha\left|x^{(n)} - \bar{x}\right|^2 , \tag{A1.9}$$

where $\varphi(x) = x - f(x)/f'(x)$. The value of the function $\varphi(x)$ in the point $x = x^{(n)}$ was obtained from the Taylor series of the function $\varphi(x)$ in the point $x = \bar{x}$.

Moreover, the evident equalities were used,

$$\bar{x} = \varphi(\bar{x})$$
$$\varphi'(\bar{x}) = 0 . \tag{A1.10}$$

When Newton's method converges, i.e. for large values of n, quantities of the order of $\left|x^{(n)} - \bar{x}\right|$ can be neglected in comparison to unity in the relationship (A1.9), then it will converge rapidly in the vicinity of the solution. Towards the end of the iteration process, the number of valid digits is practically doubled with every iteration step.

It is practically impossible to find a general procedure for constructing the first approximation $x^{(0)}$ such as to assure convergence of Newton's method. It may happen, that the value of $f'(x^{(0)})$ is low enough and therefore the quotient $f(x^{(0)})/f'(x^{(0)})$ high enough for the next approximation x_1 to achieve a senseless value. One of the most successful approaches to the problem of convergence of Newton's method is the use of a reduction parameter $\eta \in (0, 1\rangle$, controlling not the sign, but the absolute magnitude of the increment $-f(x^{(n)})/f'(x^{(n)})$, i.e.

$$x^{(n+1)} = x^{(n)} - \eta^{(n)} \frac{f(x^{(n)})}{f'(x^{(n)})} \qquad n = 0, 1, \ldots \qquad \text{(A1.11)}$$

A suitable selection of the reduction parameter then guarantees that the new approximation will not differ too much from the preceding one (e.g. by not more than 5%), or that it will not exceed the definition range of the function $f(x)$. At the end of the numerical process, the reduction parameter will obviously be equal to one. When, in the interval I, where the condition $\bar{x} \in I$ applies, the function $f(x)$ is monotonous. Newton's method with reduction parameter will converge to the solution $\bar{x}$. When, in the interval $(\bar{x}, x^{(0)})$ the function $f(x)$ has an extreme, even Newton's method with reduction parameter will not converge. In cases where the calculation of the derivative $f'(x)$ is very difficult, the so-called modified Newton's method may be applied

$$x^{(n+1)} = x^{(n)} - \frac{f(x^{(n)})}{f'(x^{(0)})} \qquad n = 0, 1, 2, \ldots, \qquad \text{(A1.12)}$$

which is a first-order method.

Appendix 2

SOLUTION OF A SET OF NON-LINEAR EQUATIONS

A set of non-linear equations

$$f_1(x_1, x_2, \ldots, x_s) = 0$$
$$f_2(x_1, x_2, \ldots, x_s) = 0$$
$$\vdots$$
$$f_s(x_1, x_2, \ldots, x_s) = 0 \tag{A2.1}$$

is mostly treated by Newton's method or a combination of one of the non-derivative methods (network method[95], Probe algorithm[95], Monte Carlo method[43,78,100], simplex method[63,95,110] etc.) and Newton's method. Convergence of Newton's method is subject to a considerable uncertainty factor and, therefore, it is preferable to obtain the first approximation to the solution of the set (A2.1) by means of one of the non-derivative methods which converge more slowly, but the certainty of convergence is far greater. The order of the method may be defined similarly as in Appendix 1, with the difference that the norm $\| \cdot \|$ is used in the relation (A1.3) instead of the absolute value $| \cdot |$.

1. Newton's method

The principle of the method will be described on the basis of a set of two non-linear equations

$$f_1(x_1, x_2) = 0$$
$$f_2(x_1, x_2) = 0 . \tag{A2.2}$$

Let us use the symbol $x^{(0)} = (x_1^{(0)}, x_2^{(0)})$ to denote the first approximation of the solution. Developing the left-hand sides of the two equations into Taylor series in the point $x^{(0)}$ and neglecting second- and higher-order terms, we obtain a set of two linear equations

$$d_{11} \Delta x_1 + d_{12} \Delta x_2 = -f_1(x^{(0)})$$
$$d_{21} \Delta x_1 + d_{22} \Delta x_2 = -f_2(x^{(0)}) , \tag{A2.3}$$

where $\Delta x_i = x_i - x_i^0$ and $d_{ij} = \partial f_i / \partial x_j$ in the point $x = x^{(0)}$. Evidently, there applies Cramer's rule

$$\Delta x_1 = \left(-f_1 d_{22} + d_{12} f_2\right)/\det$$

$$\Delta x_2 = \left(-d_{11} f_2 + f_1 d_{21}\right)/\det, \qquad (A2.4)$$

where $\det = d_{11} d_{22} - d_{12} d_{21}$. The new approximation $x^{(1)}$ is obtained thus:

$$x_1^{(1)} = x_1^{(0)} + \Delta x_1$$

$$x_2^{(1)} = x_2^{(0)} + \Delta x_2 . \qquad (A2.5)$$

For similar reasons as with the solution of one non-linear equation, the reduction parameter $\eta \in (0, 1\rangle$ may be employed, i.e.

$$x_1^{(1)} = x_1^{(0)} + \eta \, \Delta x_1$$

$$x_2^{(1)} = x_2^{(0)} + \eta \, \Delta x_2 . \qquad (A2.6)$$

Selection of $\eta < 1$ alters neither the sign, nor the mutual ratio of the increments Δx_1 and Δx_2, only their absolute magnitude being influenced. Repetition of the iteration step leads to a new approximation $x^{(2)}$, $x^{(3)}$, ... etc. The iteration process is stopped as soon as the condition

$$\left\| x^{(n+1)} - x^{(n)} \right\| < \varepsilon \qquad (A2.7)$$

is satisfied, $\| \, . \, \|$ being the norm of the vector and ε the required accuracy.

This procedure is easily extended to a generalized case of s unknown variables. A set similar to (A2.3) will take the form of

$$\sum_{j=1}^{s} d_{ij} \Delta x_j = -f_i(x^{(0)}) \quad i = 1, 2, ..., s . \qquad (A2.8)$$

The set of linear equations (A2.8) can be solved by e.g. the Gaussian elimination method. The procedure for finding a new approximation $x^{(1)}$ will be, for the generalized case,

$$x_i^{(1)} = x_i^{(0)} + \Delta x_1 \quad i = 1, 2, ..., s \qquad (A2.9)$$

or,

$$x_i^{(1)} = x_i^{(0)} + \eta \, \Delta x_i \quad i = 1, 2, ..., s \qquad (A2.10)$$

where $\eta \in (0, 1\rangle$ is the reduction parameter. The iteration process is stopped when the condition (A2.7) is satisfied.

2. Non-derivative methods

Non-derivative methods, which converge reliably but more slowly (first-order methods) than Newton's methods (second-order method) are generally employed to find the initial approximation for use in Newton's method. Non-derivative methods utilize the following properties of the solution of the set (A2.1). The absolute minimum of the function

$$\Phi(x_1, x_2, \ldots, x_s) = \sum_{i=1}^{s} f_i^2(x_1, x_2, \ldots, x_s) \tag{A2.11}$$

is achieved in the point $\bar{x} = (\bar{x}_1, \bar{x}_2, \ldots, \bar{x}_s)$, which is the solution of the set (A2.1), and conversely. The function Φ can also have a relative minimum point, in which the functional value of the function Φ is not equal to zero, and where therefore a true solution of the set (A2.1) is not achieved. This is the main disadvantage of methods based on looking for the minimum of the function Φ. There is a large number of methods of this kind. We shall now describe the simplest one, called network method, demonstrating its principle on the set (A2.2). We assume, that the solution $\bar{x}$ is known to lie within the rectangle $\bar{x} = (\bar{x}_1, \bar{x}_2) \in \langle a, b \rangle \times \langle c, d \rangle$. Let us cover the rectangle by a network of points $(x_1^{(i)}, x_2^{(j)})$

$$x_1^{(i)} = a + ih \quad i = 0, 1, \ldots, n$$

$$x_2^{(j)} = c + jk \quad j = 0, 1, \ldots, m, \tag{A2.12}$$

where h and k are distances between two meshes of the network in the x_1 and x_2 directions, resp., and $h = (b - a)/n$, $k = (d - c)/m$. We determine the value of the function Φ in every mesh of the network and remember that mesh, in which the function Φ attained its lowest value. We now construct a finer network in the vicinity of this mesh and repeat the process. It is useful to test before constructing each new network, whether Newton's method would now converge under the new conditions, i.e. whether suitably chosen parameter η will cause the value of the function Φ to decrease. In the negative case we continue applying the network method.

More rapid non-derivative methods in which the minimum of the function Φ is searched for in the direction of decreasing values, are e.g. the Probe algorithm and Rosenbrock methods. The increased speed, however, is achieved at the expense of reliable convergence.

Derivative methods also exist, in which the set (A2.1) is solved by searching for the minimum of the function Φ. These include the gradient method and Marquardt's method, which have been discussed in section 5.5.2 from the point of view of chemical equilibrium calculations.

It should finally be noted, that the problem of solving sets of non-linear equations is a difficult one, and has not yet been solved in a satisfactory manner[135]. A numerical procedure which is satisfactory for a certain type of sets, may totally fail when applied to a different type, and vice versa. Selection and design of a numerical procedure for a given type of set of non-linear equations is largely dependent on the user's mathematical experience.

Appendix 3

PROOF OF THE INEQUALITY

$$\sum_{i=1}^{N} \frac{v_i^2}{n_i} - \frac{v^2}{n} > 0$$

Let us first prove the validity of the expression

$$\sum_{i=1}^{N} \frac{\alpha_i^2}{n_i} - \frac{\alpha^2}{n} \geqq 0 , \tag{A3.1}$$

where $\alpha_i \; i = 1, 2, \ldots, N$ are arbitrary real numbers, $n_i > 0 \; i = 1, 2, \ldots, N$. Furthermore,

$$\alpha = \sum_{i=1}^{N} \alpha_i$$

$$n = \sum_{i=1}^{N} n_i . \tag{A3.2}$$

Introducing the notation

$$F(\alpha_1, \alpha_2, \ldots, \alpha_N, \; n_1, n_2, \ldots, n_N) = \sum_{i=1}^{N} \frac{\alpha_i^2}{n_i} - \frac{\alpha^2}{n} , \tag{A3.3}$$

for the left-hand side of the inequality $(A3.1)$, it will also be true that

$$\frac{\partial F}{\partial n_k} = - \frac{\alpha_k^2}{n_k^2} + \frac{\alpha^2}{n^2} \quad k = 1, 2, \ldots, N$$

$$\frac{\partial F}{\partial \alpha_k} = \frac{2\alpha_k}{n_k} - \frac{2\alpha}{n} . \tag{A3.4}$$

The function F achieves a minimum value in the point in which all first derivatives are equal to zero. There follows from $(A3.4)$, that the minimum of the function F is achieved in a point, for which there also applies

$$\alpha_k = \alpha \frac{n_k}{n} \quad k = 1, 2, \ldots, N . \tag{A3.5}$$

Substitution of relation (A3.5) into (A3.3) will easily show, that

$$F\left(\alpha\,\frac{n_1}{n}, \alpha\,\frac{n_2}{n}, \ldots, \alpha\,\frac{n_N}{n}, n_1, n_2, \ldots, n_N\right) \equiv 0,\qquad\text{(A3.6)}$$

by means of which the inequality (A3.1) is proved to be true.

From relation (A3.5) follows, that the equality in relation (A3.1) can only hold true when all values of α_k $k = 1, 2, \ldots, N$ all have the same sign. The series of stoichiometric coefficients v_i will, however, always include at least one positive and one negative number, since at least one initial substance and one product must always exist. Therefore, the sharp inequality

$$\sum_{i=1}^{N} \frac{v_i^2}{n_i} - \frac{v^2}{n} > 0,\qquad\text{(A3.7)}$$

where

$$v = \sum_{i=1}^{N} v_i$$

$$n = \sum_{i=1}^{N} n_i.\qquad\text{(A3.8)}$$

Appendix 4

PROOF OF THE ASYMPTOTIC RELATIONSHIP

$$K_a \approx \left(n_2^o\right)^{v_2 - v} \frac{\xi}{n_1^o + v_1 \xi} \quad \text{for} \quad n_2^o \to \infty \,.$$

From Chapter 4 follows, that the equilibrium condition for a system in which one chemical reaction is taking place, is

$$K_a = \left(\frac{P}{n}\right)^v \prod_{i=1}^{N} n_i^{v_i} \,, \tag{A4.1}$$

where

$$n = \sum_{i=1}^{N} n_i$$

$$n_i = n_i^o + v_i \xi \quad i = 1, 2, \ldots, N \,. \tag{A4.2}$$

Thus, the relationship (A4.1) may be rewritten in the equivalent form of

$$K_a = \left(\frac{P}{n^o + v\xi}\right)^v \times \frac{\xi^\alpha \displaystyle\prod_{i=3}^{N} v_i^{v_i}}{\left(n_1^o + v_1\xi\right)^{|v_1|} \left(n_2^o + v_2\xi\right)^{|v_2|}} \,, \tag{A4.3}$$

where

$$n_i^o = 0 \quad i = 3, 4, \ldots$$

$$n^o = n_1^o + n_2^o$$

$$\alpha = \sum_{i=3}^{N} v_i$$

$$v = \sum_{i=1}^{N} v_i \,. \tag{A4.4}$$

We assume that the system includes two initial compounds $(v_1 < 0, v_2 < 0)$ and that the initial number of moles of products is zero. For large values of n_2^o the value

244

of the reaction coordinate ξ is limited from above by conversion of the first initial compound, i.e.

$$0 < \xi < \frac{n_1^o}{|v_1|} . \tag{A4.5}$$

With a fixed selected value of n_1^o the behaviour of individual terms on the right-hand side of relation (A4.3) can be expressed for $n_2^o \to \infty$ thus:

$$n^o + v\xi \approx n_2^o$$
$$n_2^o + v_2\xi \approx n_2^o \tag{A4.6}$$

and the following therefore holds for $n_2^o \to \infty$

$$K_a \approx \beta(n_2^o)^{v_2 - v} \frac{\xi^\alpha}{(n_1^o + v_1\xi)^{|v_1|}} , \tag{A4.7}$$

where β is a non-zero constant of the form of

$$\beta = P^v \times \prod_{i=3}^{N} v_i^{v_i} . \tag{A4.8}$$

Only validity of the inequality $\alpha > 0$, and not the value of the constant α is important for a study of the relationship $\xi = \xi(n_2^o)$, as was shown in Chapter 4; therefore, we may take $\alpha = 1$. The equality $\alpha = 1$ may obviously be achieved by multiplying the stoichiometric coefficients with a suitable positive number. Similarly it is not the actual value of the value of the constant β, but only the fact that it differs from zero which is important, and therefore we may take $\beta = 1$. The same applies to the assumption of $|v_1| = 1$. Substitution of these values into the relation (A4.7) then leads to the relation that was to be proved.

Appendix 5

MAXIMUM YIELD OF A REACTION

Let us consider a system, in which one chemical reaction is taking place. This system includes two initial compounds, the initial number of moles of which is n_1^o and n_2^o. The initial number of moles of products is zero. Let us assume that the overall number of moles at the beginning of the reaction $n^o = n_1^o + n_2^o$ is a fixed selected number. The question now will be, for which ratio of n_1^o and n_2^o values, at a fixed temperature and pressure, the yield of the reaction will be maximum, i.e. what is the maximum value of the reaction coordinate ξ. In logarithmic form, the equilibrium condition

$$K_a = \left(\frac{P}{n}\right)^v \prod_{i=1}^{N} n_i^{v_i}, \tag{A5.1}$$

will be

$$\ln K_a = v \ln P - v \ln \left(n^o + v\xi\right) + \sum_{i=1}^{N} v_i \ln \left(n_i^o + v_i\xi\right), \tag{A5.2}$$

where $n_i^o = 0$ $i = 3, 4, \ldots, N$. The reaction coordinate ξ is here understood as a function $\xi = \xi(n_1^o, n_2^o \, (n_1^o))$. Differentiating the left- and right-hand sides of equation (A5.2) with respect to n_1^o, and considering the linking condition $n_2^o = n^o - n_1^o$ we obtain the equation

$$0 = \frac{d\xi}{dn_1^o} \left(\sum_{i=1}^{N} \frac{v_i^2}{n_i^o + v_i\xi} - \frac{v^2}{n^o + v\xi} \right) + \frac{v_1}{n_1^o + v_1\xi} - \frac{v_2}{n_2^o + v_2\xi}. \tag{A5.3}$$

The expression linked with the term $d\xi/dn_1^o$ is always positive (see Appendix 3) and, therefore, the derivative of $d\xi/dn_1^o$ can be zero only if

$$\frac{v_1}{n_1^o + v_1\xi} = - \frac{v_2}{n_2^o + v_2\xi}. \tag{A5.4}$$

Multiplication of the left- and right-hand sides of the equation (A5.4) by the expres-

246

sion $\left(n_1^o + v_1 \xi\right)\left(n_2^o + v_2 \xi\right)$ gives the maximum yield condition

$$\frac{n_1^o}{n_2^o} = \frac{v_1}{v_2}.$$
(A5.5)

This proof also shows that an equal result will be obtained with arbitrary but fixed values of n_i^o $i = 3, 4, \ldots, N$. The linking condition will then be $n_2^o = n^o - n_1^o$, where

$$n^o = \sum_{i=1}^{N} n_i^o$$
(A5.6)

is a fixed selected number.

Appendix 6

PROOF OF THE EXISTENCE AND UNAMBIGUOUS
NATURE OF THE SOLUTION TO THE PROBLEM
OF CHEMICAL EQUILIBRIUM IN IDEAL GAS SYSTEMS

Let us consider a closed ideal system including N compounds composed of M elements, in which R linearly independent reactions are taking place. At constant temperature and pressure, chemical equlibrium will be achieved in this system in the minimum point of the function

$$\frac{G}{RT} \equiv Q = \sum_{i=1}^{N} n_i(c_i + \ln n_i - \ln n), \tag{A6.1}$$

where

$$n = \sum_{i=1}^{N} n_i$$

$$n_i = n_i^o + \sum_{r=1}^{R} v_{ri}\xi_r \quad i = 1, 2, \ldots, N \tag{A6.2}$$

are mass balance equations, which are equivalent to the expression of the mass balance conditions by means of constitution coefficients

$$\sum_{i=1}^{N} a_{ij}n_i = b_j \quad j = 1, 2, \ldots, M. \tag{A6.3}$$

Let us study the properties of the function $Q = Q(\xi_1, \xi_2, \ldots, \xi_R)$ on a set Ω, defined as a set of all such R-membered groups, $(\xi_1, \xi_2, \ldots, \xi_R)$ for which the following applies:

$$n_i^o + \sum_{r=1}^{R} v_{ri}\xi_r > 0 \quad i = 1, 2, \ldots, N. \tag{A6.4}$$

Let us first prove that the function Q is convex on the set Ω, i.e. that the second-power form of the second differential is positively definite, which fact may be expressed by the inequality

$$\sum_{s=1}^{R} \sum_{k=1}^{R} q_{sk}h_s h_k > 0, \tag{A6.5}$$

248

where

$$q_{sk} = \frac{\partial^2 Q}{\partial \xi_s \, \partial \xi_k} \qquad \begin{matrix} s = 1, 2, ..., R \\ k = 1, 2, ..., R \, . \end{matrix} \qquad (A6.6)$$

The inequality (A6.5) must apply to an arbitrary vector $h = (h_1, h_2, ..., h_R)$ such, that $\|h\| \neq 0$. In practical cases, the symbol $d\xi_i$ is usually used instead of h_i. The relation (A6.5) is equivalent to the well-known Sylvester conditions, i.e. that the main minors of the matrix $\{q_{sk}\}$ are positive. From relations (A6.1) and (A6.2) follows

$$q_{sk} = \sum_{i=1}^{N} \frac{v_{si} v_{ki}}{n_i} - \frac{v_s v_k}{n} \qquad \begin{matrix} s = 1, 2, ..., R \\ k = 1, 2, ..., R \, , \end{matrix} \qquad (A6.7)$$

where

$$v_r = \sum_{j=1}^{N} v_{rj} \qquad r = 1, 2, ..., R \, . \qquad (A6.8)$$

Substitution of relation (A6.7) into the inequality (A6.5), the validity of which is to be proved, leads to

$$\sum_{i=1}^{N} \frac{\alpha_i^2}{n_i} - \frac{\alpha^2}{n} > 0 \, , \qquad (A6.9)$$

where

$$\alpha = \sum_{i=1}^{N} \alpha_i$$

$$\alpha_i = \sum_{r=1}^{R} v_{ri} h_r \qquad i = 1, 2, ..., N \, . \qquad (A6.10)$$

When we wish to prove the validity of the inequality (A6.9) and thus also of the inequality (A6.5), we must guarantee (see Appendix 3) that, with an arbitrary choice of $n_i > 0$ $i = 1, 2, ..., N$, there will exist no vector $h(\|h\| \neq 0)$ such, as to make valid the relation

$$\alpha_i = \alpha \frac{n_i}{n} \qquad i = 1, 2, ..., N \, . \qquad (A6.11)$$

Substitution of the relation (A6.10) into (A6.11) leads to a set of N linear equations for R unknown variables $h_1, h_2, ..., h_R$

$$\sum_{r=1}^{R} \left(v_{ri} - \frac{n_i}{n} v_r \right) h_r = 0 \qquad i = 1, 2, ..., N \, . \qquad (A6.12)$$

The set (A6.12) will have a trivial solution $h_1 = h_2 ... = h_R = 0$ only, provided that the rank of the matrix of set (A6.12) is R. The matrix of the set (A6.12) is obtained by subtracting the sum of all rows of the matrix, multiplied by the value of n_i/n, from the i-th row of the matrix of stoichiometric coefficients. Since the rank of the matrix of stoichiometric coefficients is R, the rank of the matrix of the set (A6.12) will likewise be R. Therefore there truly exists no vector $h(\|h\| \neq 0)$ such as to satisfy the relation (A6.11), whereby the validity of the inequalities (A6.9) and (A6.5) is proved. The impossibility of satisfying the relation (A6.11) can also be proved from the physical point of view. From relations (A6.2) and (A6.10) follows, that the quantity α_i represents an increment in the number of moles of the i-th compound. In a closed system, however, all values of α_i $i = 1, 2, ..., N$ cannot have the same sign, which fact disagrees with relation (A6.11).

We have proved herewith that the overall free enthalpy of the system is convex throughout the set Ω. Hence follows that the function G can have no more than one minimum on the set Ω. It remains to prove, that the function G will always have a minimum within the set Ω. It must be considered, that convexity only proves the unambiguous nature of the problem, not existence of a minimum. For example, the function $\exp(x)$ is convex in the interval $(-\infty, \infty)$, but does not have a minimum in this set. From equations (A6.1) and (A6.2) follows

$$\frac{\partial Q}{\partial \xi_r} = \sum_{i=1}^{N} v_{ri}(c_i + \ln n_i - \ln n) \quad r = 1, 2, ..., R \,. \tag{A6.13}$$

For a given value of s, i.e. for the s-th reaction, there always exists at least one positive and at least one negative stoichiometric coefficient. Imagine that the r-th reaction is running from left to right: then there will exist at least one value of $n_i \to 0_+$ and, therefore, $\partial Q/\partial \xi_r \to \infty$, since $v_{ri} < 0$. Similarly, with the reaction running completely from right to left, there will be a different value of $n_i \to 0_+$ and, therefore $\partial Q/\partial \xi_r \to -\infty$, since $v_{ri} > 0$. Thus, the existence and unambiguous nature of the minimum point of the function G has been proved on the set Ω.

The set of equations

$$\sum_{i=1}^{N} v_{ri}(c_i + \ln n_i - \ln n) = 0 \quad r = 1, 2, ..., R \,, \tag{A6.14}$$

has precisely one solution $(\xi_1, \xi_2, ..., \xi_R)$ in Ω, as follows from relation (A6.13) and from the properties of the Gibbs function which were proved above. It can easily be shown, that the set of equations

$$\left(\frac{P}{n}\right)^{v_r} \prod_{i=1}^{N} n_i^{v_{ri}} = (K_a)_r \quad r = 1, 2, ..., R \,, \tag{A6.15}$$

also has precisely one solution in Ω. $(K_a)_r$ is the equilibrium constant of the r-th reaction). This can be shown by taking the logarithm of the equation (A6.15) and employing the thermodynamic relationship

$$RT \ln (K_a)_r = - \sum_{i=1}^{N} v_{ri} G_i^{\circ} \,, \tag{A6.16}$$

which leads to a set of equations, identical with the set (A6.14).

The above-described results can obviously be extended to the case of an ideal mixture of real gases.

Appendix 7

VALUES OF $- (G° - H_0^0/T)$ OF THE MOST FREQUENTLY ENCOUNTERED COMPOUNDS

IN THE RANGE OF 298.15 TO 1200 K, AND VALUES OF H_0^0 IN CAL K^{-1} MOLE^{-1} AND CAL MOLE^{-1}, RESP.

T (K)	298.15	300	400	500	600	700	800	900	1000	1100	1200	H_0^o
Constituent												
Hydrogen	24.423	24.465	26.422	27.950	29.203	30.265	31.186	32.004	32.738	33.402	34.012	0
Nitrogen	38.817	38.859	40.861	42.415	43.688	44.769	45.711	46.550	47.306	47.994	48.629	0
Oxygen	42.061	42.106	44.112	45.675	46.968	48.071	49.044	49.911	50.697	51.415	52.077	0
Hydroxyl	36.824	36.859	38.904	40.483	41.772	42.860	43.804	44.637	45.385	46.063	46.686	10 000
Water	37.172	37.221	39.508	41.295	42.768	44.026	45.131	46.120	47.018	47.842	48.605	$-$57 104.3
Carbon monoxide	40.350	40.391	42.393	43.947	45.222	46.308	47.254	48.097	48.860	49.554	50.586	$-$27 201.9
Carbon dioxide	43.555	43.601	45.828	47.663	49.239	50.634	51.895	53.047	54.109	55.096	56.018	$-$93 968.6
Methane	36.46	36.51	38.86	40.75	42.39	43.86	45.21	46.47	47.65	48.78	49.86	$-$15 987
Ethane	45.27	45.33	48.24	50.77	53.08	55.25	57.29	59.24	61.11	62.90	64.63	$-$16 517
Propane	52.73	52.80	56.48	59.81	62.93	65.90	68.74	71.47	74.10	76.63	79.07	$-$19 482
Butane	58.52	58.62	63.49	67.93	72.05	75.95	79.60	83.28	86.73	90.03	93.20	$-$23 332
2-Methylpropane	56.08	56.16	60.72	64.95	68.95	72.78	76.45	79.98	83.38	86.65	89.80	$-$24 602
Pentane	64.26	64.37	70.33	75.76	80.84	85.66	90.26	94.68	98.92	102.98	106.88	$-$27 270
2-Methylbutane	64.74	64.84	70.35	75.52	80.40	85.13	89.67	94.01	98.21	102.24	106.12	$-$28 660
2,2-Dimethylpropane	56.36	56.46	61.93	67.04	71.96	76.70	81.27	85.67	89.90	93.98	97.92	$-$31 300
Hexane	70.01	70.15	77.20	83.65	89.68	95.41	100.89	106.14	111.18	116.00	120.64	$-$30 980
2-Methylpentane	70.20	70.34	76.9	83.0	88.8	94.4	99.8	105.0	110.0	114.7	119.3	$-$32 080
3-Methylpentane	70.15	70.29	76.9	83.1	89.0	94.6	100.1	105.3	110.3	115.1	119.7	$-$31 490
2,2-Dimethylbutane	65.81	65.93	72.4	78.4	84.2	89.8	95.2	100.4	105.4	110.2	114.8	$-$34 610
2,3-Dimethylbutane	66.97	67.08	73.6	79.8	85.7	91.2	96.6	101.8	106.8	111.6	116.0	$-$32 880
Heptane	75.77	75.92	84.06	91.53	98.53	105.17	111.51	117.60	123.63	129.03	134.40	$-$34 650
2-Methylhexane	75.75	75.90	83.7	90.9	97.7	104.2	110.4	116.5	122.3			$-$35 770
3-Methylhexane	78.05	78.19	85.8	92.9	99.6	106.0	112.2	118.2	124.0			$-$34 960
3-Ethylpentane	75.95	76.09	83.5	90.4	96.9	103.1	109.2	115.1	120.9			$-$34 100
2,2-Dimethylpentane	71.25	71.39	78.8	85.7	92.3	98.6	104.8	110.8	116.7			$-$38 000
2,3-Dimethylpentane	76.75	76.89	84.4	91.2	97.8	104.0	110.1	116.1	121.8			$-$36 290
2,4-Dimethylpentane	72.55	72.69	80.2	87.1	93.7	100.0	106.1	112.2	118.0			$-$36 980
3,3-Dimethylpentane	73.05	73.19	80.8	87.7	94.4	100.7	106.9	113.0	118.8			$-$36 920

T (K)	298.15	300	400	500	600	700	800	900	1100	1100	1200	H_0^o
Constituent												
2,2,3-Trimethylbutane	70.45	70.59	78.1	85.0	91.6	97.9	104.1	110.1	116.0			− 37 570
Octane	81.52	81.70	90.93	99.42	107.37	114.92	122.14	129.06	135.69	142.05	148.16	− 38 330
2-Methylheptane	81.49	81.66	90.5	97.8	106.5	113.9	121.1	128.1	134.6			− 39 420
3-Methylheptane	83.33	83.50	92.3	100.4	108.1	115.5	122.7	129.6	136.1			− 38 640
4-Methylheptane	81.63	81.79	90.5	98.6	106.2	113.6	120.7	127.6	134.1			− 38 430
3-Ethylhexane	84.23	84.39	92.8	100.5	107.8	114.9	121.9	128.7	135.1			− 37 710
2,2-Dimethylhexane	77.08	77.24	85.9	93.9	101.4	108.7	115.9	122.8	129.3			− 41 230
2,3-Dimethylhexane	79.76	79.92	88.7	96.9	104.6	112.0	119.2	126.2	132.8			− 38 760
2,4-Dimethylhexane	81.27	81.43	89.9	97.8	105.3	112.5	119.6	126.4	133.0			− 39 740
2,5-Dimethylhexane	79.34	79.51	88.1	96.0	103.5	110.7	117.8	124.6	131.2			− 40 610
3,3-Dimethylhexane	79.49	79.65	88.1	95.9	103.4	110.6	117.7	124.6	131.2			− 39 900
3,4-Dimethylhexane	78.10	78.26	87.0	95.1	102.8	110.2	117.4	124.4	130.9			− 38 520
2-Methyl-3-ethylpentane	79.58	79.74	88.3	96.3	103.9	111.3	118.4	125.3	131.8			− 37 960
3-Methyl-3-ethylpentane	78.24	78.40	86.9	94.8	102.2	109.5	116.6	123.6	130.1			− 38 680
2,2,3-Trimethylpentane	76.85	77.00	85.4	93.2	100.6	107.8	114.9	121.8	128.4			− 39 770
2,2,4-Trimethylpentane	76.85	77.00	85.4	93.2	100.6	107.8	114.9	121.8	128.4			− 40 730
2,3,3-Trimethylpentane	77.97	78.12	86.7	94.5	102.0	109.3	116.4	123.3	129.9			− 39 010
2,3,4-Trimethylpentane	78.25	78.40	86.9	94.7	102.2	109.4	116.5	123.3	129.9			− 39 120
2,2,3,3-Tetramethylbutane	69.77	69.92	78.3	86.2	93.7	101.1	108.3	115.2	121.7			− 41 090
Ethylene	43.98	44.03	46.61	48.74	50.70	52.50	54.19	55.78	57.29	58.74	60.12	14 522
Propene	52.95	53.02	56.39	59.32	62.05	64.61	67.04	69.36	71.57	73.69	75.73	8 468
1-Butene	59.32	59.41	63.87	67.84	71.56	75.08	78.42	81.61	84.66	87.58	90.39	5 158
2-Butene, cis	58.67	58.75	62.89	66.51	69.94	73.19	76.30	79.29	82.17	84.95	87.62	3 794
2-Butene, trans	56.80	56.89	61.31	65.19	68.84	72.27	75.53	78.64	81.62	84.47	87.22	2 506
2-Methylpropene	56.47	56.56	60.90	64.77	68.42	71.88	75.15	78.29	81.29	84.17	86.94	1 676
1-Pentene	65.11	65.23	70.88	75.96	80.68	85.13	89.37	93.42	97.29	100.98	104.52	1 019
2-Pentene, cis	66.51	66.50	71.73	76.30	80.64	84.80	88.76	92.59	96.27	99.82	103.22	− 178
2-Pentene, trans	64.54	64.65	70.13	75.01	79.60	83.97	88.12	92.07	95.87	99.51	103.02	− 1 362

T (K)	298.15	300	400	500	600	700	800	900	1000	1100	1200	H_0^o
Constituent												
2-Methyl-1-butene	64.96	65.06	70.41	75.23	79.80	84.14	88.27	92.24	96.04	99.69	103.18	— 2 303
3-Methyl-1-butene	62.47	62.57	67.99	73.12	77.87	82.34	86.59	90.65	94.51	98.23	101.77	— 681
2-Methyl-2-butene	64.52	64.63	69.84	74.48	78.88	83.09	87.09	90.93	94.62	98.18	101.60	— 3 677
1-Hexene	70.85	70.98	77.7	83.8	89.5	94.8	99.9	104.8	109.5	113.9	118.2	— 2 690
2-Hexene, cis	72.30	72.42	78.7	84.4	89.8	94.9	99.7	104.4	108.9			— 3 890
2-Hexene, trans	70.33	70.47	77.1	83.1	88.7	94.0	99.1	103.9	108.5			— 5 190
3-Hexene, cis	71.44	71.54	77.7	83.6	88.4	93.4	98.2	102.9	107.3			— 3 660
3-Hexene, trans	69.53	69.66	76.2	82.1	87.6	92.9	98.0	102.8	107.4			— 5 020
2-Methyl-1-pentene	70.75	70.88	77.4	83.4	88.9	94.2	99.2	104.1	108.7			— 6 040
3-Methyl-1-pentene	70.32	70.44	76.8	82.9	88.5	93.9	99.0	103.9	108.5			— 3 370
4-Methyl-1-pentene	69.95	70.07	76.4	82.2	87.7	92.9	98.0	102.8	107.4			— 3 860
2-Methyl-2-pentene	71.40	71.51	77.6	83.1	88.3	93.4	98.2	102.8	107.3			— 6 990
3-Methyl-2-pentene, cis	71.40	71.51	77.6	83.1	88.3	93.4	98.2	102.8	107.3			— 6 350
3-Methyl-2-pentene, trans	72.21	72.32	78.4	83.9	89.2	94.2	99.0	103.6	108.1			— 6 350
4-Methyl-2-pentene, cis	69.72	69.83	76.0	81.8	87.2	92.4	97.3	102.1	106.6			— 5 420
4-Methyl-2-pentene, trans	67.69	67.81	74.3	80.3	85.9	91.2	96.3	101.1	105.7			— 6 670
2-Ethyl-1-butene	70.20	70.31	76.7	82.4	87.9	93.1	98.1	102.8	107.4			— 5 170
2,3-Dimethyl-1-butene	67.36	67.48	73.9	79.9	85.6	90.9	96.0	100.9	105.5			— 7 100
3,3-Dimethyl-1-butene	65.08	65.19	71.4	77.1	82.4	87.5	92.4	97.2	101.9			— 6 180
2,3-Dimethyl-2-butene	67.56	67.68	73.8	79.3	84.5	89.5	94.3	98.9	103.3			— 7 960
1-Heptene	76.60	76.76	84.6	91.7	98.3	104.6	110.6	116.3	121.7	127.0	132.0	— 6 370
Propadiene	48.18	48.24	51.35	54.08	56.55	58.85	61.00	63.02	64.94	66.76	68.50	47 700
1,2-Butadiene	57.11	57.19	61.21	64.79	68.10	71.19	74.11	76.88	79.52	82.04	84.47	42 780
1,3-Butadiene	54.46	54.54	58.38	61.89	65.18	68.29	71.24	74.05	76.72	79.28	81.73	30 200
1,2-Pentadiene	63.5	63.6	68.7	73.3	77.6	81.7	85.5	89.1	92.6	95.9	99.1	39 320
1,3-Pentadiene, cis	62.9	63.0	67.6	71.8	75.8	79.6	83.3	86.7	90.1	93.3	96.4	23 730
1,3-Pentadiene, trans	61.1	61.2	66.1	70.5	74.7	78.7	82.5	86.1	89.5	92.8	96.0	23 390
1,4-Pentadiene	63.2	63.3	68.5	73.1	77.4	81.5	85.4	89.0	92.5	95.8	99.0	29 630

T (K)	298.15	300	400	500	600	700	800	900	1000	1100	1200	H_0^o
Constituent												
2,3-Pentadiene	61.9	62.0	66.9	71.4	75.5	79.4	83.1	86.6	90.0	93.2	96.3	37 770
2-Methyl-1,3-butadiene	60.4	60.5	65.3	69.8	74.0	78.0	81.8	85.4	88.9	92.2	95.4	22 980
2-Methyl-2,3-butadiene	60.6	60.7	65.7	70.2	74.5	78.5	82.2	85.8	89.2	92.5	95.7	35 640
Acetylene (Ethylene)	39.976	40.025	42.451	44.508	46.313	47.930	49.400	50.752	52.005	53.175	54.275	54 329
Propyne	48.89	48.95	52.14	54.92	57.44	59.76	61.91	63.94	65.86	67.68	69.42	46 017
1-Butyne	56.70	56.78	60.78	64.38	67.70	70.81	73.74	76.51	79.16	81.69	84.11	42 960
2-Butyne	54.43	54.51	58.59	62.18	65.44	68.48	71.35	74.06	76.65	79.12	81.50	38 491
1-Pentyne	62.49	62.60	67.8	72.5	76.8	80.9	84.7	88.3	91.8	95.1	98.2	38 900
2-Pentyne	63.62	63.72	68.6	73.0	77.1	80.9	84.5	88.0	91.3	94.5	97.6	35 480
2-Methyl-3-butyne	60.86	60.95	65.8	70.3	74.5	78.5	82.2	85.8	89.2	92.5	95.6	37 370
1-Hexyne	68.23	68.35	74.6	80.3	85.6	90.6	95.3	99.7	104.0	108.0	111.9	35 200
Vinylacetylene	55.11	55.18	58.82	62.06	65.03	67.79	70.37	72.82	75.10			74 350
Cyclopropane	47.69	47.73	50.60	53.22	55.72	58.15	60.47	62.71	64.90			16 830
Cyclobutane	52.37	52.44	55.95	59.25	62.43	65.54	68.56	71.50	74.36			12 180
Cyclopentane	57.93	58.00	61.88	65.62	69.30	72.95	76.52	80.04	83.48	86.84	90.13	− 10 680
Methylcyclopentane	65.23	65.33	70.45	75.33	80.07	84.72	89.26	93.68	97.97	102.13	106.18	− 16 620
1,1-Dimethylcyclopentane	67.21	67.36	73.36	79.14	84.72	90.28	95.75	101.02	106.13	111.05	115.87	− 22 690
1,2-Dimethylcyclopentane, cis	68.65	68.81	74.91	80.71	86.35	91.91	97.36	102.65	107.76	112.70	117.50	− 20 660
1,2-Dimethylcyclopentane, trans	68.73	68.87	75.01	80.81	86.45	92.03	97.54	102.79	107.90	112.84	117.64	− 22 390
1,3-Dimethylcyclopentane, cis	68.73	68.87	75.01	80.81	86.45	92.03	97.54	102.79	107.90	112.84	117.64	− 21 650
1,3-Dimethylcyclopentane, trans	68.73	68.87	75.01	80.81	86.45	92.03	97.54	102.79	107.90	112.84	117.64	− 22 190
Ethylcyclopentane	71.71	71.83	77.90	83.72	89.34	94.89	100.30	105.53	110.62	115.54	120.32	− 20 080
Cyclohexane	57.07	57.16	61.80	66.39	70.96	75.50	79.98	84.40	88.74	93.0	97.1	− 20 010
Methylcyclohexane	64.51	64.62	70.38	76.06	81.68	87.24	92.70	98.04	103.24	108.3	113.2	− 26 300

T (K)	298.15	300	400	500	600	700	800	900	1000	1100	1200	H_0°
Constituent												
Ethylcyclohexane	70.99	71.12	77.8	84.5	91.0	97.4	103.7	109.9	115.9	121.7	127.3	−28 940
1,1-Dimethylcyclohexane	67.52	67.67	74.1	80.6	87.0	93.3	99.6	105.7	111.7	117.5	123.1	−30 930
1,2-Dimethylcyclohexane, cis	69.35	69.49	76.1	82.6	89.1	95.5	101.8	107.9	113.9	119.7	125.3	−28 950
1,2-Dimethylcyclohexane, trans	68.21	68.36	75.1	81.7	88.3	94.7	101.1	107.3	113.3	119.1	124.8	−30 910
1,3-Dimethylcyclohexane, cis	68.19	68.34	75.0	81.6	88.1	94.5	100.8	107.0	113.0	118.8	124.4	−32 020
1,3-Dimethylcyclohexane, trans	69.57	69.72	76.4	83.0	89.4	95.8	102.1	108.3	**114.2	120.0	125.6	−30 060
1,4-Dimethylcyclohexane, cis	68.19	68.34	75.0	81.6	88.1	94.5	100.8	106.9	112.8	118.6	124.2	−30 080
1,4-Dimethycyclohexane, trans	66.81	66.96	73.6	80.2	86.7	93.2	99.6	105.7	111.8	117.6	123.2	−31 990
Propylcyclohexane	76.45	76.62	84.4	91.9	99.4	106.7	113.9	120.9	127.7	134.2	140.5	−32 790
Butylcyclohexane	82.20	82.40	91.3	99.8	108.2	116.5	124.5	132.4	140.0	147.2	154.3	−36 290
Cyclopentene	57.62	57.69	61.37	64.82	68.18	71.46	74.67	77.80	80.85	83.81	86.70	7 050
Cyclohexene	60.29	60.38	64.96	69.43	73.84	78.16	82.38	86.49	90.48	94.36	98.12	− 2 240
Benzene	52.93	53.00	56.69	60.24	63.70	67.06	70.34	73.50	76.57	79.54	82.40	24 000
Toluene	61.98	62.07	66.74	71.20	75.52	79.72	83.79	87.72	91.53	95.21	98.77	17 500
Ethylbenzene	68.26	68.37	74.14	79.64	84.94	90.08	95.05	99.84	104.47	108.94	113.25	13 917
o-Xylene	65.51	65.62	71.67	77.35	82.74	87.97	92.98	97.76	102.40	106.87	111.19	11 057
m-Xylene	67.63	67.74	73.50	78.95	84.20	89.28	94.18	98.91	103.48	107.89	112.15	10 926
p-Xylene	66.26	66.37	72.15	77.59	82.83	87.89	92.76	97.48	102.02	106.42	110.66	11 064
Propylbenzene	74.05	74.19	81.2	87.8	94.1	100.1	106.0	111,6	117.1	122.3	127.4	9 810
Isopropylbenzene	72.42	72.54	79.2	85.6	91.8	97.8	103.6	109.2	114.6	119.8	124.8	9 250
1,2,3-Trimethylbenzene	69.53	69.66	76.80	83.49	89.85	95 99	101.79	107.43	112.83	118.12	123.05	5 480
1,2,4-Trimethylbenzene	72.49	72.62	79.72	86.25	92.51	98.50	104.27	109.84	115.17	120.38	125.39	4 503

T (K)	298.15	300	400	500	600	700	800	900	1000	1100	1200	H_0°
Constituent												
1,3,5-Trimethylbenzene	70.93	71.06	77.66	84.04	90.18	96.08	101.77	107.26	112.56	117.68	122.62	4 241
1-Methyl-2-ethylbenzene	73.27	73.41	80.5	87.2	93.6	99.8	105.6	111.3	116.8	122.0	127.1	8 092
1-Methyl-3-ethylbenzene	75.29	75.42	82.3	88.8	95.0	101.0	106.8	112.4	117.8	123.0	128.0	7 593
1-Methyl-4-ethylbenzene	73.92	74.05	80.9	87.4	93.6	99.6	105.4	111.0	116.3	121.5	126.5	7 241
1,2,3,4-Tetramethylbenzene	73.33	73.48	81.92	89.82	97.33	104.59	111.33	117.91	124.20			— 1 155
1,2,4,5-Tetramethylbenzene	73.31	73.46	82.00	89.89	97.35	104.40	111.20	117.72	123.93			— 2 104
1,2,3,5-Tetramethylbenzene	74.98	75.13	83.48	91.27	98.68	105.80	112.51	119.01	125.24			— 1 782
Pentamethylbenzene	76.69	76.86	86.37	95.31	103 85	112.14	119.74	127.21	134.34			— 7 608
n-Butylbenzene	79.79	79.94	88.0	95.6	102.8	109.8	116.6	123.0	129.3	135.3	141.1	5 890
Styrene	65.76	65.86	71.28	76.44	81.42	86.21	90.82	95.26	99.54	103.66	107.61	40 340
1-Phenyl-1-propene, cis	71.4	71.5	78.1	84.2	90.1	95.8	101.2	106.4	111.4	116.3	120.9	35 330
1-Phenyl-1-propene, trans	71.0	71.1	77.6	83.7	89.6	95.3	100.7	105.9	111.0	115.8	120.5	34 450
2-Phenyl-1-propene	71.4	71 5	78.1	84.2	90.1	95.8	101.2	106.4	111.4	116.3	120.6	33 330
1-Phenyl-2-ethenylbenzene	71.4	71.5	78.1	84.2	90.1	95.8	101.2	106.4	111.4	116.3	120.9	34 630
1-Methyl-3-ethenylbenzene	72.8	72.9	79.5	85.6	91.5	97.2	102.6	107.8	112.8	117.7	122.3	33 930
1-Methyl-4-ethenylbenzene	71.4	71.5	78.1	84.2	90.1	95.8	101.2	106.4	111.4	116.3	120.9	33 730
Cyclooktatetraene	61.34	61.44	66.78	71.88	76.92	81.65	86.31	90.78	95.12			76 330
1,2,3,4-Tetrahydro-naphthalene	65.2	65.3	71.7	78.0	84.2	90.3	96.6	102.4	108.1			21 320
Naphtalene	63.83	63.93	69.39	74.80	80.15	85.40	90.53	95.45	100.32	104.98	109.46	40 290
Methanol	47.73	47.78	50.42	52.56	54.60	56.41	58.11	59.66	61.15			—45 820
Ethanol	52.27	54.34	58.11	61.42	64.43	67.22	69.84	72.31	74.65	76.88	79.02	—52 260
1-Propanol	62.73	62.83	73.36	77.70	81.78	85.59	89.18	92.57	95.80			56 660
1-Butanol	71.61	71.71	77.37	82.69	87.63	92.32	96.70	100.93	104.97			61 040
Ethylene Glycol	61.90	62.03	66.80	71.01	74.98	78.58	81.68	84.75	87.53			—91 060
Dimethylether	52.285	52.352	55.887	58.928	61.720	64.290	66.867	69.109	71.291			—40 107
Furan	53.95	54.01	57.16	60.10	62.92	65.63	68.24	70.75	73.15	75.47	77.69	— 5 157
Formaldehyde	44.254	44.304	46.642	48.539	50.115	51.531	52.813	54.004	55.111	56.145	57.129	—27 700

T (K)	298.15	300	400	500	600	700	800	900	1000	1100	1200	H_0°
Constituent												
Acetaldehyde	52.85	52.91	56.03	58.67	61.05	63.24	65.28	67.20	69.03			—37 160
Acetone	57.45	57.53	61.52	65.03	68.22	71.19	74.00	76.67	79.22	81.68	84.02	—47 740
Formic Acid	51.029	51.086	53.829	56.158	58.220	60.091	61.811	63.410	64.907	66.314	67.646	—88 400
Acetic Acid (monomer)	56.50	56.58	60.29		66.19		71.44		75.92		79.96	—101 200
Acetic Acid (dimer)	73.91	74.05	81.12		93.33		103.95		113.40		122.01	—217 740
Methanethiol	51.16	51.23	54.09	56.60	58.76	60.73	62.55	64.25	65.85	67.37	68.80	—15 900
Ethanethiol	58.64	58.72	62.46	65.76	68.78	71.58	74.21	76.69	79.07			—22 420
Thiacyclopropane	51.84	51.90	54.73	57.25	59.58	61.77	63.84	65.81	67.68			6 820
1-Propanethiol	64.86	64.95	69.76	74.05	78.00	81.69	85.18	88.51	91.67			—26 170
2-Propanethiol	62.72	62.82	67.45	71.65	75.56	79.22	82.71	86.03	89.16			—27 950
Methyl Sulphide	55.71	55.79	59.66	63.04	66.11	68.93	71.63	74.10	76.47			—20 520
Methyl Disulfide	64.64	64.72	69.58	73.81	77.56	81.02	84.25	87.27	90.13			— 3 200
Thiacyclobutane	56.85	56.92	60.44	63.64	66.67	69.55	72.31	74.97	77.52			3 910
2-Methyl-2-propanthiol	63.41	63.52	69.08	74.19	79.04	83.62	88.00	92.16	96.14			—34 770
3-Methyl-2-thiabutane	67.66	67.76	73.49	78.71	83.58	88.15	92.49	96.63	100.58			—30 060
3-Thiapentane	69.62	69.74	75.48	80.65	85.43	89.94	94.20	98.24	102.1			—28 600
3,4-Dithiahexane	75.90	76.04	83.19	89.55	95.35	100.7	105.7	110.5	115.0			—13 600
Thiacyclopentane	60.73	60.81	65.04	69.02	72.84	76.52	80.07	83.49	86.80			—17 300
Thiacyclohexane	62.46	62.56	67.37	72.03	76.63	81.16	85.61	89.95	94.20			—22 350
Thiophene	55.98	56.05	59.45	62.62	65.64		71.27		76.42			15 080
3-Methylthiophene	62.93	63.01	67.44	71.55	75.46	79.19	82.75	86.17	89.43			8 810
2-Methylthiophene	62.49	62.58	67.07	71.23	75.17	78.93	82.51	85.96	89.25			8 930
Benzenethiol	65.60	65.70	70.49	75.00	79.30	83.43	87.39	91.17	94.82			15 440
Cyanogen	47.71	47.77	50.92	53.47	55.71	57.70	59.47	61.15	62.66			73 380
Hydrogen Cyanide	40.811	40.856	43.046	44.832	46.356	47.697	48.898	49.986	50.993			31 300
Acetonitrile	48.39	48.45	51.36	53.83	56.04	58.05	59.92	61.68	63.30			22 700
Acrylonitrile	54.68	54.75	58.07	60.97	63.61	66.05	68.31	70.46	72.49			45 650
Pyridine	56.57	56.65	60.39	63.95	67.36	70.69	73.85	76.92	79.87			37 530

Appendix 8

C_p° VALUES OF THE MOST FREQUENTLY OCCURING
COMPOUNDS IN THE 298.15 TO 1200 K RANGE,
IN CAL K^{-1} MOLE^{-1}

T (K)	298.15	300	400	500	600	700	800	900	1000	1100	1200	$(\Delta H_f^{\circ})_{298,15}$
Constituent												
Hydrogen	6.892	6.895	6.974	6.993	7.008	7.035	7.078	7.139	7.217	7.308	7.404	0
Nitrogen	6.960	6.961	6.991	7.070	7.197	7.351	7.512	7.671	7.816	7.947	8.063	0
Oxygen	7.017	7.019	7.194	7.429	7.670	7.885	8.064	8.212	8.335	8.440	8.530	0
Hydroxyl	7.141	7.139	7.074	7.048	7.053	7.087	7.150	7.234	7.333	7.440	7.551	10 060
Water	8.025	8.026	8.185	8.415	8.677	8.959	9.254	9.559	9.869	10.172	10.467	−57 797.9
Carbon Monoxide	6.965	6.965	7.013	7.120	7.276	7.451	7.624	7.787	7.932	8.058	8.167	−26 415.7
Carbon Dioxide	8.874	8.894	9.871	10.662	11.311	11.850	12.300	12.678	12.995	13.26	13.49	−94 051.8
Methane	8.536	8.552	9.736	11.133	12.546	13.88	15.10	16.21	17.21	18.09	18.88	−17 889
Ethane	12.585	12.648	15.68	18.66	21.35	23.72	25.83	27.69	29.33	30.77	32.02	−20 236
Propane	17.57	17.66	22.54	27.04	30.88	34.20	37.08	39.61	41.83	43.75	45.42	−24 820
Butane	23.61	23.77	29.80	35.54	40.42	44.61	48.23	51.42	54.20	56.60	58.72	−29 812
2-Methylpropane	23.14	23.25	29.77	35.62	40.62	44.85	48.49	51.65	54.40	56.81	58.89	−31 452
Pentane	29.30	29.51	36.91	43.96	49.88	54.98	59.37	63.21	66.57	69.46	72.01	−35 000
2-Methylbutane	28.83	28.97	37.05	44.23	50.28	55.41	59.80	63.60	66.90	69.80	72.29	−36 920
2,2-Dimethylpropane	29.07	29.21	37.55	45.00	51.21	56.40	60.78	64.55	67.80	70.62	73.04	−39 670
Hexane	35.06	35.32	44.04	52.39	59.38	65.35	70.51	75.01	78.04	78.94	82.32	−39 960
2-Methylpentane	34.46	34.63	44.0	52.5	59.6	65.7	70.8	75.3	79.2			−41 660
3-Methylpentane	35.14	35.31	44.6	52.9	59.9	65.9	71.0	75.4	79.3			−41 020
2,2-Dimethylbutane	34.25	34.43	44.2	53.0	60.4	66.5	71.7	75.9	79.9			−44 350
2,3-Dimethylbutane	34.64	34.84	44.3	52.8	60.0	65.9	71.1	75.5	79.4			−42 490
Heptane	40.82	41.13	51.18	60.83	68.83	75.71	81.65	86.80	91.31	95.18	98.59	−44 890
2-Methylhexane	40.82	41.13	51.18	60.83	68.83	75.71	81.65	86.80	91.31			−46 600
3-Methylhexane	40.82	41.13	51.18	60.83	68.83	75.71	81.65	86.80	91.31			−45 960
3-Ethylpentane	40.82	41.13	51.18	60.83	68.83	75.71	81.65	86.80	91.31			−45 340
2,2-Dimethylpentane	40.82	41.13	51.18	60.83	68.83	75.71	81.65	86.80	91.31			−49 290
2,3-Dimethylpentane	40.82	41.13	51.18	60.83	68.83	75.71	81.65	86.80	91.31			−47 620
2,4-Dimethylpentane	40.82	41.13	51.18	60.83	68.83	75.71	81.65	86.80	91.31			−43 300

T (K)	298.15	300	400	500	600	700	800	900	1000	1100	1200	$(\Delta H_f^o)_{298,15}$
Constituent												
3,3-Dimethylpentane	40.82	41.13	51.18	60.83	68.83	75.71	81.65	86.80	91.31			— 48 170
2,2,3-Trimethylbutane	40.82	41.13	51.18	60.83	68.83	75.71	81.65	86.80	91.31			— 48 960
Octane	46.58	46.96	58.31	69.27	78.28	86.08	92.79	98.59	103.86	108.04	111.88	— 49 820
2-Methylheptane	46.58	46.94	58.31	69.27	78.28	86.08	92.79	98.59	103.68			— 51 500
3-Methylheptane	46.58	46.94	58.31	69.27	78.28	86.08	92.79	98.59	103.68			— 50 820
4-Methylheptane	46.58	46.94	58.31	69.27	78.28	86.08	92.79	98.59	103.68			— 50 690
3-Ethylhexane	46.58	46.94	58.31	69.27	78.28	86.08	92.79	98.59	103.68			— 50 400
2,2-Dimethylhexane	46.58	46.94	58.31	69.27	78.28	86.08	92.79	98.59	103.68			— 53 710
2,3-Dimethylhexane	46.58	46.94	58.31	69.27	78.28	86.08	92.79	98.59	103.68			— 51 130
2,4-Dimethylhexane	46.58	46.94	58.31	69.27	78.28	86.08	92.79	98.59	103.68			— 52 440
2,5-Dimethylhexane	46.58	46.94	58.31	69.27	78.28	86.08	92.79	98.59	103.68			— 53 210
3,3-Dimethylhexane	46.58	46.94	58.31	69.27	78.28	86.08	92.79	98.59	103.68			— 52 610
3,4-Dimethylhexane	46.58	46.94	58.31	69.27	78.28	86.08	92.79	98.59	103.68			— 50 910
2-Methyl-3-ethylpentane	46.58	46.94	58.31	69.27	78.28	86.08	92.79	98.59	103.68			— 50 480
3-Methyl-3-ethylpentane	46.58	46.94	58.31	69.27	78.28	86.08	92.79	98.59	103.68			— 51 380
2,2,3-Trimethylpentane	46.58	46.94	58.31	69.27	78.28	86.08	92.79	98.59	103.68			— 52 610
2,2,4-Trimethylpentane	46.58	46.94	58.31	69.27	78.28	86.08	92.79	98.59	103.68			— 53 570
2,3,3-Trimethylpentane	46.58	46.94	58.31	69.27	78.28	86.08	92.79	98.59	103.68			— 51 730
2,3,4-Trimethylpentane	46.58	46.94	58.31	69.27	78.28	86.08	92.79	98.59	103.68			— 51 970
2,2,3,3-Tetramethyl- butane	46.58	46.94	58.31	69.27	78.28	86.08	92.79	98.59	103.68			— 53 990
Ethylene	10.41	10.45	12.90	15.16	17.10	18.76	20.20	21.46	22.57	23.54	24.39	12 496
Propene	15.27	15.34	19.10	22.62	25.70	28.37	30.68	32.70	34.46	35.99	37.32	4 879
1-Butene	21.35	21.45	26.94	31.75	35.82	39.31	42.33	44.95	47.24	49.23	50.96	280
2-Butene, cis	18.86	18.96	24.33	29.39	33.80	37.60	40.87	43.70	46.15	48.28	50.13	— 1 362
2-Butene, trans	20.99	21.08	26.02	30.68	34.80	38.38	41.50	44.20	46.58	48.65	50.44	— 2 405
2-Methylpropene	21.30	21.39	26.57	31.24	35.30	38.81	41.86	44.53	46.85	48.88	50.63	— 3 343

T (K)	298.15	300	400	500	600	700	800	900	1000	1100	1200	$(\Delta H_f^o)_{298,15}$
Constituent												
1-Pentene	27.39	27.56	34.20	40.25	45.36	49.72	53.48	56.76	59.61	62.08	64.26	— 5 000
2-Pentene, cis	24.32	24.45	31.57	38.05	43.62	48.25	52.29	55.76	58.78	61.38	63.66	— 6 710
2-Pentene, trans	26.80	26.92	33.57	39.57	44.70	49.14	52.98	56.31	59.23	61.75	63.96	— 7 590
2-Methyl-1-butene	26.69	26.82	33.71	39.81	44.97	49.40	53.23	56.54	59.44	61.93	64.13	— 8 680
3-Methyl-1-butene	28.35	28.47	35.26	40.97	45.90	50.15	53.85	57.03	59.83	62.28	64.42	— 6 920
2-Methyl-2-butene	25.49	25.62	32.22	38.33	43.64	48.23	52.22	55.67	58.68	61.28	63.56	—10 170
1-Hexene	33.08	33.30	41.3	48.7	54.8	60.1	64.6	68.6	72.0	74.9	77.6	— 9 960
2-Hexene, cis	30.36	30.56	38.8	46.6	53.2	58.7	63.4	67.6	71.2			—11 560
2-Hexene, trans	32.84	33.03	40.8	48.1	54.2	59.6	64.1	68.1	71.6			—12 560
3-Hexene, cis	29.55	29.71	38.5	46.4	53.2	58.7	63.5	67.6	71.2			—11 560
3-Hexene, trans	32.63	32.78	41.1	48.4	54.6	59.8	64.4	68.3	71.8			—12 560
2-Methyl-1-pentene	32.73	32.93	41.0	48.3	54.5	59.8	64.4	68.4	71.8			—13 560
3-Methyl-1-pentene	34.04	34.19	42.5	49.6	55.6	60.7	65.2	69.0	72.3			—11 020
4-Methyl-1-pentene	32.61	32.76	41.5	48.9	55.2	60.5	65.1	68.9	72.3			—11 660
2-Methyl-2-pentene	30.26	30.42	39.0	46.6	53.2	58.6	63.4	67.5	71.1			—14 960
3-Methyl-2-pentene, cis	30.26	30.42	39.0	46.6	53.2	58.6	63.4	67.5	71.1			—14 320
3-Methyl-2-pentene, trans	30.26	30.42	39.0	46.6	53.2	58.6	63.4	67.5	71.1			—14 320
4-Methyl-2-penetne, cis	31.92	32.07	40.5	47.8	54.1	59.4	64.0	68.0	71.5			—13 260
4-Methyl-2-pentene, trans	33.80	33.94	41.9	48.8	54.8	60.0	64.5	68.4	71.8			—14 260
2-Ethyl-1-butene	31.92	32.08	40.7	48.2	54.5	59.8	64.4	68.4	71.9			—12 920
2,3-Dimethyl-1-butene	34.29	34.44	42.6	49.5	55.4	60.5	65.0	68.8	72.2			—14 780
3,3-Dimethyl-1-butene	31.72	31.87	40.6	48.4	55.0	60.6	65.3	69.2	72.7			—14 250
2,3-Dimethyl-2-butene	30.48	30.63	38.6	46.1	52.5	58.0	62.9	67.0	70.7			—14 890
1-Heptene	38.84	39.11	48.4	57.1	64.3	70.5	75.8	80.4	84.4	87.8	90.8	—14 890
Propadiene	14.10	14.16	17.21	19.82	22.00	23.84	25.42	26.80	28.00	29.04	29.96	45 920
1,2-Butadiene	19.15	19.23	23.54	27.39	30.72	33.54	36.01	38.16	40.02	41.62	43.02	39 550
1,3-Butadiene	19.01	19.11	24.29	28.52	31.84	34.55	36.84	38.81	40.52	42.02	43.32	26 750

T (K)	298.15	300	400	500	600	700	800	900	1000	1100	1200	$(\Delta H^{\circ}_{f})_{298.15}$
Constituent												
1,2-Pentadiene	25.2	25.3	31.4	36.5	40.8	44.5	47.7	50.4	52.8	54.9	56.7	34 800
1,3-Pentadiene, cis	22.6	22.7	29.5	35.3	39.9	43.8	47.0	49.8	52.2	54.3	56.1	18 700
1,3-Pentadiene, trans	24.7	24.9	31.2	38.6	40.9	44.6	47.7	50.3	52.6	54.7	56.4	18 600
1,4-Pentadiene	25.1	25.2	31.3	36.5	40.8	44.4	47.6	50.3	52.7	54.7	56.5	25 200
2,3-Pentadiene	24.2	24.3	29.9	35.0	39.4	43.2	46.6	49.5	52.0	54.2	56.1	33 100
2-Methyl-1,3-butadiene	25.0	25.2	31.8	37.1	41.4	45.0	48.0	50.6	52.9	54.9	56.6	18 100
2-Methyl-2,3-butadiene	25.2	25.3	31.0	36.0	40.3	44.0	47.2	50.0	52.4	54.5	56.3	31 000
Acetylene	10.499	10.532	11.973	12.967	13.728	14.366	14.933	15.449	15.922	16.353	16.744	54 194
Propyne	14.50	14.55	17.33	19.74	21.80	23.58	25.14	26.51	27.71	28.77	29.69	44 319
1-Butyne	19.46	19.54	23.87	27.63	30.83	33.57	35.95	38.02	39.84	41.42	42.80	39 700
2-Butyne	18.63	18.70	22.62	26.36	29.68	32.59	35.14	37.36	39.29	40.98	42.44	35 374
1-Pentyne	25.50	25.65	31.1	36.1	40.4	44.0	47.1	49.8	52.2	54.3	56.1	34 500
2-Pentyne	23.59	23.69	29.2	34.3	38.7	42.6	45.9	48.9	51.4	53.6	55.6	30 800
2-Methyl-3-butyne	25.02	25.13	31.1	36.2	40.6	44.2	47.4	50.1	52.4	54.5	56.3	32 600
1-Hexyne	31.19	31.39	38.2	44.6	49.8	54.4	58.2	61.6	64.6	67.1	69.4	29 550
Vinylacetylene	17.49	17.57	21.26	24.25	26.67	28.68	30.40	31.87	33.16			72 800
Cyclopropane	13.35	13.44	18.33	22.64	26.14	29.04	31.45	33.57	35.37			12 740
Cyclobutane	17.26	17.37	23.89	29.86	34.76	38.89	42.42	45.41	47.96			17 980
Cyclopentane	19.82	19.98	28.24	35.86	42.36	47.81	52.44	56.37	59.75	62.68	65.18	$-18\ 460$
Methylcyclopentane	26.24	26.46	36.11	44.94	52.43	58.68	64.00	68.53	72.44	75.82	78.72	$-25\ 500$
1,1-Dimethylcyclopentane	31.86	32.16	43.55	54.01	62.78	70.08	76.18	81.38	85.83	89.62	92.89	$-33\ 050$
1,2-Dimethylcyclo-pentane, cis	32.06	32.34	43.67	54.03	62.72	69.92	75.98	81.14	85.57	89.38	92.65	$-30\ 960$
1,2-Dimethylcyclo-pentane, trans	32.14	32.44	43.71	54.03	62.66	69.88	75.84	80.98	85.43	89.24	92.51	$-32\ 670$
1,3-Dimethylcyclo-pentane, cis	32.14	32.44	43.71	54.03	62.66	69.88	75.84	80.98	85.43	89.24	92.51	$-31\ 930$

T (K)	298.15	300	400	500	600	700	800	900	1000	1100	1200	$(\Delta H_f^o)_{298.15}$
Constituent												
1,3-Dimethylcyclo-pentane, trans	32.14	32.44	43.71	54.03	62.66	69.88	75.84	80.98	85.43	89.24	92.511	— 31 470
Ethylcyclopentane	31.93	32.18	43.39	53.55	62.09	69.24	75.31	80.48	84.94	88.81	92.12	— 30 370
Cyclohexane	25.40	25.58	35.82	45.47	53.83	60.87	66.76	71.68	75.80	79.3	82.2	— 29 430
Methylcyclohexane	32.27	32.51	44.35	55.21	64.46	72.23	78.74	84.20	88.79	92.7	96.0	— 36 990
Ethylcyclohexane	37.9	38.2	51.6	63.8	74.1	82.8	90.1	96.2	101.3	105.7	109.4	— 41 050
1,1-Dimethylcyclohexane	36.9	37.2	50.7	63.3	74.1	83.2	90.7	97.0	102.2	106.6	110.3	— 43 260
1,2-Dimethylcyclo-hexane, cis	37.4	37.7	51.1	63.5	74.0	82.8	90.1	96.3	101.4	105.8	109.5	— 41 150
1,2-Dimethylcyclo-hexane, trans	38.0	38.3	51.9	62.2	74.6	83.3	90.5	96.6	101.7	106.0	109.7	— 43 020
1,3-Dimethylcyclo-hexane, cis	37.6	37.9	51.2	63.6	74.2	83.1	90.5	96.7	102.0	106.4	110.1	— 44 160
1,3-Dimethylcyclo-hexane, trans	37.6	37.9	51.1	63.4	73.8	82.5	89.8	95.9	101.1	105.5	109.2	— 42 200
1,4-Dimethylcyclo-hexane, cis	37.6	37.9	51.1	63.4	73.8	82.5	89.8	95.9	101.1	105.5	109.2	— 42 200
1,4-Dimethylcyclo-hexane, trans	37.7	38.0	51.6	64.0	74.6	83.3	90.6	96.8	101.9	106.2	109.9	— 44 120
n-Propylcyclohexane	43.59	43.89	58.6	72.1	83.5	93.1	101.1	107.9	113.6	118.6	122.7	— 46 200
n-Butylcyclohexane	49.35	49.70	65.7	80.5	93.0	103.5	112.2	119.7	126.0	131.4	136.0	— 50 950
Cyclopentene	17.95	18.8	25.08	31.62	37.19	41.86	45.78	49.11	51.94	54.37	56.45	1 160
Cyclohexene	25.10	25.28	34.64	42.78	49.45	54.92	59.49	63.34	66.62	69.43	71.85	— 9 700
Benzene	19.52	19.65	26.74	32.80	37.74	41.75	45.06	47.83	50.16	52.16	53.86	19 820
Toluene	24.80	24.95	33.25	40.54	46.58	51.57	55.72	59.22	62.19	64.73	66.90	11 820
Ethylbenzene	30.49	30.88	40.76	49.35	56.44	62.28	67.15	71.27	74.77	77.77	80.35	7 120
o-Xylene	31.85	32.02	41.03	49.11	55.98	61.76	66.64	70.80	74.35	77.40	80.02	4 540

T (K)	298.15	300	400	500	600	700	800	900	1000	1100	1200	$(\Delta H_f^o)_{298.15}$
Constituent												
m-Xylene	30.49	30.66	40.03	48.43	55.51	61.43	66.41	70.63	74.23	77.31	79.95	4 120
p-Xylene	30.32	30.49	39.70	48.06	55.16	61.12	66.14	70.39	74.02	77.13	79.80	4 290
n-Propylbenzene	36.73	36.99	48.0	57.8	66.0	72.7	78.3	83.1	87.1	90.6	93.6	1 870
Isopropylbenzene	36.26	36.47	48.0	57.9	66.2	72.9	78.6	83.3	87.3	90.8	93.8	940
1,2,3-Trimethylbenzene	36.85	37.04	36.9	56.1	64.0	70.9	76.7	81.6	85.9	89.5	92.7	— 2 290
1,2,4-Trimethylbenzene	37.10	37.28	47.1	56.2	64.2	71.0	76.8	81.7	86.0	89.6	92.8	— 3 330
1,3,5-Trimethylbenzene	35.91	36.10	46.41	55.92	64.08	70.99	76.84	81.81	86.07	89.72	92.86	— 3 840
1-Methyl-2-ethylbenzene	37.74	37.94	48.5	57.9	65.8	72.5	78.1	82.8	86.9	90.4	93.5	290
1-Methyl-3-ethylbenzene	36.38	36.59	47.5	57.2	65.4	72.1	77.8	82.7	86.8	90.4	93.4	— 460
1-Methyl-4-ethylbenzene	36.22	36.42	47.2	56.9	65.0	71.8	77.6	82.4	86.6	90.2	93.2	— 780
1,2,3,4-Tetramethyl-benzene	45.31	45.50	56.81	67.01	75.68	83.13	89.42	94.82	99.47			—10 020
1,2,4,5-Tetramethyl-benzene	44.58	44.77	55.50	65.62	74.38	81.97	88.41	93.94	98.71			—10 820
1,2,3,5-Tetramethyl-benzene	44.39	44.57	55.76	66.03	74.81	82.39	88.79	94.27	99.01			—10 710
Pentamethylbenzene	51.74	51.99	65.00	76.43	86.08	94.27	101.29	107.20	112.33			—17 800
n-Butylbenzene	42.42	42.73	55.1	66,3	75 4	83.1	89.4	94.9	99.5	103.5	106.9	— 3 300
Styrene	29.18	29.35	38.32	45.94	52.14	57.21	61.40	64.93	67.92	70.48	72.66	35 220
1-Phenyl-1-propene, cis	34.7	34.9	44.8	53.5	60.7	66.8	71.8	76.1	79.8	82.9	85.6	29 000
1-Phenyl-1-propene, trans	34.9	35.1	45.2	54.0	61.2	67.2	72.2	76.4	80.0	83.1	85.8	28 000
2-Phenyl-1-propene	34.7	34.9	44.8	53.5	60.7	66.8	71.8	76.1	79.8	82.9	85.6	27 000
1-Methyl-2-ethylbenzene	34.7	34.9	44.8	53.5	60.7	66.8	71.8	76.1	79,8	82.9	85.6	28 300
1-Methyl-3-ethylbenzene	34.7	34.9	44.8	53.5	60.7	66.8	71.8	76.1	79.8	82.9	85.6	27 600
1-Methyl-4-ethylbenzene	54.7	34.9	44.8	53.5	60.7	66.8	71.8	76.1	79.8	82.9	85.6	27 400
Cyclooktatetraene	29.16	29.34	38.45	46.38	52.77	57.90	62.23	65.75	68.88			71 120
1,2,3,4-Tetrahydro-naphthalene	37.41	37.66	50.55	61.81	71.45	79.50	85.85	90.62	93.76			12 330

T (K)	298.15	300	400	500	600	700	800	900	1000	1100	1200	$(\Delta H_{\mathrm{f}}^{\mathrm{o}})_{298.15}$
Constituent												
Naphthalene	31.03	31.25	42.81	52.46	60.16	66.31	71.32	75.43	78.87	81.76	84.22	34 590
Methanol	10.8	10.8	12.7	14.5	16.3	17.8	19.2	20.4	21.6			− 48 490
Ethanol	17.59	17.66	21.00	24.09	26.81	29.18	31.25	33.07	34.66	36.06	37.28	− 56 240
1-Propanol	20.82	20.91	25.86	30.51	34.56	38.03	41.04	43.65	45.93			− 61 550
1-Butanol	26.29	26.41	32.80	38.76	43.90	48.31	52.11	55.40	58.26			− 65 590
Ethylene Glycol	23.20	23.28	27.06	30.10	32.72	35.00	36.90	38.00	39.88			− 93 050
Dimethylether	15.77	15.83	19.17	22.48	25.41	27.83	30.24	32.25	33.95			− 44 300
Furan	15.64	15.75	21.20	25.73	29.31	32.13	34.41	36.30	37.89	39.23	40.37	− 8 293
Formaldehyde	8.45	8.47	9.37	10.44	11.50	12.48	13.36	14.13	14.80	15.38	15.88	− 28 620
Acetaldehyde	13.06	13.11	15.73	18.27	20.52	22.50	24.20	25.68	26.96			− 39 670
Acetone	17.90	17.97	22.00	25.89	29.34	32.34	34.93	37.19	39.15	40.85	42.32	− 51 720
Formic Acid	11.636	11.674	13.613	15.274	16.648	17.785	18.738	19.546	20.238	20.834	21.351	− 90 031
Acetic Acid (monomer)	15.90	15.97	19.52	25.15	29.08	31.99	34.14					− 104 530
Acetic Acid (dimer)	34.99	35.13	42.32		53.89		62.45		68.06		72.47	− 224 260
Methanethiol	12.12	12.15	14.11	15.96	17.62	19.08	20.38	21.53	22.54	23.43	24.21	− 18 380
Ethanethiol	17.37	17.44	21.08	24.36	27.21	29.68	31.83	33.71	35.38			− 26 450
Thiacyclopropane	12.83	12.90	16.53	19.56	21.99	23.96	25.61	27.03	28.21			3 930
1-Propanthiol	22.65	22.75	27.86	32.56	36.72	40.37	43.60	46.47	49.01			− 31 460
2-Propanthiol	22.94	23.04	28.35	33.06	37.02	40.38	43.26	45.74	47.92			− 33 460
Dimethylsulfide	17.71	17.77	21.12	24.24	27.01	29.44	31.59	33.50	35.17			− 24 400
Dimethyldisulfide	21.97	22.04	25.81	29.23	32.20	34.75	36.96	38.86	40.53			− 7 200
Thiacyklobutane	16.57	16.66	21.89	26.56	30.45	33.69	36.40	38.70	40.67			− 620
2-Methyl-2-propanthiol	28.91	29.04	36.13	42.39	47.60	51.92	55.53	58.61	61.24			− 41 790
3-Methyl-2-thiabutane	28.70	28.82	35.41	41.36	46.50	50.97	54.89	58.32	61.35			− 36 380
3-Thiapentane	27.95	28.07	34.51	40.33	45.31	49.57	53.20	56.39	59.17			− 35 344
3,4-Dithiahexane	34.24	34.37	40.97	46.74	51.66	55.89	59.54	62.73	65.52			− 20 000
Thiacyklopentane	21.72	21.85	28.95	35.04	40.05	44.20	47.66	50.60	53.14			− 23 500

T (K)	298.15	300	400	500	600	700	800	900	1000	1100	1200	$(\Delta H_f^o)_{298.15}$
Constituent												
Thiacyklohexane	25.86	26.03	35.71	44.73	52.37	58.71	64.00	68.49	72.34			$-30\,380$
Thiofen	17.32	17.43	22.81	27.10	30.46		35.27		38.56			12 130
3-Methylthiofen	22.67	22.80	29.38	34.89	39.34	42.95	45.95	48.47	50.59			4 540
2-Methylthiofen	22.80	22.92	29.43	35.01	39.57	43.32	46.43	49.06	51.30			4 740
Benzenethiol	25.07	25.22	32.76	39.07	44.13	48.23	51.59	54.41	56.79			11 240
Cyanogen	13.60	13.63	14.80	15.64	16.32	16.91	17.43	17.87	18.25			73 840
Hydrogen Cyanide	8.58	8.59	9.38	9.98	10.47	10.91	11.30	11.65	11.97			31 200
Acetonitrile	12.40	12.44	14.59	16.61	18.36	19.94	21.31	22.52	23.53			21 000
Acrylonitrile	15.24	15.30	18.36	20.95	23.11	24.90	26.43	27.75	28.88			44 040
Pyridine	19.72	19.85	26.30	31.68	35.99	39.50	42.35	44.74	46.76			33 630

Appendix 9

CRITICAL DATA OF SELECTED ORGANIC SUBSTANCES

This table lists values of critical constants of substances, the thermochemical data of which are given in Appendices 7 and 8. Data taken from: Kudchadker A. P., Alani G. H., Zwolinski B. J.: Chem. Rev. *68*, 659 (1968).

Constituent	T_c (K)	P_c (atm)	v_c (l mol^{-1})
Hydrogen	39.25	16.50	0.0603
Nitrogen	126.26	33.98	0.09008
Oxygen	154.74	50.14	0.074
Water	647.30	218.3	0.0572
Carbon Monoxide	132.92	34.53	0.0931
Carbon Dioxide	304.19	72.85	0.0940
Methane	190.55	45.44	0.099
Ethane	305.43	48.16	0.148
Propane	369.82	41.94	0.203
Butane	425.16	37.47	0.255
2-Methylpropane	408.13	36.00	0.263
Pentane	469.6	33.25	0.304
2-Methylbutane	460.39	33.37	0.306
2,2-Dimethylpropane	433.75	31.57	0.303
Hexane	507.4	29.3	0.370
2-Methylpentane	497.45	29.71	0.367
3-Methylpentane	504.4	30.83	0.367
2,2-Dimethylbutane	488.73	30.40	0.359
2.3-Dimethylbutane	499.93	30.86	0.358
Heptane	540.2	27.00	0.432
2-Methylhexane	530.31	26.98	0.421
3-Methylhexane	535.19	27.77	0.404
3-Ethylpentane	540.57	28.53	0.416
2,2-Dimethylpentane	520.44	27.37	0.416
2,3-Dimethylpentane	537.29	28.70	0.393

Constituent	T_c (K)	P_c (atm)	v_c (l mol^{-1})
2,4-Dimethylpentane	519.73	27.01	0.418
3,3-Dimethylpentane	536.34	29.07	0.414
2,2,3-Trimethylbutane	531.11	29.15	0.398
Octane	568.76	24.54	0.492
2-Methylheptane	595.57	24.52	0.488
3-Methylheptane	563.60	25.13	0.644
4-Methylheptane	561.67	25.09	0.476
3-Ethylhexane	565.42	25.74	0.455
2,2-Dimethylhexane	549.80	24.96	0.478
2,3-Dimethylhexane	563.42	25.94	0.468
2,4-Dimethylhexane	553.45	25.23	0.472
2,5-Dimethylhexane	549.99	24.54	0.482
3,3-Dimethylhexane	561.95	26.19	0.443
3,4-Dimethylhexane	568.78	26.57	0.466
2-Methyl-3-ethylpentane	567.02	26.65	0.443
3-Methyl-3-ethylpentane	576.51	27.71	0.455
2,2,3-Trimethylpentane	563.43	26.94	0.436
2,2,4-Trimethylpentane	543.89	25.34	0.468
2,3,3-Trimethylpentane	573.49	27.83	0.455
2,3,4-Trimethylpentane	566.34	26.94	0.461
2,2,3,3-Tetramethylbutane	567.8	28.3	0.461
Ethylene	282.36	49.66	0.129
Propene	365.0	45.6	0.181
1-Butene	419.6	39.7	0.240
2-Butene, cis	435.55	41.5	0.234
2-Butene, trans	428.61	40.5	0.238
2-Methylpropene	417.89	39.48	0.239
1-Pentene	464.74	40	
2-Pentene, cis	476.0	36	
2-Pentene, trans	475.0	36	
2-Methyl-1-butene	465.0	34	
2-Methyl-2-butene	470	34	
1-Hexene	503.98		
1-Heptene	537.23		
Acetylene	308.33	60.59	0.113
Propyne	402.38	55.54	0.164
1-Butyne	463.6		
2-Butyne	488.6		
1-Pentyne	493.4		
Propadiene	393		
1,3-Butadiene	425	42.7	0.221
Cyclopropane	397.80	54.23	
Cyclopentane	511.6	44.49	0.26
Methylcyclopentane	532.73	37.35	0.319

Constituent	T_c (K)	P_c (atm)	v_c (l mol^{-1})
Ethylcyclopentane	569.45	33.53	0.375
Cyclohexane	553.4	40.2	0.308
Methylcyclohexane	572.12	34.26	0.368
Cyclopentene	506.0		
Cyclohexene	560.41		
Benzene	562.09	48.34	0.259
Toluene	591.72	40.55	0.316
Ethylbenzene	617.09	35.62	0.374
1,2-Dimethylbenzene	630.2	36.84	0.369
1,3-Dimethylbenzene	616.97	34.95	0.376
1,4-Dimethylbenzene	616.2	34.65	0.379
n-Propylbenzene	638.30	31.58	0.440
Isopropylbenzene	631.0	31.67	
1,2,3-Trimethylbenzene	664.45	34.09	
1,2,4-Trimethylbenzene	649.05	31.90	
1,3,5-Trimethylbenzene	637.28	30.86	
n-Butylbenzene	660.4	28.49	0.497
Naphtalene	748.4	39.98	0.41
Methanol	512.58	79.9	0.118
Ethanol	516.2	63.0	0.167
1-Propanol	536.71	51.02	0.218
1-Butanol	562.93	43.55	0.274
Dimethylether	400.00	53.0	0.190
Furan	490.2	54.3	0.218
Acetaldehyde	461		
Acetone	508.2	46.4	0.209
Acetic Acid	594.45	57.1	0.171
Methanthiol	470.0	71.4	0.145
Ethanthiol	499.0	54.2	0.207
Dimethylsulfide	503.0	54.6	0.309
Diethylsulfide	557.0	39.1	0.318
Thiophen	579.4	56.2	0.219
Acetonitril	548	47.7	0.173
Pyridine	620.0	55.6	0.254

Appendix 10

VALUES OF CONSTANTS OF THE BEATTIE-BRIDGMAN EQUATION

Constituent	A_0	B_0	a	b	$10^{-4} \times c$
Helium	0.0216	0.01400	0.05984	0.0	0.0040
Neon	0.2125	0.02060	0.02196	0.0	0.101
Argon	1.2907	0.03913	0.02328	0.0	5.99
Hydrogen	0.1975	0.02096	−0.00506	−0.04359	0.0504
Nitrogen	1.3445	0.05046	+0.02617	−0.00691	4.20
Oxygen	1.4911	0.04624	0.02562	+0.004208	4.80
Air	1.3012	0.04611	0.01931	−0.01101	4.34
Carbon dioxide	5.0065	0.10476	0.07132	+0.07235	66.00
Ammonia	2.3930	0.03415	0.17031	+0.19112	476.87
Methane	2.2769	0.05587	0.01855	−0.01587	12.83
Ethane	5.8800	0.09400	0.05861	+0.01915	90.00
Ethylene	6.1520	0.12156	0.04964	+0.03597	22.68
Propane	11.9200	0.18100	0.07321	+0.04293	120.00
Butane	17.7940	0.24620	0.12161	+0.09423	350.00
2-Methylpropane	16.6037	0.23540	0.11171	+0.07697	300.00
1-Butene	16.6979	0.24046	0.11988	+0.10690	300.00
2-Methylpropene	16.9600	0.24200	0.10860	+0.08750	250.00
Pentane	28.2600	0.39400	0.15099	+0.13960	400.00
2,2-Dimethylpropane	23.3300	0.33560	0.15174	+0.13358	400.00
Heptane	54.520	0.70816	0.20066	+0.19179	400.00
Methanol	33.309	0.60362	0.09246	+0.09929	32.03
Diethylether	31.278	0.45446	0.12426	+0.11954	33.33

Appendix 11

VALUES OF CONSTANTS OF THE BENEDICT-WEBB--RUBIN EQUATION

P atm, volume l/mole, temperature in K, $R = 0.08205$

From: Novák J. P., Malijevský A., Matouš L., Šobr J.: Plyny a plynné směsi, Stavové chování a termodynamické vlastnosti (Gases and gas mixtures, Behaviour of state and thermodynamic properties), published by the Publishing House of the Czechoslovak Academy of Sciences, Prague 1972, in Czech.

References

1. Anderson J. W., Beyer G. H., Watson K. M.: Natl. Petroleum News Tech. Sec., *36*, R 476 (1944); cit. Hougen O. A., Watson K. M.: Chemical Process Principles, str. 758, Wiley, New York 1958.
2. Anthony R. G., Himmelblau D. M.: J. Phys. Chem. *67*, 1083 (1963).
3. Aris R., Mah R. H. S.: Ind. Eng. Chem. Fundamentals *2* (2), 90 (1963).
4. Aston J. G.: Chem. Rev. *27*, 63 (1940).
5. Azbel I. J., Kilštedt K. K., Makajev T. A.: Proizvodstvo ammiaka (Ammonia Production) Goschimizdat, Moskva 1954.
6. Barnhard P., Hawkins A. W.: Proc. 2nd Conf. High Temp. Systems (1963).
7. Barrar R. B., Loeb H. L.: Numer. Math. *15*, 382 (1970).
8. Beattie J. A., Bridgman O. C.: J. Am. Chem. Soc. *50*, 3133 (1928).
9. Beattie J. A.: Proc. Natl. Acad. Sci. U. S., *16*, 14 (1930).
10. Beattie J. A., Stockmayer W. H., Ingersoll H. G.: J. Chem. Phys. *9*, 871 (1941).
11. Beattie J. A.: Chem. Rev. *44*, 141 (1949).
12. Benedict M., Webb G. W., Rubin L. C.: J. Chem. Phys. *8*, 334 (1940).
13. Bennewitz K., Rossner W.: Z. Physik Chem. *38 B*, 126 (1938).
14. Bremner J. G. M., Thomas G. D.: Trans. Faraday Soc. *44*, 230 (1948).
15. Brinkley S. R. Jr.: J. Chem. Phys. *14*, 563 (1946).
16. Brinkley S. R. Jr.: J. Chem. Phys. *15*, 107 (1947).
17. Brinkley S. R. Jr.: Ind. Eng. Chem. *43*, 2471 (1951).
18. Brown F. W.: Ind. Eng. Chem. *33*, 1536 (1941).
19. Brown K. M.: Communications of the ACM *10* (11) 728 (1967).
20. Browne H. Jr., Williams M. M., Cruise D. R.: NAVWEPS Report 7043 (1960), NOTS-Tech. Publ. 2434.
21. Bruins P. F., Czarnecki J. D.: Ind. Eng. Chem. *33*, 201 (1941).
22. Canjar L. N., Smith R. F., Volianitis E., Galluzzo J. F., Carboreos M.: Ind. Eng. Chem. *47*, 1028 (1955).
23. Cappelli A.: Chim. ind. (Milan) *48*, (12) 1271 (1966).
24. Clasen R. J.: Rand Corp. Memo 4345-PR, AD-609 904 (1965).
25. Collatz L.: Funktionalanalysis und numerische mathematik chapter 2, Springer Verlag, Berlin—Göttingen—Heidelberg 1964.
26. Collected Works of Willard Gibbs, Vol. 1., Longmans Green 1931.
27. Corner J.: Trans. Faraday Soc.: *37*, 358 (1941).
28. Craine G. D., Thompson H. V.: Trans. Faraday Soc. *49*, 1273 (1953).
29. Crawford B. L., Parr R. G.: J. Chem. Phys. *16*, 233 (1948).
30. Cruise D. R.: J. Phys. Chem. *68*, 3797 (1964).
31. Černý Č.: Chem. listy *55*, 119 (1961).
32. Černý Č., Erdös E.: Chem. listy *47*, 1742 (1953).

33. Černý Č., Erdös E.: Collection Czechoslov. Chem. Commun. *31*, 1915 (1966).
34. Černý Č., Erdös E.: Collection Czechoslov. Chem. Commun. *31*, 1934 (1966).
35. Černý Č.: Collection Czechoslov. Chem. Commun. *31*, 4487 (1966).
36. Černý Č.: Collection Czechoslov. Chem. Commun. *32*, 219 (1967).
37. Damköhler G., Edse R.: Z. Elektrochem. *49*, 178 (1943).
38. Dantzig G. B., De Haven J. C.: J. Chem. Phys. *36*, 2620 (1962).
39. Dantzig G. B.: Linear programming and extensions, Princeton University Press, Princeton N. J., 1963 (Czech translation SVTL, Bratislava 1967, p. 546).
40. Darin S. R., Stuart E. B., Coull J.: Oil Gas J. *56*, (24) 105 (1958).
41. David H. G., Hamann S. D.: Proceedings of the Conference on Thermodynamics and Transport Properties of Fluids, p. 74, London 1957.
42. Demidovich B. P., Maron I. A.: Osnovy vychislitelnoi matematiky (Fundamentals of numerical mathematics) (Czech translation SNTL, Prague 1966, chapter 4 and 13) Fizmatgiz, Moscow 1960.
43. Demidovich B. P., Maron I. A.: Osnovy vychislitelnoi matematiky (Fundamentals of numerical mathematics) (Czech translation SNTL, Prague 1966, chapter 17) Fizmatgiz, Moscow 1960.
44. Demidovich B. P., Maron I. A.: Osnovy vychislitelnoi matematiky (Fundamentals of numerical mathematics) (Czech translation SNTL, Prague 1966, p. 125) Fizmatgiz, Moscow 1960.
45. Dyatkina M. E.: Zhur. fiz. khim. *28*, 377 (1954).
46. Dobratz C. J.: Ind. Eng. Chem. *33*, 759 (1941).
47. Dodge B. F.: Trans. Am. Inst. Chem. Engrs. *34*, 540 (1938).
48. Dorn W. S.: J. Chem. Phys. *32*, 1490 (1960).
49. Dorn W. S.: IBM Journal of Res. *4* (4) 407 (1960).
50. Dowel R. S. Mc., Kruse F. H.: J. Chem. Eng. Data *8* (4) 547 (1963).
51. Falkovskii V. B.: Zhur. obschei khim. *18*, 1639 (1948).
52. Ferigle S. L., Cleveland F. F., Meister A. G.: J. Chem. Phys. *20* (3) 526 (1952).
53. Ferigle S. L., Weber A.: J. Chem. Phys. *20* (10) 1657 (1952).
54. Fichtengolc G. M.: Kurs diferencionalnovo i integralnovo ischislenia, (A course of differential and integral calculation) I, p. 539, Gostěchizdat, Moscow 1951.
55. Franklin J. L.: Ind. Eng. Chem. *41*, 1070 (1949).
56. Fuchs W., Glaser F.: Chem. Ing. Tech. *28*, 689 (1956).
57. Gamson B. W., Watson K. M.: Natl. Petroleum News. Tech. Sec. *36*, R 623 (1944).
58. Goldwasser S. R.: Ind. Eng. Chem. *51*, 595 (1959).
59. Gordon J. S.: J. Chem. Eng. Data *8* (3) 294 (1963).
60. Gordon S., Zeleznik F. J., Huff V. N.: NASA Tech. Note D-132, 1959.
61. Griskey R. G., Beyer H. H.: AICHE J. *9* (4) 507 (1963).
62. Groot S. E. de: Thermodynamics of Irreversible Processes, North Holland Publ. Co., Amsterdam 1961.
63. Hadley G.: Nonlinear and Dynamic Programming, Addison-Wesley, Reading 1964.
64. Hála E., Reiser A.: Fysikální chemie 1 (Physical Chemistry 1) p. 137, Academia, Prague 1960.
65. Hála E., Reiser A.: Fysikální chemie 1 (Physical Chemistry 1) p. 328, Academie, Prague 1960.
66. Hála E., Boublík T.: Úvod do statistické termodynamiky (Introduction to Statistical Thermodynamics) p. 62, Academia, Prague 1969.
67. Hirschfelder J. O., Roseveare W. E.: J. Phys. Chem. *43*, 15 (1939).
68. Hirschfelder J. O., Curtiss Ch. F., Bird B. R.: Molecular Theory of Gases and Liquids, p. 144, John Wiley & Sons, New York 1954.

69. Hirschfelder J. O., Curtiss Ch. F., Bird B. R.: Molecular Theory of Gases and Liquids, p. 198, John Wiley & Sons, New York 1954.

70. Holub R.: Chem. listy *62*, 87 (1968).

71. Holub R.: Habilitation Thesis, Institute of Chemical Technology, Prague 1968.

72. Holub R.: Chem. listy *64*, 639 (1970).

73. Hooke R., Jeeves T. A.: J. Assn. Comp. Mach. *8*, 212 (1961).

74. Horn F., Schüller W.: Dechema Monogr. *29*, 143 (1957).

75. Horn F., Troltenier U.: Chem. Ing. Tech. *34*, 551 (1962).

76. Hougen O. A., Watson K. M.: Chemical Process Principles, p. 765, John Wiley & Sons, New York 1948.

77. Hougen O. A., Watson K. M., Ragatz R. A.: Chemical Process Principles, Part II, p. 1050, John Wiley & Sons, New York 1959.

78. Householder A. S.: Principles of Numerical Analysis, chapter *8*, McGraw Hill Company, New York 1953.

79. Huff V. N., Gordon S., Morell V. E.: NACA Report 1037 (1951).

80. Huff V. N., Gordon S., Zeleznik F. J.: NASA Report D 132 (1959).

81. Hughes E. M., Howland E. H., Grootenhuis P., Moore N. P. W.: Chem. and Ind. (London) (5) 189 (1963).

82. Hutchinson H. P.: Chem. Eng. Sci. *17*, 703 (1962).

83. Ivash E. V., Li J. C. M., Pitzer K. S.: J. Chem. Phys. *23*, 1814 (1955).

84. Janz G. J.: Estimation of Thermodynamic Properties of Organic Compounds, Academic Press, New York 1958.

85. Joffe J.: Chem. Eng. Progr. *45*, 160 (1949).

86. Kaeppeler J. H., Baumann G.: Astronaut. Acta (Vienna) *3*, 28 (1957).

87. Kamptner H. K., Krause W. R., Schilkem H. P.: Chem. Eng. *73*, (5) 93 (1966).

88. Kandiner H. J., Brinkley S. R. Jr.: Ind. Eng. Chem. *42*, 850 (1950).

89. Kandiner H. J., Brinkley S. R. Jr.: Ind. Eng. Chem. *42*, 1526 (1950).

90. Keyes F. G.: J. Am. Chem. Soc. *60*, 1761 (1938).

91. Kobe K. A., Lynn R. E.: Chem. Rev. *52*, 117 (1953).

92. Krevelen D. W. van. Chermin H. A. G.: Chem. Eng. Sci. *1*, 238 (1952).

93. Krieger F. J., White W. B.: J. Chem. Phys. *16*, 358 (1948).

94. Kubíček M., Hlaváček V.: Numerické metody a optimalisace v chemickém inženýrství (Numerical methods and optimation in chemical engineering, Part I, Chapter 4), SNTL, Praha 1972.

95. Kubíček M., Hlaváček V.: Numerické metody a optimalisace v chemickém inženýrství (Numerical methods and optimation in chemical engineering, Part II, Chapters 9—12), SNTL, Praha 1972.

96. Kunz K. S.: Numerical Analysis, McGraw Hill, New York 1957.

97. Kühl H.: VDI — Forschungsheft *6*, 373 (1935).

98. Le Chatelier M. H.: Compt. rend. *100*, 441 (1885).

99. Leroux P. J., Mathieu P. M.: Chem. Eng. Progr. *57* (11) 54 (1961).

100. Luus R., Jaakola T. H. I.: AICHE J. Vol. *19*, No. 4, 760 (1973).

101. Marek J., Holub R.: Collection Czechoslov. Chem. Commun. *29*, 1085 (1964).

102. Marquardt D. W.: J. Soc. Industr. Appl. Math. *11* (2) 431 (1963).

103. Martin F. J., Yachter M.: Ind. Eng. Chem. *43*, 2446 (1951).

104. McGlashan M. L., Potter D. J. B.: Proceedings of the Conference on Thermodynamics and Transport Properties of Fluids, p. 60, London 1957.

105. Meghreblian R. V.: J. Am. Rocket Soc. (9) *128* (1951).

106. Mentz R. M.: Proc. 1st Conf. High Temp. Systems (1960).

107. Michels H. H., Schneiderman S. B.: Proc. 2nd Conf. High Temp. Systems (1963).

108. Naphtali L.: J. Chem. Phys. *31*, 263 (1960).
109. Naphtali L.: Ind. Eng. Chem. *53*, 387 (1961).
110. Nelder L., Mead A.: Computer J. *7*, 308 (1965).
111. Newitt D. M., Byrne B. J., Strong H. W.: Proc. Roy. Soc. (London) A *123*, 236 (1929).
112. Nikolskii S. S.: Theor. Exp. Chem. *2*, 343 (1966).
113. Novák J. P.: Chem. listy *58*, 1471 (1964).
114. Novák J. P.: Chem. listy *60*, 385 (1966).
115. Novák J. P.: Thesis, VŠCHT, Praha 1966.
116. Novák J. P., Pick J.: Collection Czechoslov. Chem. Commun. *35*, 2068 (1970).
117. Novák J. P., Malijevský A., Matouš J., Šobr J.: Plyny a plynné směsi, Stavové chování a termodynamické vlastnosti (Gases and Gas Mixtures, Behaviour of State and Thermodynamic Properties) Academia, Prague 1972.
118. Opfell J. B., Sage B. H., Pitzer K. S.: Ind. Eng. Chem. *48*, 2069 (1956).
119. Offtermatt W. F.: Thesis, Pasadena 1953, cit. Reid R. C., Sherwood T. K.: The Properties of Gases and Liquids, McGraw Hill, New York 1966.
120. Pankiewicz W.: Communications of the ACM *13*, (4) 259 (1970).
121. Pennington R. E., Kobe K. A.: J. Am. Chem. Soc. *79*, 300 (1957).
122. Peters K., Kappelmacher E., Voetter H.: Gas, Wasser, Wärme *6*, 189 (1952).
123. Pereyra V.: SIAM J. Numer. Anal. *4* (1) 27 (1967).
124. Pichler H.: Chem. Eng. Progr. *61*, (8) 63 (1965).
125. Pings C. J.: Chem. Eng. Sci. *17*, 573 (1962).
126. Pitzer K. S., Weltner W.: J. Am. Chem. Soc. *71*, 2842 (1949).
127. Pitzer K. S., Curl R. F.: Proceedings of the Conference on Thermodynamics and Transport Properties of Fluids, p. 1, London 1957.
128. Pitzer K. M.: Quantum Chemistry, Prentice Hall, New York 1953.
129. Powell H. N., Sarner S. F.: General Electric Co., Report R 59/FPD 796 (1959).
130. Putnam W. E., Kilpatrick J. E.: J. Chem. Phys. *21*, (5) 951 (1953).
131. Ralston A.: A First Course in Numerical Analysis chapter 8, McGraw-Hill Book Company New York, St. Louis, San Francisco, Toronto, London, Sydney 1965.
132. Redlich O., Kwong J. N. S.: Chem. Rev. *44*, 233 (1949).
133. Rektorys K.: Přehled matematiky (Survey of Mathematics) p. 96, SNTL, Prague 1963.
134. Rideal E. K.: Proc. Roy. Soc. (London) A *99*, 133 (1921).
135. Robert H. Fariss and Victor J. Law: Practical Tactics for Overcoming Difficulties in Non-Linear Regression and Equation Solving (research report), Monsanto Company, St. Louis, Missouri.
136. Rossini F. D.: Chemical Thermodynamics, John Wiley & Sons, New York 1950.
137. Rossini F. D., Pitzer K. S., Arnett R. I., Braun R. M., Pimentel G. C.: Selected Values of Physical and Thermodynamic Properties of Hydrocarbons and Related Compounds, Carnegie Press, Pittsburg 1953.
138. Samuels M. R., Elliassen J. D.: Ind. Eng. Chem. Process Des. Dev. *11* (3), 383 (1972).
139. Satoh S.: J. Sci. Research Inst. (Tokio) *43*, 79 (1948).
140. Scarborough J. B.: Numerical Mathematical Analysis, 2. Ed., p. 203, John Hopkins, Baltimore 1950.
141. Scott D. W., McCullough J. P., Messerly J. F., Pennington R. E., Hossenlopp I. A., Finke H. L., Waddington G.: J. Am. Chem. Soc. *80*, 55 (1958).
142. Scully D. B.: Chem. Eng. Sci. *17*, 977 (1962).
143. Souders M., Matthews C. S., Hurd C. O.: Ind. Eng. Chem. *41*, 1037 (1949).
144. Souders M., Matthews C. S., Hurd C. O.: Ind. Eng. Chem. *41*, 1048 (1949).
145. Stamm R. F., Halverson F., Whalen J. J.: J. Chem. Phys. *17* (1) 104 (1949).
146. Stein M.: Forsch. Gebiete Ingenieurw. *14*, 113 (1943).

147. Stein M., Voetter H.: Z. Elektrochem. *57*, 119 (1953).
148. Storey S. H., van Zeggeren F.: Canad. J. Chem. Engrs. *42* (4) 54 (1964).
149. Stull D. R., Mayfield F. D.: Ind. Eng. Chem. *35*, 639 (1943).
150. Stull D. R., Westrum E. F. Jr., Sinke G. C.: The Chemical Thermodynamics of Organic Compounds, p. 90, John Wiley & Sons, New York 1969.
151. Su G. J., Viswanath D. S.: AICHE J.: *11* (2) 205 (1965).
152. Su G. J., Chang C. H.: J. Am. Chem. Soc. *68*, 1080 (1946).
153. Su G. J., Chang C. H.: Ind. Eng. Chem. *38*, 802 (1946).
154. Šeha Z., Habada M.: Chem. zvesti *10*, 25 (1956).
155. Šeha Z.: Chem. listy *49*, 1569 (1955).
156. Šeha Z.: Collection Czechoslov. Chem. Commun. *25*, 1688 (1960).
157. Šeha Z., Holub R.: Chem. listy *52*, 2216 (1958).
158. Taylor W. J., Wagmann D. D., Williams M. G., Pitzer K. S., Rossini F. D.: J. Research Natl. Bur. Standards *37*, 95 (1946).
159. Taylor H. S., Glasstone S.: A Treatise of Physical Chemistry, Vol. II., States of Matter, D. van Nostrand Co., Amsterdam 1951.
160. Tiggelen A. van, Nenquin G.: Ind. chim. Belge *24*, 350 (1959).
161. Traustel S.: Z. Ver. dent. Ing. *88*, 688 (1944).
162. Tsao C. C., Wiederhold G.: Proc. 2nd Conf. High Temp. Systems (1963).
163. Verma K. K., Doraiswamy L. K.: Ind. Eng. Chem. Fundamentals *4*, 389 (1965).
164. Villars D. S.: J. Phys. Chem. *63*, 521 (1959).
165. Voňka P., Holub R.: Collection Czechoslov. Chem. Commun. *36*, 2446 (1971); *40*, 931 (1975).
166. Vvedenskii A. A.: Termodinamicheskie raschoty v neftekhimicheskikh processakh (Thermodynamic calculations of petrochemical processes) Gostoptekhizdat, Leningrad 1960.
167. Vysoký A., Fauser G.: Chem. prům. *40*, (12) 713 (1965).
168. White W. B.: J. Chem. Phys. *46*, 4171 (1967).
169. White W. B., Johnson S. M., Dantzig G. B.: J. Chem. Phys. *28*, 751 (1958).
170. Wilkins, R. L.: Proc. 1st Conf. High Temp. Systems, *123* (1960).
171. Van Zeggeren F., Storey S. H.: The Computation of Chemical Equilibria, Cambridge University Press 1970.
172. Zeleznik F. J.: Thesis, Case Inst. of Technology, 1959.
173. Zeleznik F. J., Gordon S.: NASA, Tech. Note D-473 (1960).
174. Zeleznik F. J., Gordon S.: NASA, Tech. Note D-1454 (1962).
175. IUPAC, New values of universal constants, Chem. listy *64*, 930 (1970).

Index